Introduction to Finite Element Analysis
Using SOLIDWORKS Simulation® 2022

Randy H. Shih
Oregon Institute of Technology

PUBLICATIONS

SDC Publications
P.O. Box 1334
Mission, KS 66222
913-262-2664
www.SDCpublications.com
Publisher: Stephen Schroff

ISBN-13: 978-1-63057-484-0
ISBN-10: 1-63057-484-8

Printed and bound in the United States of America.

Preface

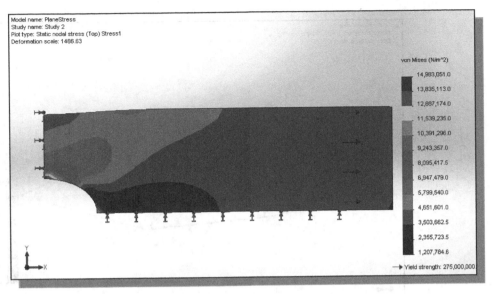

The primary goal of ***Introduction to Finite Element Analysis Using SOLIDWORKS Simulation*** is to introduce the aspects of finite element analysis (**FEA**) that are important to engineers and designers. Theoretical aspects of finite element analysis are also introduced as they are needed to help gain a better understanding of the operation. The primary emphasis of the text is placed on the practical concepts and procedures for using SOLIDWORKS Simulation in performing *Linear Static Stress Analysis* and basic *Modal Analysis*. This text is intended to be used as a training guide for students and professionals. This text covers SOLIDWORKS Simulation, and the lessons proceed in a pedagogical fashion to guide you from constructing basic truss elements to generating three-dimensional solid elements from solid models. This text takes a hands-on, exercise-intensive approach to all the important finite element analysis techniques and concepts. This textbook contains a series of thirteen tutorial style lessons designed to introduce beginning FEA users to SOLIDWORKS Simulation. This text is also helpful to SOLIDWORKS Simulation users upgrading from a previous release of the software. The finite element analysis techniques and concepts discussed in this text are also applicable to other FEA packages. The basic premise of this book is that the more designs you create using SOLIDWORKS Simulation, the better you learn the software. With this in mind, each lesson introduces a new set of commands and concepts building on previous lessons. This book does not attempt to cover all of SOLIDWORKS Simulation's features, only to provide an introduction to the software. It is intended to help you establish a good basis for exploring and growing in the exciting field of Computer Aided Engineering.

Acknowledgments

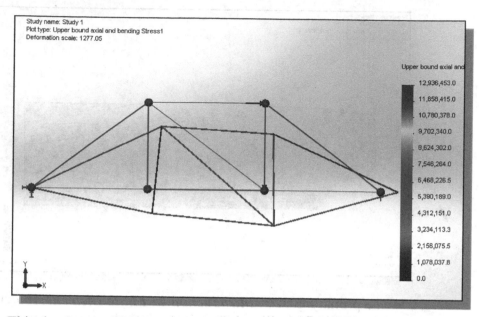

This book would not have been possible without a great deal of support. First, special thanks to two great teachers, Prof. George R. Schade of University of Nebraska-Lincoln and Mr. Denwu Lee, who taught me the fundamentals, the intrigue, and the sheer fun of Computer Aided Engineering.

The effort and support of the editorial and production staff of SDC Publications is gratefully acknowledged. I would especially like to thank Stephen Schroff for his support and helpful suggestions during this project.

I am grateful that the Mechanical and Manufacturing Engineering Department of Oregon Institute of Technology has provided me with an excellent environment in which to pursue my interests in teaching and research.

Finally, truly unbounded thanks are due to my wife Hsiu-Ling and our daughter Casandra for their understanding and encouragement throughout this project.

Randy H. Shih
Klamath Falls, Oregon
Winter, 2022

Table of Contents

Chapter 1
The Direct Stiffness Method

Chapter 2
Truss Elements in Two-Dimensional Spaces

Chapter 3
2D Trusses in MS Excel and Truss Solver

Chapter 4
Truss Elements in SOLIDWORKS Simulation

Chapter 5
SOLIDWORKS Simulation Two-Dimensional Truss Analysis

Chapter 6
Three-Dimensional Truss Analysis

Chapter 7
Basic Beam Analysis

Chapter 8
Beam Analysis Tools

Chapter 9
Statically Indeterminate Structures

Chapter 10
Two-Dimensional Surface Analysis

Chapter 11
Three-Dimensional Solid Elements

Chapter 12
3D Thin Shell Analysis

Appendix

Index

Notes:

Introduction

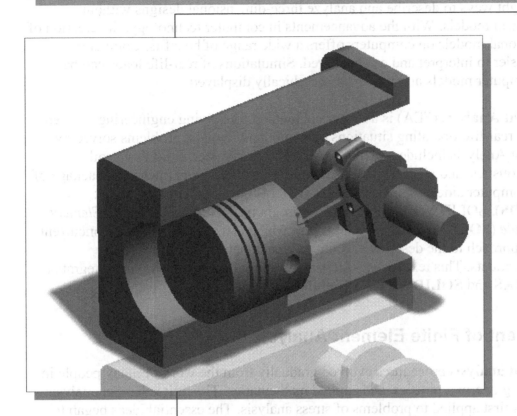

Learning Objectives

- ♦ **Development of Finite Element Analysis**
- ♦ **FEA Modeling Considerations**
- ♦ **Finite Element Analysis Procedure**
- ♦ **Getting Started with SOLIDWORKS**
- ♦ **Startup Options**
- ♦ **SOLIDWORKS Screen Layout**
- ♦ **Mouse Buttons**
- ♦ **SOLIDWORKS Online Help**

Introduction

Design includes all activities involved from the original concept to the finished product. Design is the process by which products are created and modified. For many years, designers sought ways to describe and analyze three-dimensional designs without building physical models. With the advancements in computer technology, the creation of three-dimensional models on computers offers a wide range of benefits. Computer models are easier to interpret and easily altered. Simulations of real-life loads can be applied to computer models and the results graphically displayed.

Finite Element Analysis (FEA) is a numerical method for solving engineering problems by simulating real-life-operating situations on computers. Typical problems solved by Finite Element Analysis include structural analysis, heat transfer, fluid flow, soil mechanics, acoustics, and electromagnetism. SOLIDWORKS is an integrated package of mechanical computer aided engineering software tools developed by **Dassault Systèmes** (DS). SOLIDWORKS is a suite of programs, including the *Finite Element Analysis module* (**SOLIDWORKS Simulation**), which is used to facilitate a concurrent engineering approach to the design, analysis, and manufacturing of mechanical engineering products. This text focuses on basic structural analysis using the integrated **SOLIDWORKS** and **SOLIDWORKS Simulation**.

Development of Finite Element Analysis

Finite element analysis procedures evolved gradually from the work of many people in the fields of engineering, physics, and applied mathematics. The finite element analysis procedure was first applied to problems of stress analysis. The essential ideas began to appear in publications during the 1940s. In 1941, Hrenikoff proposed that the elastic behavior of a physically continuous plate would be similar to a framework of one-dimensional rods and beams, connected together at discrete points. The problem could then be handled by familiar methods for trusses and frameworks. In 1943, Courant's paper detailed an approach to solving the torsion problem in elasticity. Courant described the use of piecewise linear polynomials over a triangularized region. Courant's work was not noticed and soon forgotten, since the procedure was impractical to solve by hand.

In the early 1950s, with the developments in digital computers, Argyris and Kelsey converted the well-established "framework-analysis" procedure into matrix format. In 1956, Turner, Clough, Matin, and Topp derived stiffness matrices for truss elements, beam elements and two-dimensional triangular and rectangular elements in plane stress. Clough introduced the first use of the phrase "finite element" in 1960. In 1961, Melosh developed a flat, rectangular-plate bending-element, followed by development of the curved-shell bending-element by Grafton and Strome in 1963. Martin developed the first three-dimensional element in 1961 followed by Gallagher, Padlog and Bijlaard in 1962 and Melosh in 1964.

From the mid-1960s to the end of the 1970s, finite element analysis procedures spread beyond structural analysis into many other fields of application. Large general purpose

FEA software began to appear. By the late 1980s, FEA software became available on microcomputers, complete with automatic mesh-generation, interactive graphics, and pre-processing and post-processing capabilities.

In this text, we will follow a logical order, parallel to the historical development of finite element analysis procedures, in learning the fundamental concepts and commands for performing finite element analysis using SOLIDWORKS and SOLIDWORKS Simulation. We will begin with the one-dimensional truss element, beam element, and move toward the more advanced features of SOLIDWORKS Simulation. This text also covers the general procedures of performing two-dimensional and three-dimensional solid FE analyses. The concepts and techniques presented in this text are also applicable to other FEA packages. Throughout the text, many of the classic strength of materials and machine design problems are used as examples and exercises, which hopefully will help build up the user's confidence in performing FEA analyses.

FEA Modeling Considerations

The analysis of an engineering problem requires the idealization of the problem into a mathematical model. It is clear that we can only analyze the selected mathematical model and that all the assumptions in this model will be reflected in the predicted results. We cannot expect any more information in the prediction than the information contained in the model. Therefore, it is crucial to select an appropriate mathematical model that will most closely represent the actual situation. It is also important to realize that we cannot predict the response exactly because it is impossible to formulate a mathematical model that will represent all the information contained in an actual system.

As a general rule, finite element modeling should start with a simple model. Once a mathematical model has been solved accurately and the results have been interpreted, it is feasible to consider a more refined model in order to increase the accuracy of the prediction of the actual system. For example, in a structural analysis, the formulation of the actual loads into appropriate models can drastically change the results of the analysis. The results from the simple model, combined with an understanding of the behavior of the system, will assist us in deciding whether and at which part of the model we want to use further refinements. Clearly, the more complicated model will include more complex response effects, but it will also be more costly and sometimes more difficult to interpret the solutions.

Modeling requires that the physical action of the problem be understood well enough to choose suitable kinds of analyses. We want to avoid the waste of time and computer resources caused by over-refinement and badly shaped elements. Once the results have been calculated, we must check them to see if they are reasonable. Checking is very important because it is easy to make mistakes when we rely upon the FEA software to solve complicated systems.

Types of Finite Elements

The finite element analysis method is a numerical solution technique that finds an approximate solution by dividing a region into small sub-regions. The solution within each sub-region that satisfies the governing equations can be reached much more simply than that required for the entire region. The sub-regions are called *elements*, and the elements are assembled through interconnecting a finite number of points on each element called *nodes*. Numerous types of finite elements can be found in commercial FEA software, and new types of elements are being developed as research is done worldwide. Depending on the dimensions, finite elements can be divided into three categories:

1. **One-dimensional** line elements: Truss, beam and boundary elements.

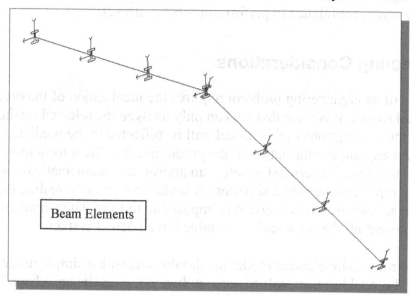

Beam Elements

2. **Two-dimensional** plane elements: Plane stress, plane strain, axisymmetric, membrane and shell elements.

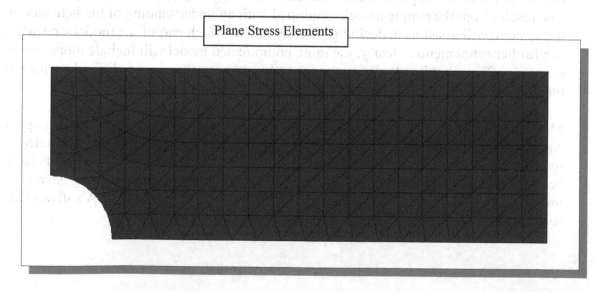

Plane Stress Elements

3. **Three-dimensional** volume elements: Tetrahedral, hexahedral, and brick
   elements.

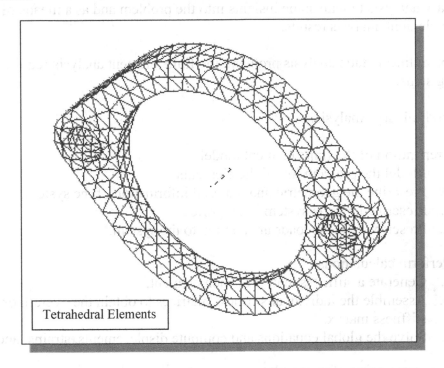

Tetrahedral Elements

Typically, finite element solutions using one-dimensional line elements are as accurate as
solutions obtained using conventional truss and beam theories. It is usually easier to get
FEA results than doing hand calculations using conventional theories. However, very few
closed form solutions exist for two-dimensional elements and almost none exist for three-
dimensional solid elements.

In theory, all designs could be modeled with three-dimensional volume elements.
However, this is not practical since many designs can be simplified with reasonable
assumptions to obtain adequate FEA results without any loss of accuracy. Using
simplified models greatly reduces the time and effort in reaching FEA solutions.

Finite Element Analysis Procedure

Prior to carrying out the finite element analysis, it is important to do an approximate preliminary analysis to gain some insights into the problem and as a means of checking the finite element analysis results.

For a typical linear static analysis problem, the finite element analysis requires the following steps:

1. Preliminary Analysis.

2. Preparation of the finite element model:
    a. Model the problem into finite elements.
    b. Prescribe the geometric and material information of the system.
    c. Prescribe how the system is supported.
    d. Prescribe how the loads are applied to the system.

3. Perform calculations:
    a. Generate a stiffness matrix of each element.
    b. Assemble the individual stiffness matrices to obtain the overall, or global, stiffness matrix.
    c. Solve the global equations and compute displacements, strains, and stresses.

4. Post-processing of the results:
    a. Viewing the stress contours and the displaced shape.
    b. Checking any discrepancy between the preliminary analysis results and the FEA results.

Matrix Definitions

The use of vectors and matrices is of fundamental importance in engineering analysis because it is with the use of these quantities that complex procedures can be expressed in a compact and elegant manner. One need not understand vectors or matrices in order to use FEA software. However, by studying the matrix structural analysis, we develop an understanding of the procedures common to the implementation of structural analysis as well as the general finite element analysis. The objective of this section is to present the fundamentals of matrices, with emphasis on those aspects that are important in finite element analysis. In the next chapter we will introduce the derivation of matrix structural analysis, the stiffness matrix method. *MATRIX Algebra* is a powerful tool for use in programming the FEA methods for electronic digital computers. Matrix notation represents a simple and easy-to-use notation for writing and solving sets of simultaneous algebraic equations.

- **Matrix** A **matrix** is a rectangular array of elements arranged in rows and columns. Applications in this text deal only with matrices whose elements are real numbers. For example,

$$[A] = \begin{bmatrix} A_{11} & A_{12} & A_{13} \\ A_{21} & A_{22} & A_{23} \end{bmatrix}$$

[A] is a rectangular array of two rows and three columns, thus called a 2×3 matrix. The element A_{ij} is the element in the ith row and j^{th} column.

- **Column Matrix (Row Matrix)** A **column (row) matrix** is a matrix having one column (row). A single-column array is commonly called a column matrix or *vector*. For example:

$$\{F\} = \begin{Bmatrix} F_1 \\ F_2 \\ F_3 \end{Bmatrix}$$

- **Square Matrix** A **square matrix** is a matrix having equal numbers of rows and columns.

- **Diagonal Matrix** A **diagonal matrix** is a square matrix with nonzero elements only along the diagonal of the matrix.

- **Addition** The **addition of matrices** involves the summation of elements having the same "address" in each matrix. The matrices to be summed must have identical dimensions. The addition of matrices of different dimensions is not defined. Matrix addition is associative and commutative.

For example:

$$[A] + [B] = \begin{bmatrix} ① & 2 \\ 3 & \boxed{4} \end{bmatrix} + \begin{bmatrix} ② & 4 \\ 6 & \boxed{8} \end{bmatrix} = \begin{bmatrix} ③ & 6 \\ 9 & \boxed{12} \end{bmatrix}$$

- **Multiplication by a Constant** If a matrix is to be multiplied by a constant, every element in the matrix is multiplied by that constant. Also, if a constant is factored out of a matrix, it is factored out of each element. For example:

$$3 \times [A] = 3 \times \begin{bmatrix} 1 & 2 \\ 3 & 4 \end{bmatrix} = \begin{bmatrix} 3 & 6 \\ 9 & 12 \end{bmatrix}$$

- **Multiplication of Two Matrices**   Assume that [C] = [A][B], where [A], [B], and [C] are matrices. Element C_{ij} in matrix [C] is defined as follows:

$$C_{ij} = A_{i1} \times B_{1j} + A_{i2} \times B_{2j} + \cdots + A_{ik} \times B_{kj}$$

For example:

$$[\,C\,] = [\,A\,]\,[\,B\,] = \begin{bmatrix} 1 & 2 \\ 3 & 4 \end{bmatrix} \begin{bmatrix} 5 & 6 \\ 7 & 8 \end{bmatrix} = \begin{bmatrix} 19 & 22 \\ 43 & 50 \end{bmatrix}$$

$C_{11} = 1 \times 5 + 2 \times 7 = 19, \quad C_{12} = 1 \times 6 + 2 \times 8 = 22$

$C_{21} = 3 \times 5 + 4 \times 7 = 43, \quad C_{22} = 3 \times 6 + 4 \times 8 = 50$

- **Identity Matrix**   An **identity matrix** is a diagonal matrix with each diagonal element equal to unity.

For example:

$$[\,I\,] = \begin{bmatrix} 1 & 0 & 0 & 0 \\ 0 & 1 & 0 & 0 \\ 0 & 0 & 1 & 0 \\ 0 & 0 & 0 & 1 \end{bmatrix}$$

- **Transpose of a Matrix**   The **transpose of a matrix** is a matrix obtained by interchanging rows and columns. Every matrix has a transpose. The transpose of a column matrix (vector) is a row matrix; the transpose of a row matrix is a column matrix.

For example:

$$[A] = \begin{bmatrix} 1 & 2 & 3 \\ 4 & 5 & 6 \end{bmatrix} \qquad [A]^T = \begin{bmatrix} 1 & 4 \\ 2 & 5 \\ 3 & 6 \end{bmatrix}$$

- **Inverse of a Square Matrix**   A square matrix *may* have an inverse. The product of a matrix and its inverse matrix yields the identity matrix.

$$[A]\,[A]^{-1} = [A]^{-1}\,[A] = [\,I\,]$$

The reader is referred to the following techniques for matrix inversion:
  1. Gauss-Jordan elimination method
  2. Gauss-Seidel iteration method

These techniques are popular and are discussed in most texts on numerical techniques.

Getting Started with SOLIDWORKS

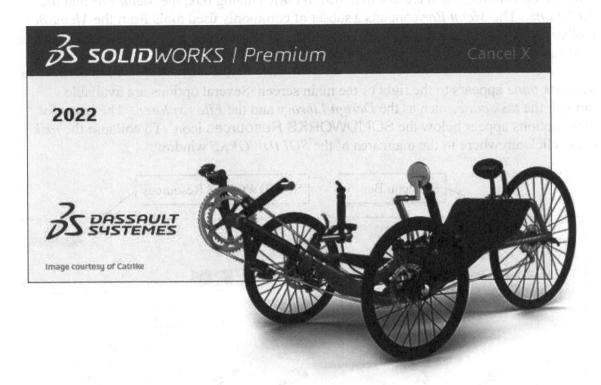

Image courtesy of Catrike

- SOLIDWORKS is composed of several application software modules (these modules are called *applications*) all sharing a common database. In this text, the main concentration is placed on the solid modeling modules used for part design. The general procedures required in creating solid models, engineering drawings, and assemblies are illustrated.

Starting SOLIDWORKS

1. How to start SOLIDWORKS depends on the type of workstation and the particular software configuration you are using. With most *Windows* systems, you may select **SOLIDWORKS** on the *Start* menu or select the **SOLIDWORKS** icon on the desktop. Consult your instructor or technical support personnel if you have difficulty starting the software.

❖ The program takes a while to load, so be patient. The tutorials in this text are based on the assumption that you are using SOLIDWORKS' default settings. If your system has been customized for other uses, some of the settings may appear differently and not work with the step-by-step instructions in the tutorials. Contact your instructor and/or technical support personnel to restore the default software configuration.

Once the program is loaded into memory, the *SOLIDWORKS* program window appears. This screen contains the *Welcome to SOLIDWORKS* dialog box, the *Menu Bar* and the *Task Pane*. The *Menu Bar* contains a subset of commonly used tools from the *Menu Bar* toolbar (**New**, **Open**, **Save**, etc.), the *SOLIDWORKS* menus, the *SOLIDWORKS Search* oval, and a flyout menu of *Help* options.

The *task pane* appears to the right of the main screen. Several options are available through the *task pane*, such as the *Design Library* and the *File Explorer*. The icons for these options appear below the **SOLIDWORKS Resources** icon. To collapse the *task pane*, click anywhere in the main area of the *SOLIDWORKS* window.

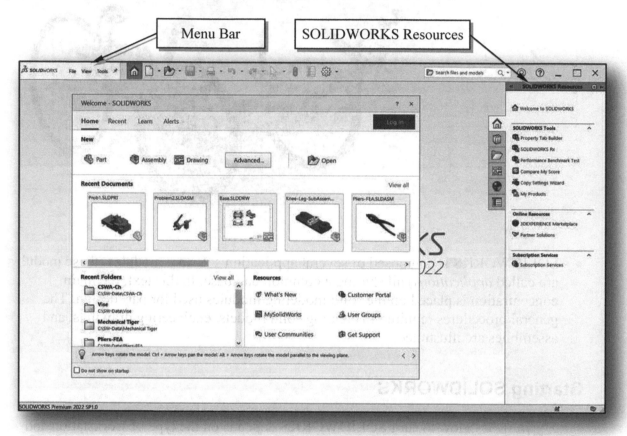

To the left side of the main window is the *Welcome to SOLIDWORKS* dialog box, which contains four tabs, *Home*, *Recent*, *Learn* and *Alert*. The Home tab can be used to quickly start a new document or open recently modified files. Note that this dialog box can be toggled on/off by hitting [Ctrl + F2].

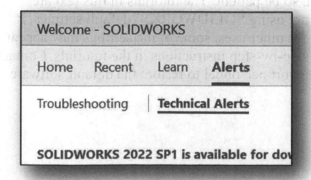

2. Click on the *Alerts* tab to display technical alerts, such as new updates and known issues, from SOLIDWORKS.

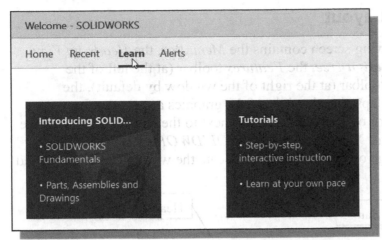

3. Click on the **Learn** tab to display the built-in SOLIDWORKS learning tools.

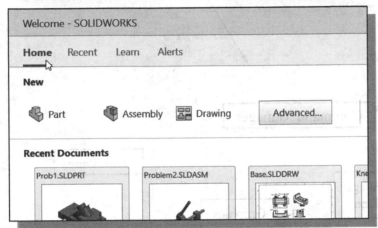

4. Switch back to the Home tab and notice the top section in the *Home* tab is the **New** file option, which allows us to start a new *SOLIDWORKS* PART file, a new ASSEMBLY file, or a new DRAWING file.

A part is a single three-dimensional (3D) solid model. Parts are the basic building blocks in modeling with *SOLIDWORKS*. An assembly is a 3D arrangement of parts (components) and/or other assemblies (subassemblies). A drawing is a 2D representation of a part or an assembly.

5. Select the **Part** icon as shown. Click **OK** in the *New SOLIDWORKS Document* dialog box to open a new part file.

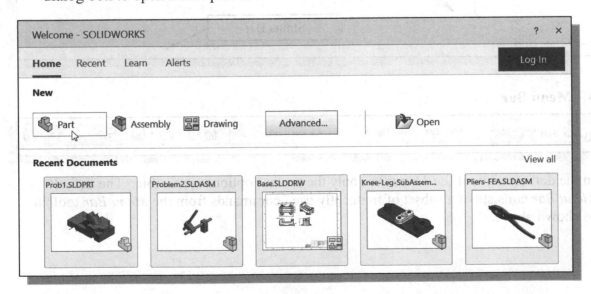

SOLIDWORKS Screen Layout

The default *SOLIDWORKS* drawing screen contains the *Menu Bar*, the *Heads-up View* toolbar, the *Feature Manager Design Tree*, the *Features* toolbar (at the left of the window by default), the *Sketch* toolbar (at the right of the window by default), the graphics area, the *task pane* (collapsed to the right of the graphics area in the figure below), and the *Status Bar*. A line of quick text appears next to the icon as you move the *mouse cursor* over different icons. You may resize the *SOLIDWORKS* drawing window by clicking and dragging the edge of the window, or relocate the window by clicking and dragging the *window title* area.

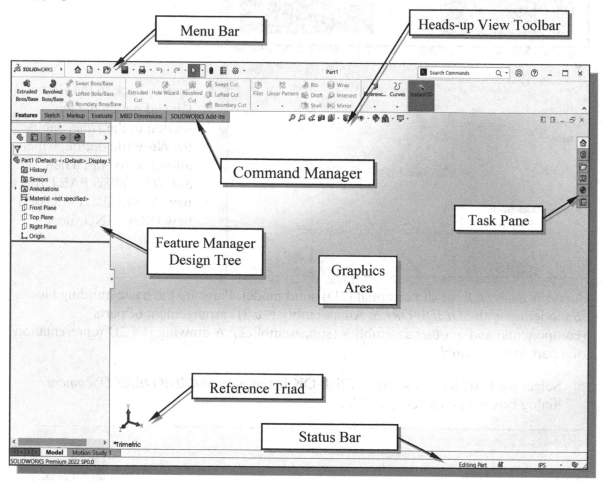

- **Menu Bar**

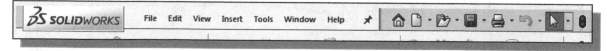

In the default view of the *Menu Bar*, only the toolbar options are visible. The default *Menu Bar* consists of a subset of frequently used commands from the *Menu Bar* toolbar as shown above.

- **Menu Bar Pull-down Menus**

To display the *pull-down* menus, move the cursor over or click the *SOLIDWORKS* logo. The pull-down menus contain operations that you can use for all modes of the system.

- **Heads-up View Toolbar**

The *Heads-up View* toolbar allows us quick access to frequently used view-related commands, such as **Zoom, Pan** and **Rotate**. Note: You cannot hide or customize the *Heads-up View* toolbar.

- **Features Toolbar**

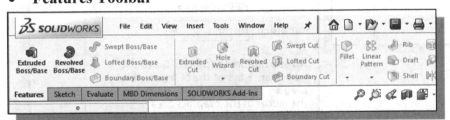

By default, the *Features* toolbar is displayed at the top of the *SOLIDWORKS* window. The *Features* toolbar allows us quick access to frequently used feature-related commands, such as **Extruded Boss/Base**, **Extruded Cut**, and **Revolved Boss/Base**.

- **Sketch Toolbar**

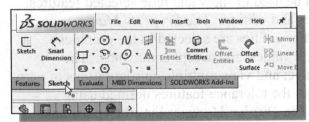

By default, the *Sketch* toolbar can be accessed by clicking on the **Sketch** tab in the *Ribbon* toolbar area. The *Sketch* toolbar provides tools for creating the basic geometry that can be used to create features and parts.

- **Feature Manager-Design Tree/Property Manager/Configuration Manager/ DimXpert Manager/Display Manager**

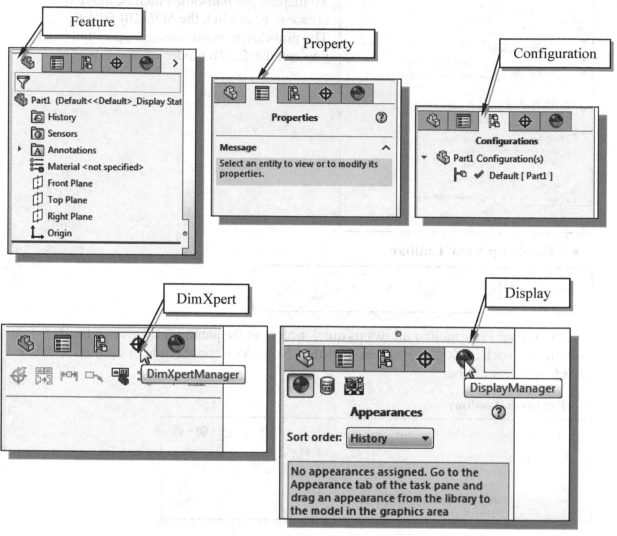

The left panel of the *SOLIDWORKS* window is used to display the *Feature Manager-Design Tree*, the *Property Manager*, the *Configuration Manager*, and the *DimXpert Manager*. These options can be chosen by selecting the appropriate tab at the top of the panel. The *Feature Manager Design Tree* provides an overview of the active part, drawing, or assembly in outline form. It can be used to show and hide selected features, filter contents, and manage access to features and editing. The *Property Manager* opens automatically when commands are executed or entities are selected in the graphics window, and is used to make selections, enter values, and accept commands. The *Configuration Manager* is used to create, select and view multiple configurations of parts and assemblies. The *DimXpert Manager* lists the tolerance features defined using *SOLIDWORKS* 'DimXpert for parts' tools. The *Display Manager* lists the appearance, decals, lights, scene, and cameras applied to the current model. From the *Display Manager*, we can view applied content, and add, edit, or delete items. The *Display Manager* also provides access to *Photo View* options if the module is available.

- **Graphics Area**

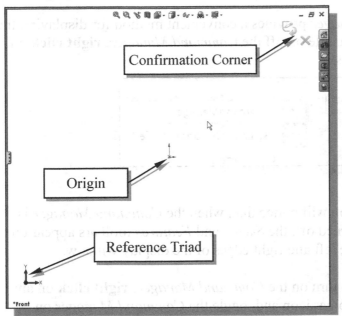

The graphics area is the area where models and drawings are displayed.

- **Reference Triad**

The *Reference Triad* appears in the graphics area of part and assembly documents. The triad is shown to help orient the user when viewing models and is for reference only.

- **Origin**

The *Origin* represents the (0,0,0) coordinate in a model or sketch. A model origin appears blue; a sketch origin appears red.

- **Confirmation Corner**

The *Confirmation Corner* offers an alternate way to accept features.

- **Graphics Cursor or Crosshairs**

The *graphics cursor* shows the location of the pointing device in the graphics window. During geometric construction, the coordinate of the cursor is displayed in the *Status Bar* area, located at the bottom of the screen. The cursor's appearance depends on the selected command or option.

- **Message and Status Bar**

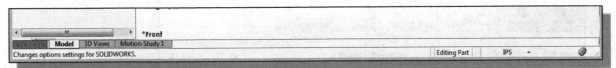

The *Message and Status Bar* area displays a single-line description of a command when the cursor is on top of a command icon. This area also displays information pertinent to the active operation. In the figure above, the cursor coordinates are displayed while in the *Sketch* mode.

Using the *SOLIDWORKS* Command Manager

The *SOLIDWORKS Command Manager* provides a convenient method for displaying the most commonly used toolbars. To toggle on/off the *Command Manager*, **right click** on any toolbar icon, and select in the list.

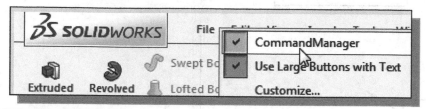

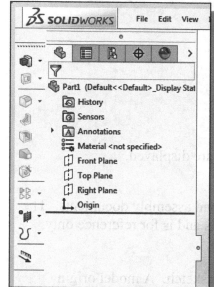

You will notice that, when the *Command Manager* is turned off, the *Sketch* and *Features* toolbars appear on the left and right edges of the display window.

To turn on the *Command Manager*, **right click** on any toolbar icon and toggle the *Command Manager* on by selecting it at the top of the pop-up menu.

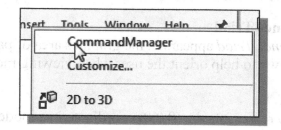

The *Command Manager* is a context-sensitive toolbar that dynamically updates based on the user's selection. When you click a tab below the *Command Manager*, it updates to display the corresponding toolbar. For example, if you click the **Sketches** tab, the *Sketch* toolbar appears. By default, the *Command Manager* has toolbars embedded in it based on the document type.

The default display of the *Command Manager* is shown below. You will notice that, when the *Command Manager* is used, the *Sketch* and *Features* toolbars no longer appear on the left and right edges of the display window.

Important Note: The illustrations in this text use the *Command Manager*. If a user prefers to use the standard display of toolbars, the only change is that it may be necessary to activate the appropriate toolbars prior to selecting a command.

Mouse Buttons

SOLIDWORKS utilizes the mouse buttons extensively. In learning SOLIDWORKS' interactive environment, it is important to understand the basic functions of the mouse buttons.

- **Left mouse button**
  The **left-mouse-button** is used for most operations, such as selecting menus and icons, or picking graphic entities. One click of the button is used to select icons, menus and form entries, and to pick graphic items.

- **Right mouse button**
  The **right-mouse-button** is used to bring up additionally available options in a context-sensitive pop-up menu. These menus provide shortcuts to frequently used commands.

- **Middle mouse button/wheel**
  The middle mouse button/wheel can be used to Rotate (hold down the wheel button and drag the mouse), Pan (hold down the wheel button and drag the mouse while holding down the [**Ctrl**] key), or Zoom (hold down the wheel button and drag the mouse while holding down the [**Shift**] key) real time. Spinning the wheel allows zooming to the position of the cursor.

[Esc] – Canceling Commands

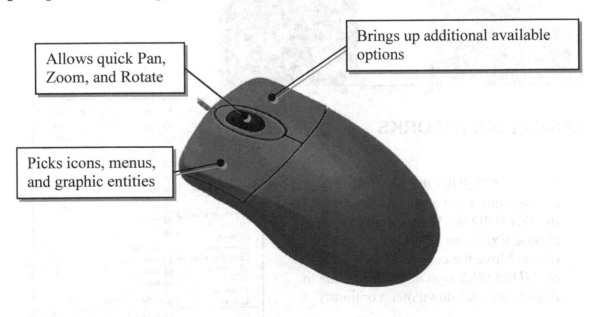

Brings up additional available options

Allows quick Pan, Zoom, and Rotate

Picks icons, menus, and graphic entities

The [**Esc**] key is used to cancel a command in SOLIDWORKS. The [**Esc**] key is located near the top left corner of the keyboard. Sometimes, it may be necessary to press the [**Esc**] key twice to cancel a command; it depends on where we are in the command sequence. For some commands, the [**Esc**] key is used to exit the command.

SOLIDWORKS Help System

❖ *SOLIDWORKS* provides on-line help functions, available at any time during a *SOLIDWORKS* session.

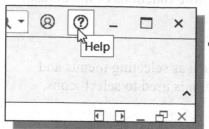

• The **SOLIDWORKS Help** option can be accessed by clicking on the **Help** icon at the right end of the *Menu Bar*.

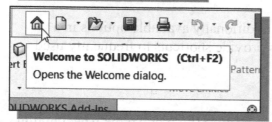

• The SOLIDWORKS Tutorials can also be accessed from the *Welcome to SOLIDWORKS dialog box* by selecting the **Learn** tab.

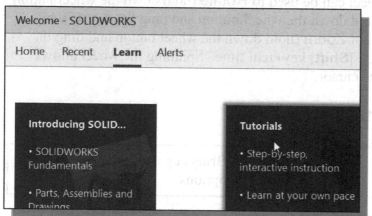

• The **SOLIDWORKS Tutorials** option provides a collection of tutorials illustrating different *SOLIDWORKS* operations.

Leaving SOLIDWORKS

➢ To leave *SOLIDWORKS*, use the left-mouse-button and click on **File** at the top of the *SOLIDWORKS* screen window, then choose **Exit** from the pull-down menu. (Note: Move the cursor over the *SOLIDWORKS* logo in the *Menu Bar* to display the pull-down menu options.)

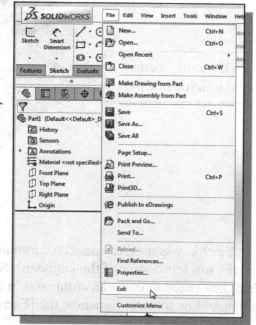

Creating a CAD Files Folder

It is a good practice to create a separate folder to store your CAD files. You should not save your CAD files in the same folder where the SOLIDWORKS application is located. It is much easier to organize and back up your project files if they are in a separate folder. Making folders within this folder for different types of projects will help you organize your CAD files even further. When creating CAD files in SOLIDWORKS, it is strongly recommended that you *save* your CAD files on the hard drive.

➢ To create a new folder in the *Windows* environment:

1. On the *desktop* or under the *My Documents* folder, in which you want to create a new folder:

2. *Right-mouse-click* once to bring up the option menu, then select **New→ Folder**.

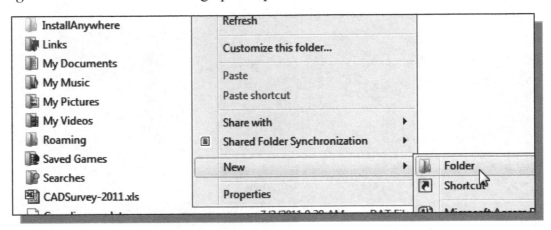

3. Type a name, such as **SOLIDWORKS Projects**, for the new folder, and then press **[ENTER]**.

Notes:

Chapter 1
The Direct Stiffness Method

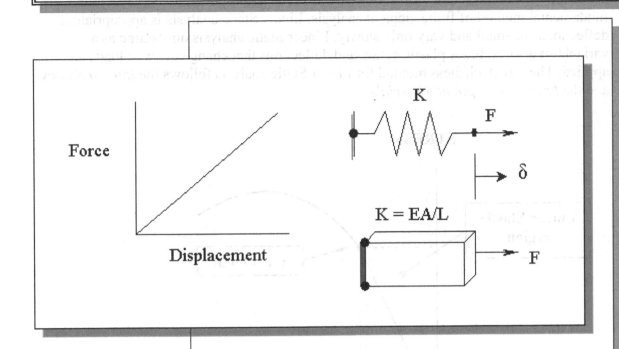

Learning Objectives

♦ **Understand system equations for truss elements.**

♦ **Understand the setup of a Stiffness Matrix.**

♦ **Apply the Direct Stiffness Method.**

♦ **Understand Units Setup.**

♦ **Create an Extruded solid model using SOLIDWORKS.**

♦ **Use the SOLIDWORKS 2D Sketch and Constraints tools.**

♦ **Use the Display Viewing commands.**

Introduction

The **direct stiffness method** is used mostly for *Linear Static analysis*. The development of the direct stiffness method originated in the 1940s and is generally considered the fundamental method of finite element analysis. Linear Static analysis is appropriate if deflections are small and vary only slowly. Linear Static analysis omits time as a variable. It also excludes plastic action and deflections that change the way loads are applied. The direct stiffness method for Linear Static analysis follows the *laws of Statics* and the *laws of Strength of Materials*.

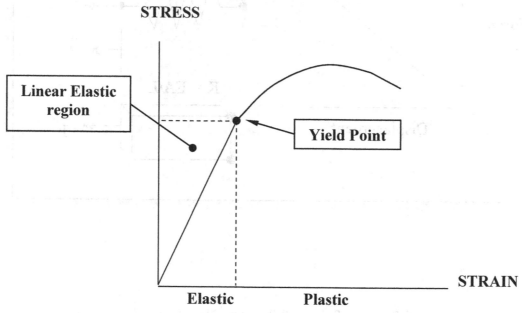

Stress-Strain diagram of typical ductile material

This chapter introduces the fundamentals of finite element analysis by illustrating an analysis of a one-dimensional truss system using the direct stiffness method. The main objective of this chapter is to present the classical procedure common to the implementation of structural analysis. The direct stiffness method utilizes *matrices* and *matrix algebra* to organize and solve the governing system equations. Matrices, which are ordered arrays of numbers that are subjected to specific rules, can be used to assist the solution process in a compact and elegant manner. Of course, only a limited discussion of the direct stiffness method is given here, but we hope that the focused practical treatment will provide a strong basis for understanding the procedure to perform finite element analysis with SOLIDWORKS *Simulation*.

The later sections of this chapter demonstrate the procedure to create a solid model using SOLIDWORKS. The step-by-step tutorial introduces the SOLIDWORKS user interface and serves as a preview to some of the basic modeling techniques demonstrated in the later chapters.

One-dimensional Truss Element

The simplest type of engineering structure is the truss structure. A truss member is a slender (the length is much larger than the cross section dimensions) **two-force** member. Members are joined by pins and only have the capability to support tensile or compressive loads axially along the length. Consider a uniform slender prismatic bar (shown below) of length L, cross-sectional area A, and elastic modulus E. The ends of the bar are identified as nodes. The nodes are the points of attachment to other elements. The nodes are also the points for which displacements are calculated. The truss element is a two-force member element; forces are applied to the nodes only, and the displacements of all nodes are confined to the axes of elements.

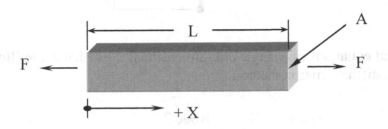

In this initial discussion of the truss element, we will consider the motion of the element to be restricted to the horizontal axis (one-dimensional). Forces are applied along the X axis and displacements of all nodes will be along the X axis.

For the analysis, we will establish the following sign conventions:

1. Forces and displacements are defined as positive when they are acting in the positive X direction as shown in the above figure.

2. The position of a node in the un-deformed condition is the finite element position for that node.

If equal and opposite forces of magnitude F are applied to the end nodes, from the elementary strength of materials, the member will undergo a change in length according to the equation:

$$\delta = \frac{FL}{EA}$$

This equation can also be written as $\delta = F/K$, which is similar to *Hooke's Law* used in a linear spring. In a linear spring, the symbol K is called the **spring constant** or **stiffness** of the spring. For a truss element, we can see that an equivalent spring element can be used to simplify the representation of the model, where the spring constant is calculated as **K=EA/L**.

Force-Displacement Curve of a Linear Spring

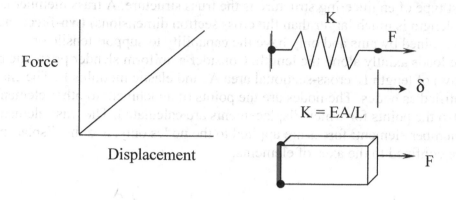

We will use the general equations of a single one-dimensional truss element to illustrate the formulation of the stiffness matrix method:

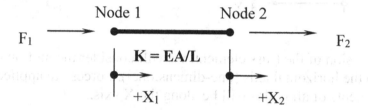

By using the *Relative Motion Analysis* method, we can derive the general expressions of the applied forces (F_1 and F_2) in terms of the displacements of the nodes (X_1 and X_2) and the stiffness constant (K).

1. Let $X_1 = 0$,

Node 1   Node 2

$F_1 \longrightarrow$ ●————————● $\longrightarrow F_2$

K = EA/L

$X_1 = 0$ $+X_2$

Based on Hooke's law and equilibrium equation:

$$\begin{cases} F_2 = K\,X_2 \\ F_1 = -\,F_2 = -\,K\,X_2 \end{cases}$$

2. Let $X_2 = 0$,

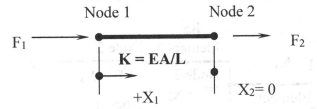

Based on *Hooke's Law* and equilibrium:

$$\begin{cases} F_1 = K\, X_1 \\ F_2 = -F_1 = -K\, X_1 \end{cases}$$

Using the *Method of Superposition,* the two sets of equations can be combined:

$$F_1 = K\, X_1 - K\, X_2$$
$$F_2 = -K\, X_1 + K\, X_2$$

The two equations can be put into matrix form as follows:

$$\left\{ \begin{matrix} F_1 \\ F_2 \end{matrix} \right\} = \begin{bmatrix} +K & -K \\ -K & +K \end{bmatrix} \left\{ \begin{matrix} X_1 \\ X_2 \end{matrix} \right\}$$

This is the general force-displacement relation for a two-force member element, and the equations can be applied to all members in an assemblage of elements. The following example illustrates a system with three elements.

Example 1.1:
Consider an assemblage of three of these two-force member elements. (Motion is restricted to one-dimension, along the X axis.)

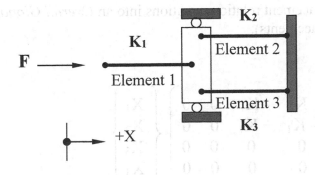

The assemblage consists of three elements and four nodes. The *Free Body Diagram* of the system with node numbers and element numbers labeled:

Consider now the application of the general force-displacement relation equations to the assemblage of the elements.

Element 1:

$$\begin{Bmatrix} F_1 \\ F_{21} \end{Bmatrix} = \begin{bmatrix} +K_1 & -K_1 \\ -K_1 & +K_1 \end{bmatrix} \begin{Bmatrix} X_1 \\ X_2 \end{Bmatrix}$$

Element 2:

$$\begin{Bmatrix} F_{22} \\ F_3 \end{Bmatrix} = \begin{bmatrix} +K_2 & -K_2 \\ -K_2 & +K_2 \end{bmatrix} \begin{Bmatrix} X_2 \\ X_3 \end{Bmatrix}$$

Element 3:

$$\begin{Bmatrix} F_{23} \\ F_4 \end{Bmatrix} = \begin{bmatrix} +K_3 & -K_3 \\ -K_3 & +K_3 \end{bmatrix} \begin{Bmatrix} X_2 \\ X_4 \end{Bmatrix}$$

Expanding the general force-displacement relation equations into an *Overall Global Matrix* (containing all nodal displacements):

Element 1:

$$\begin{Bmatrix} F_1 \\ F_{21} \\ 0 \\ 0 \end{Bmatrix} = \begin{bmatrix} +K_1 & -K_1 & 0 & 0 \\ -K_1 & +K_1 & 0 & 0 \\ 0 & 0 & 0 & 0 \\ 0 & 0 & 0 & 0 \end{bmatrix} \begin{Bmatrix} X_1 \\ X_2 \\ X_3 \\ X_4 \end{Bmatrix}$$

Element 2:

$$\begin{Bmatrix} 0 \\ F_{22} \\ F_3 \\ 0 \end{Bmatrix} = \begin{bmatrix} 0 & 0 & 0 & 0 \\ 0 & +K_2 & -K_2 & 0 \\ 0 & -K_2 & +K_2 & 0 \\ 0 & 0 & 0 & 0 \end{bmatrix} \begin{Bmatrix} X_1 \\ X_2 \\ X_3 \\ X_4 \end{Bmatrix}$$

Element 3:

$$\begin{Bmatrix} 0 \\ F_{23} \\ 0 \\ F_4 \end{Bmatrix} = \begin{bmatrix} 0 & 0 & 0 & 0 \\ 0 & +K_3 & 0 & -K_3 \\ 0 & 0 & 0 & 0 \\ 0 & -K_3 & 0 & +K_3 \end{bmatrix} \begin{Bmatrix} X_1 \\ X_2 \\ X_3 \\ X_4 \end{Bmatrix}$$

Summing the three sets of general equation: (Note $F_2 = F_{21} + F_{22} + F_{32}$)

$$\begin{Bmatrix} F_1 \\ F_2 \\ F_3 \\ F_4 \end{Bmatrix} = \begin{bmatrix} K_1 & -K_1 & 0 & 0 \\ -K_1 & (K_1+K_2+K_3) & -K_2 & -K_3 \\ 0 & -K_2 & K_2 & 0 \\ 0 & -K_3 & 0 & +K_3 \end{bmatrix} \begin{Bmatrix} X_1 \\ X_2 \\ X_3 \\ X_4 \end{Bmatrix}$$

Overall Global Stiffness Matrix

Once the *Overall Global Stiffness Matrix* is developed for the structure, the next step is to substitute boundary conditions and solve for the unknown displacements. At every node in the structure, either the externally applied load or the nodal displacement is needed as a boundary condition. We will demonstrate this procedure with the following example.

Example 1.2:

Given:

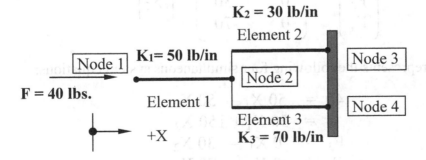

Find: Nodal displacements and reaction forces.

Solution:

From Example 1.1, the overall global force-displacement equation set:

$$\begin{Bmatrix} F_1 \\ F_2 \\ F_3 \\ F_4 \end{Bmatrix} = \begin{bmatrix} 50 & -50 & 0 & 0 \\ -50 & (50+30+70) & -30 & -70 \\ 0 & -30 & 30 & 0 \\ 0 & -70 & 0 & 70 \end{bmatrix} \begin{Bmatrix} X_1 \\ X_2 \\ X_3 \\ X_4 \end{Bmatrix}$$

Next, apply the known boundary conditions to the system: the right-ends of element 2 and element 3 are attached to the vertical wall; therefore, these two nodal displacements (X_3 and X_4) are zero.

$$\begin{Bmatrix} F_1 \\ F_2 \\ F_3 \\ F_4 \end{Bmatrix} = \begin{bmatrix} 50 & -50 & 0 & 0 \\ -50 & (50+30+70) & -30 & -70 \\ 0 & -30 & 30 & 0 \\ 0 & -70 & 0 & 70 \end{bmatrix} \begin{Bmatrix} X_1 \\ X_2 \\ 0 \\ 0 \end{Bmatrix}$$

The two displacements we need to solve the system are X_1 and X_2. Remove any unnecessary columns in the matrix:

$$\begin{Bmatrix} F_1 \\ F_2 \\ F_3 \\ F_4 \end{Bmatrix} = \begin{bmatrix} 50 & -50 \\ -50 & 150 \\ 0 & -30 \\ 0 & -70 \end{bmatrix} \begin{Bmatrix} X_1 \\ X_2 \end{Bmatrix}$$

Next, include the applied loads into the equations. The external load at *Node 1* is 40 lbs. and there is no external load at *Node 2*.

$$\begin{Bmatrix} 40 \\ 0 \\ F_3 \\ F_4 \end{Bmatrix} = \begin{bmatrix} 50 & -50 \\ -50 & 150 \\ 0 & -30 \\ 0 & -70 \end{bmatrix} \begin{Bmatrix} X_1 \\ X_2 \end{Bmatrix}$$

The Matrix represents the following four simultaneous system equations:

$$40 = 50\,X_1 - 50\,X_2$$
$$0 = -50\,X_1 + 150\,X_2$$
$$F_3 = 0\,X_1 - 30\,X_2$$
$$F_4 = 0\,X_1 - 70\,X_2$$

From the first two equations, we can solve for X_1 and X_2:

$$X_1 = 1.2 \text{ in.}$$
$$X_2 = 0.4 \text{ in.}$$

Substituting these known values into the last two equations, we can now solve for F_3 and F_4:

$$F_3 = 0 \, X_1 - 30 \, X_2 = -30 \times 0.4 = 12 \text{ lbs.}$$
$$F_4 = 0 \, X_1 - 70 \, X_2 = -70 \times 0.4 = 28 \text{ lbs.}$$

From the above analysis, we can now reconstruct the *Free Body Diagram* (*FBD*) of the system:

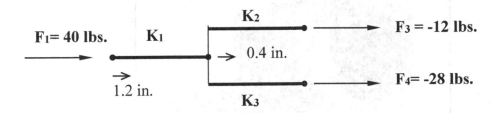

> ➢ The above sections illustrated the fundamental operation of the direct stiffness method, the classical finite element analysis procedure. As can be seen, the formulation of the global force-displacement relation equations is based on the general force-displacement equations of a single one-dimensional truss element. The two-force-member element (truss element) is the simplest type of element used in FEA. The procedure to formulate and solve the global force-displacement equations is straightforward but somewhat tedious. In real-life applications, the use of a truss element in one-dimensional space is rare and very limited. In the next chapter, we will expand the procedure to solve two-dimensional truss frameworks.

The following sections illustrate the procedure to create a solid model using SOLIDWORKS. The step-by-step tutorial introduces the basic SOLIDWORKS's user interface, and the tutorial serves as a preview to some of the basic modeling techniques demonstrated in the later chapters.

Basic Solid Modeling Using SOLIDWORKS

One of the methods to create solid models in SOLIDWORKS is to create a two-dimensional shape and then *extrude* the two-dimensional shape to define a volume in the third dimension. This is an effective way to construct three-dimensional solid models since many designs are in fact the same shape in one direction. This method also conforms to the design process that helps the designer with conceptual design along with the capability to capture the design intent. SOLIDWORKS provides many powerful modeling tools and there are many different approaches available to accomplish modeling tasks. We will start by introducing the basic two-dimensional sketching and parametric modeling tools.

The Adjuster Design

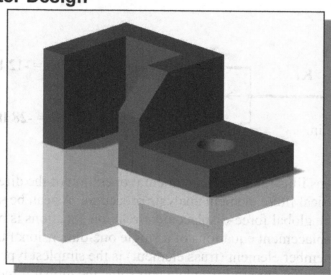

Starting SOLIDWORKS

How to start SOLIDWORKS depends on the type of workstation and the particular software configuration you are using. With most *Windows* and *Linux* systems, you may select **SOLIDWORKS** on the *Start* menu or select the **SOLIDWORKS** icon on the desktop. Consult your instructor or technical support personnel if you have difficulty starting the software.

1. Select the **SOLIDWORKS** option on the *Start* menu or select the **SOLIDWORKS** icon on the desktop to start SOLIDWORKS. The SOLIDWORKS main window will appear on the screen.

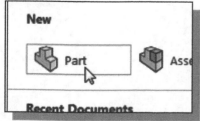

2. Select **Part** by clicking on the first icon in the *New SOLIDWORKS Document* dialog box as shown.

3. In the *Standard* toolbar area, **right-mouse-click** on any icon and activate **Command Manager** in the option list as shown.

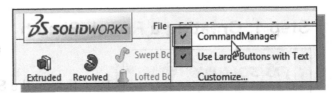

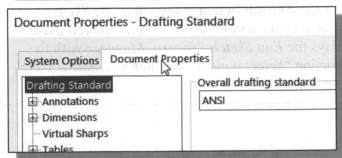

4. Select the **Options** icon from the *Menu* toolbar to open the *Options* dialog box.

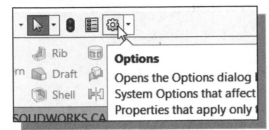

5. Select the **Document Properties** tab as shown in the figure.

6. Set the *Overall drafting standard* to **ANSI** to reset the default setting.

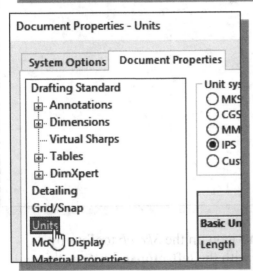

7. Click **Units** as shown in the figure.

8. Select **IPS (inch, pound, second)** under the *Unit system* options.

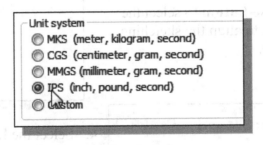

9. Click **OK** in the *Options* dialog box to accept the selected settings.

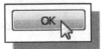

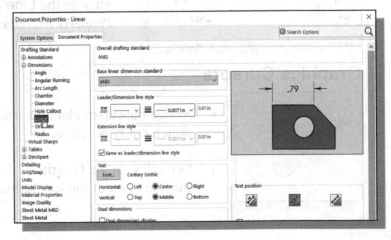

Step 1: Creating a Rough Sketch

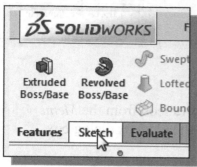

1. Click on the **Sketch** tab in the *Command Manager* to switch to the *Sketch* toolbar.

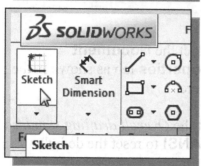

2. Select the **Sketch** button at the top of the *Sketch* toolbar to create a new sketch. Notice the left panel displays the *Edit Sketch Property Manager* with the instruction "*Select a plane on which to create a sketch for the entity.*"

3. Move the cursor over the edge of the **Front Plane** in the graphics area. When the **Front Plane** is highlighted, click once with the **left-mouse-button** to select the plane to align the sketching plane.

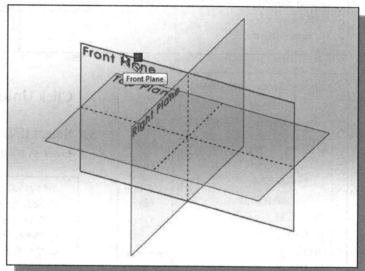

4. Select the **Line** icon on the *Sketch* toolbar by clicking once with the **left-mouse-button**; this will activate the Line command. The *Line Properties Property Manager* is displayed in the left panel.

Graphics Cursors

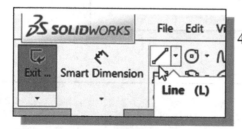

- Notice the cursor changes from an arrow to a pencil when a sketch entity is active.

1. Left click on the **Origin** of the coordinate system to place the starting point of the line segments.

2. As you move the graphics cursor, you will see a digital readout next to the cursor. This readout gives you the line length. In the *Status Bar* area at the bottom of the window, the readout gives you the cursor location. Move the cursor around and you will notice different symbols appear at different locations.

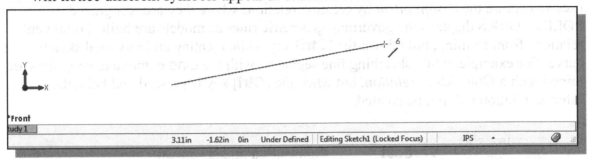

3. Move the graphics cursor toward the right side of the graphics window to create a horizontal line as shown below. Notice the geometric relation symbol displayed. When the **Horizontal** relation symbol is displayed, left-click to select **Point 2**.

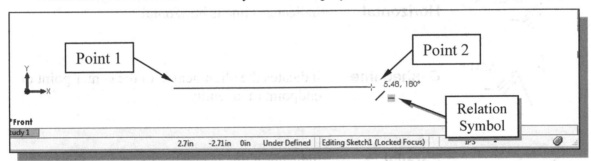

4. Complete the sketch as shown below, creating a closed region ending at the starting point (**Point 1**). Do not be overly concerned with the actual size of the sketch. Note that all line segments are sketched horizontally or vertically.

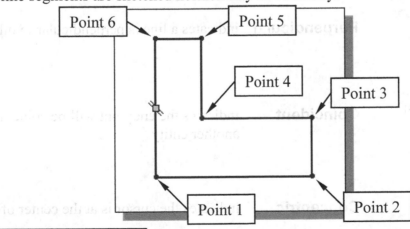

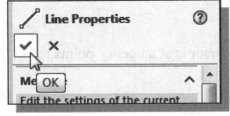

5. Click the **OK** icon (green check mark) in the *Property Manager* to end editing of the current line, then click the **OK** icon again, or hit the [**Esc**] key once, to end the **Sketch Line** command.

Geometric Relation Symbols

During sketching, SOLIDWORKS automatically displays different visual clues, or symbols, to show you alignments, perpendicularities, tangencies, etc. These relations are used to capture the *design intent* by creating relations where they are recognized. SOLIDWORKS displays the governing geometric rules as models are built. To prevent relations from forming, hold down the [**Ctrl**] key while creating an individual sketch curve. For example, while sketching line segments with the Line command, endpoints are joined with a *Coincident relation*, but when the [**Ctrl**] key is pressed and held, the inferred relation will not be created.

1.182, 90°	**Vertical**	indicates a line is vertical
1.863, 180°	**Horizontal**	indicates a line is horizontal
0.641	**Dashed line**	indicates the alignment is to the center point or endpoint of an entity
51.83	**Parallel**	indicates a line is parallel to other entities
0.338	**Perpendicular**	indicates a line is perpendicular to other entities
5.151	**Coincident**	indicates the endpoint will be coincident with another entity
	Concentric	indicates the cursor is at the center of an entity
0.689	**Tangent**	indicates the cursor is at tangency points to curves

Step 2: Apply/Modify Relations and Dimensions

As the sketch is made, SOLIDWORKS automatically applies some of the geometric constraints (such as horizontal, parallel, and perpendicular) to the sketched geometry. We can continue to modify the geometry, apply additional constraints, and/or define the size of the existing geometry. In this example, we will illustrate adding dimensions to describe the sketched entities.

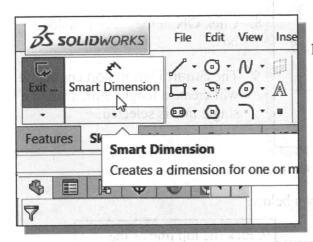

1. Move the cursor to the second icon of the *Sketch* toolbar; it is the **Smart Dimension** icon. Activate the command by left clicking once on the icon.

2. Select the bottom horizontal line by left clicking once on the line.

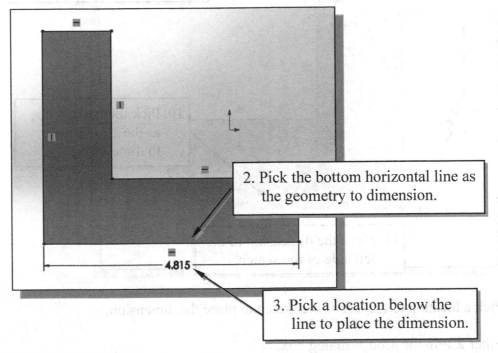

> 2. Pick the bottom horizontal line as the geometry to dimension.

4.815

> 3. Pick a location below the line to place the dimension.

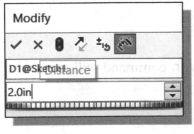

3. Pick a location below the line to place the dimension. Enter **2.0** in the *Modify* dialog box.

4. Left-click **OK** (green check mark) in the *Modify* dialog box to save the current value and exit the dialog. Notice the associated geometry is adjusted.

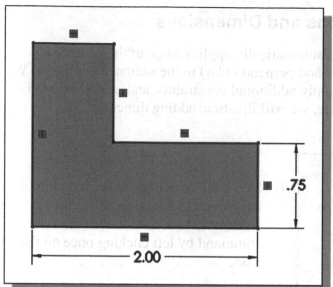

5. Select the lower right-vertical line.

6. Pick a location toward the right of the sketch to place the dimension.

7. Enter **0.75** in the *Modify* dialog box.

8. Click **OK** in the *Modify* dialog box.

❖ The **Smart Dimension** command will create a length dimension if a single line is selected.

9. Select the top-horizontal line as shown below.

10. Select the bottom-horizontal line as shown below.

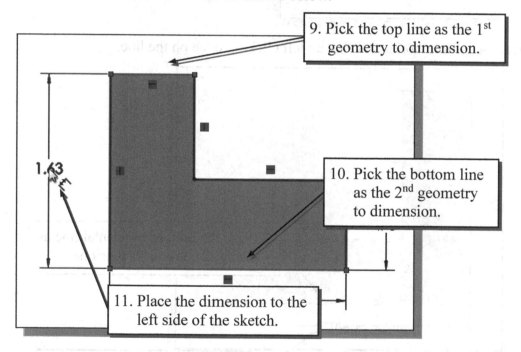

9. Pick the top line as the 1st geometry to dimension.

10. Pick the bottom line as the 2nd geometry to dimension.

11. Place the dimension to the left side of the sketch.

11. Pick a location to the left of the sketch to place the dimension.

12. Enter **2.0** in the *Modify* dialog box.

13. Click **OK** in the *Modify* dialog box.

❖ When two parallel lines are selected, the **Smart Dimension** command will create a dimension measuring the distance between them.

14. On your own, repeat the above steps and create an additional dimension for the top line. Make the dimension **0.75**.

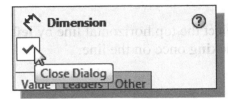

15. Click the **OK** icon in the *PropertyManager* as shown, or hit the [**Esc**] key once, to end the Smart Dimension command.

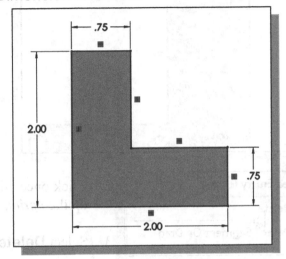

Viewing Functions – Zoom and Pan

SOLIDWORKS provides a special user interface that enables convenient viewing of the entities in the graphics window. There are many ways to perform the **Zoom** and **Pan** operations.

1. Hold the [**Ctrl**] function key down. While holding the [**Ctrl**] function key down, press the mouse wheel down and drag the mouse to **Pan** the display. This allows you to reposition the display while maintaining the same scale factor of the display.

2. Hold the [**Shift**] function key down. While holding the [**Shift**] function key down, press the mouse wheel down and drag the mouse to **Zoom** the display. Moving downward will reduce the scale of the display, making the entities display smaller on the screen. Moving upward will magnify the scale of the display.

3. Turning the mouse wheel can also adjust the scale of the display. Turn the mouse wheel forward. Notice the scale of the display is reduced, making the entities display smaller on the screen.

4. Turn the mouse wheel backward. Notice scale of the display is magnified. (**NOTE:** Turning the mouse wheel allows zooming to the position of the cursor.)

5. On your own, use the options above to change the scale and position of the display.

6. Press the **F** key on the keyboard to automatically fit the model to the screen.

Delete an Existing Geometry of the Sketch

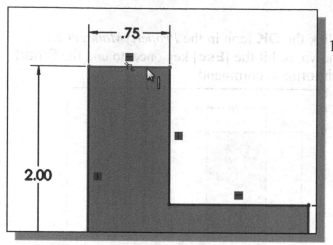

1. Select the top horizontal line by **left clicking** once on the line.

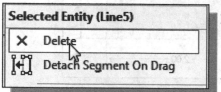

2. Click once with the **right-mouse-button** to bring up the *option menu*.

3. Select **Delete** from the option list as shown.

4. Click **Yes** to proceed with deleting the selected line.

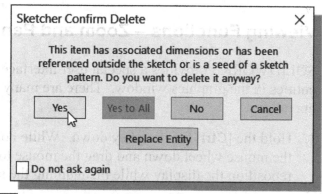

❖ Note that the height dimensions and the horizontal constraint attached to the geometry are also deleted.

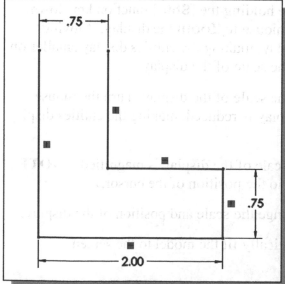

5. Activate the **Undo** command by left clicking once on the icon in the standard toolbar as shown.

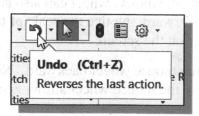

Modify the Dimensions of the Sketch

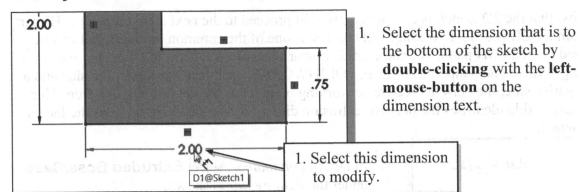

1. Select the dimension that is to the bottom of the sketch by **double-clicking** with the **left-mouse-button** on the dimension text.

1. Select this dimension to modify.

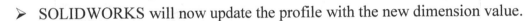

2. In the *Modify* window, the current length of the line is displayed. Enter **2.5** to reset the length of the line.

3. Click on the **OK** icon to accept the entered value.

➢ SOLIDWORKS will now update the profile with the new dimension value.

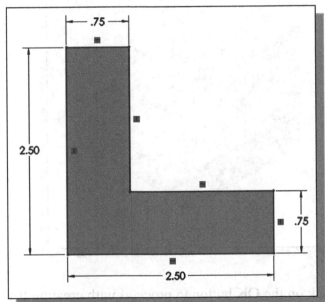

4. On your own, repeat the above steps and adjust the left vertical dimension to **2.5** so that the sketch appears as shown.

5. Press the [**Esc**] key once to exit the **Dimension** command.

6. Click once with the **left-mouse-button** on the **Exit Sketch** icon on the *Sketch* toolbar to exit the sketch.

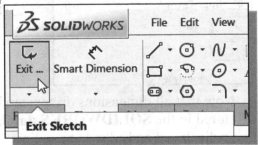

Step 3: Complete the Base Solid Feature

Now that the 2D sketch is completed, we will proceed to the next step: creating a 3D part from the 2D profile. Extruding a 2D profile is one of the common methods that can be used to create 3D parts. We can extrude planar faces along a path. We can also specify a height value and a tapered angle. In SOLIDWORKS, each face has a positive side and a negative side; the current face we're working on is set as the default positive side. This positive side identifies the positive extrusion direction, and it is referred to as the face's *normal*.

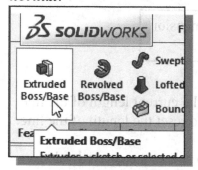

1. In the *Ribbon toolbar*, select **Extruded Boss/Base** under the *Feature tab* as shown.

2. In the *Extrude Property Manager*, enter **2.5** as the extrusion distance. Notice that the completed sketch region is automatically selected as the extrusion profile.

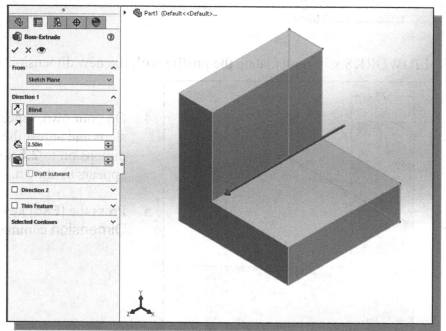

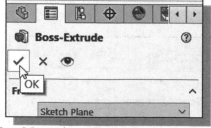

3. Click on the **OK** button to proceed with creating the 3D part.

> Note that all dimensions disappeared from the screen. All parametric definitions are stored in the **SOLIDWORKS database** and any of the parametric definitions can be re-displayed and edited at any time.

Isometric View

SOLIDWORKS provides many ways to display views of the three-dimensional design. We will first orient the model to display in the *isometric view*, by using the *View Orientation* pull-down menu on the *Heads-up View* toolbar.

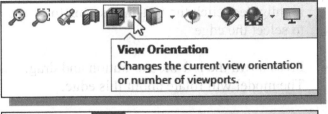

1. Select the **View Orientation** button on the *Heads-up View* toolbar by clicking once with the left-mouse-button.

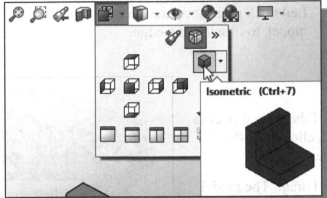

2. Select the **Isometric** icon in the *View Orientation* pull-down menu.

❖ Note that many other view-related commands are also available under the **View** *pull-down menu.*

Rotation of the 3D Model – Rotate View

The Rotate View command allows us to rotate a part or assembly in the graphics window. Rotation can be around the center mark, free in all directions, or around a selected entity (vertex, edge, or face) on the model.

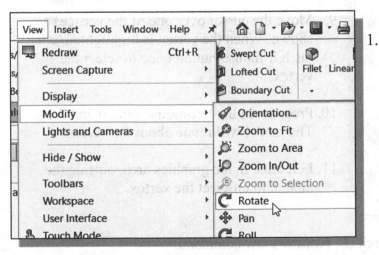

1. Move the cursor over the **SOLIDWORKS** logo to display the pull-down menus. Select **View → Modify → Rotate** from the pull-down menu as shown.

2. Move the cursor inside the graphics area. Press down the left-mouse-button and drag in an arbitrary direction; the Rotate View command allows us to freely rotate the solid model.

- The model will rotate about an axis normal to the direction of cursor movement. For example, drag the cursor horizontally across the screen and the model will rotate about a vertical axis.

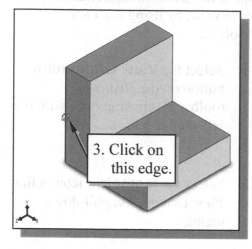

3. Click on this edge.

3. Move the cursor over one of the vertical edges of the solid model as shown. When the edge is highlighted, click the **left-mouse-button** once to select the edge.

4. Press down the left-mouse-button and drag. The model will rotate about this edge.

5. Left click in the graphics area, outside the model, to unselect the edge.

6. Move the cursor over the front face of the solid model as shown. When the face is highlighted, click the **left-mouse-button** once to select the face.

7. Press down the left-mouse-button and drag. The model will rotate about the direction normal to this face.

8. Left click in the graphics area, outside the model, to unselect the face.

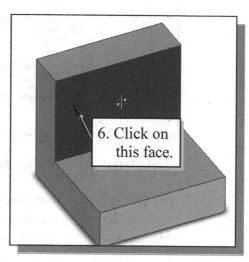

6. Click on this face.

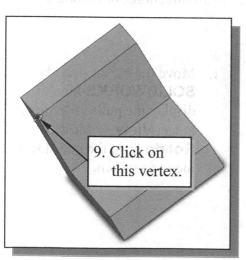

9. Click on this vertex.

9. Move the cursor over one of the vertices as shown. When the vertex is highlighted, click the left-mouse-button once to select the vertex.

10. Press down the left-mouse-button and drag. The model will rotate about the vertex.

11. Left click in the graphics area, outside the model, to unselect the vertex.

12. Press the [**Esc**] key once to exit the **Rotate View** command.

13. On your own, reset the display to the *isometric* view.

Rotation and Panning – Arrow Keys

SOLIDWORKS allows us to easily rotate a part or assembly in the graphics window using the **arrow** keys on the keyboard.

- Use the **arrow** keys to rotate the view horizontally or vertically. The **left-right** keys rotate the model about a vertical axis. The **up-down** keys rotate the model about a horizontal axis.

- Hold down the [**Alt**] key and use the **left-right arrow** keys to rotate the model about an axis normal to the screen, i.e., to rotate clockwise and counter-clockwise.

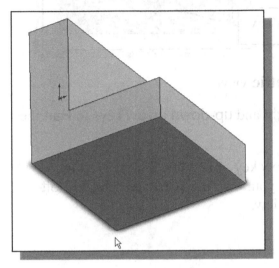

1. Hit the **left arrow** key. The model view rotates by a pre-determined increment. The default increment is 15°. (This increment can be set in the *Options* dialog box.) On your own, use the **left-right** and **up-down arrow** keys to rotate the view.

2. Hold down the [**Alt**] key and hit the **left arrow** key. The model view rotates in the clockwise direction. On your own use the **left-right** and **up-down arrow** keys, and the [**Alt**] key plus the **left-right arrow** keys, to rotate the view.

3. On your own, reset the display to the *isometric* view.

- Hold down the [**Shift**] key and use the **left-right** and **up-down arrow** keys to rotate the model in 90° increments.

4. Hold down the [**Shift**] key and hit the **right arrow** key. The view will rotate by 90°. On your own use the [**Shift**] key plus the **left-right arrow** keys to rotate the view.

5. Select the **Front** icon in the *View Orientation* pull-down menu as shown to display the **Front** view of the model.

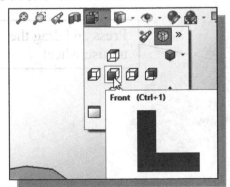

6. Hold down the [**Shift**] key and hit the **left arrow** key. The view rotates to the **Right** side view.

7. Hold down the [**Shift**] key and hit the **down arrow** key. The view rotates to the **Top** view.

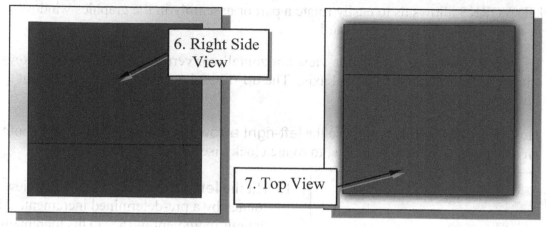

6. Right Side View

7. Top View

8. On your own, reset the display to the **Isometric** view.

• Hold down the [**Ctrl**] key and use the **left-right** and **up-down arrow** keys to **Pan** the model in increments.

9. Hold down the [**Ctrl**] key and hit the **left arrow** key. The view **Pans**, moving the model toward the left side of the screen. On your own use [**Ctrl**] key plus the **left-right** and **up-down arrow** keys to **Pan** the view.

Dynamic Viewing – Quick Keys

We can also use the function keys on the keyboard and the mouse to access the *Dynamic Viewing* functions.

❖ **Panning**

(1) Hold the Ctrl key; press and drag the mouse wheel

Hold the [**Ctrl**] function key down, and press and drag with the mouse wheel to **Pan** the display. This allows you to reposition the display while maintaining the same scale factor of the display.

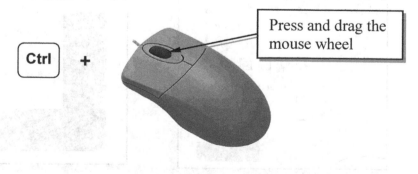

Ctrl   +

Press and drag the mouse wheel

(2) Hold the Ctrl key; use arrow keys

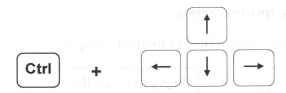

❖ Zooming

(1) Hold the Shift key; press and drag the mouse wheel

Hold the [**Shift**] function key down, and press and drag with the mouse wheel to **Zoom** the display. Moving downward will reduce the scale of the display, making the entities display smaller on the screen. Moving upward will magnify the scale of the display.

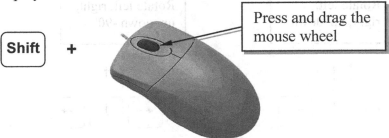

Press and drag the mouse wheel

(2) Turning the mouse wheel

Turning the mouse wheel can also adjust the scale of the display. Turning forward will reduce the scale of the display, making the entities display smaller on the screen. Turning backward will magnify the scale of the display.

- Turning the mouse wheel allows zooming to the position of the cursor.

- The cursor position, inside the graphics area, is used to determine the center of the scale of the adjustment.

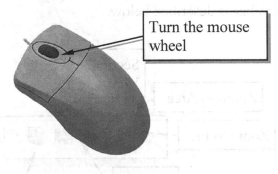

Turn the mouse wheel

(3) Z key or Shift + Z key

Pressing the [**Z**] key on the keyboard will zoom out. Holding the [**Shift**] function key and pressing the [**Z**] key will zoom in.

3D Rotation

(1) Press and drag the mouse wheel

Press and drag with the mouse wheel to rotate the display.

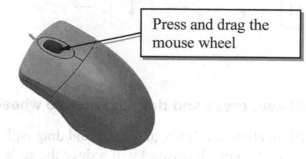

Press and drag the mouse wheel

(2) Use arrow keys

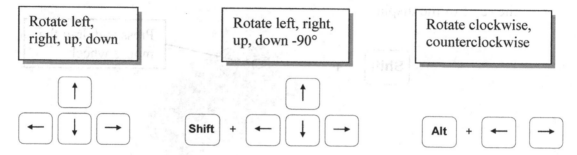

| Rotate left, right, up, down | Rotate left, right, up, down -90° | Rotate clockwise, counterclockwise |

Viewing Tools – Heads-up View Toolbar

The *Heads-up View* toolbar is a transparent toolbar which appears in each viewport and provides easy access to commonly used tools for manipulating the view. The default toolbar is described below.

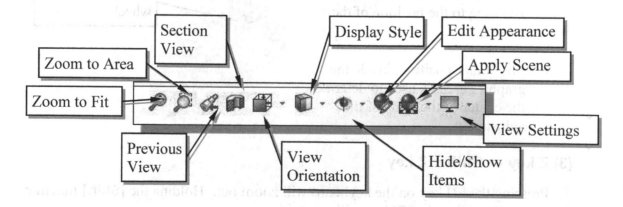

Section View — Display Style — Edit Appearance — Apply Scene — Zoom to Area — Zoom to Fit — Previous View — View Orientation — Hide/Show Items — View Settings

Zoom to Fit – Adjusts the view so that all items on the screen fit inside the graphics window.

Zoom to Area – Use the cursor to define a region for the view; the defined region is zoomed to fill the graphics window.

Previous View – Returns to the previous view.

Section View – Displays a cutaway of a part or assembly using one or more section planes.

View Orientation – This allows you to change the current view orientation or number of viewports.

Display Style – This can be used to change the display style (shaded, wireframe, etc.) for the active view.

Hide/Show Items – This pull-down menu is used to control the visibility of items (axes, sketches, relations, etc.) in the graphics area.

Edit Appearance – Modifies the appearance of entities in the model.

Apply Scene – Cycles through or applies a specific scene.

View Settings – Allows you to toggle various view settings (e.g., shadows, perspective).

View Orientation

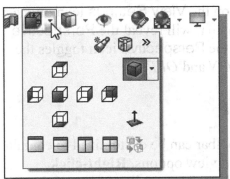

1. Click on the **View Orientation** icon on the *Heads-up View* toolbar to reveal the view orientation and number of viewports options.

❖ Standard view orientation options: **Front**, **Back**, **Left**, **Right**, **Top**, **Bottom**, **Isometric**, **Trimetric** or **Dimetric** icons can be selected to display the corresponding standard view.

Normal to – In a part or assembly, zooms and rotates the model to display the selected plane or face. You can select the element either before or after clicking the Normal to icon.

❖ The icons across the bottom of the pull-down menu allow you to display a single viewport (the default) or multiple viewports.

Display Style

1. Click on the **Display Style** icon on the *Heads-up View* toolbar to reveal the display style options.

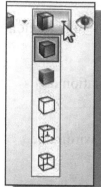

Shaded with Edges – Allows the display of a shaded view of a 3D model with its edges.

Shaded – Allows the display of a shaded view of a 3D model.

Hidden Lines Removed – Allows the display of the 3D objects using the basic wireframe representation scheme. Only those edges which are visible in the current view are displayed.

Hidden Lines Visible – Allows the display of the 3D objects using the basic wireframe representation scheme in which all the edges of the model are displayed, but edges that are hidden in the current view are displayed as dashed lines (or in a different color).

Wireframe – Allows the display of 3D objects using the basic wireframe representation scheme in which all the edges of the model are displayed.

Orthographic vs. Perspective

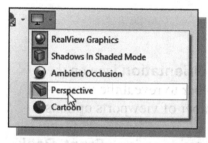

1. Besides the basic display modes, we can also choose an orthographic view or perspective view of the display. Clicking on the **View Settings** icon on the *Heads-up View* toolbar will reveal the **Perspective** icon. Clicking on the **Perspective** icon toggles the perspective view *ON* and *OFF*.

Customizing the Heads-up View Toolbar

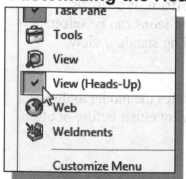

1. The *Heads-up View* toolbar can be customized to contain icons for user-preferred view options. **Right-click** anywhere on the *Heads-up View* toolbar to reveal the display menu option list. Click on the **View (Heads-up)** icon to turn *OFF* the display of the toolbar. Notice the **Customize** option is also available to add/remove different icons.

➤ On your own, use the different options described in the above sections to familiarize yourself with the 3D viewing/display commands. Reset the display to the standard **Isometric view** before continuing to the next section.

Step 4-1: Adding an Extruded Boss Feature

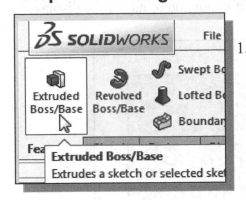

1. In the *Features* toolbar select the **Extruded Boss/ Base** command by clicking once with the left-mouse-button on the icon.

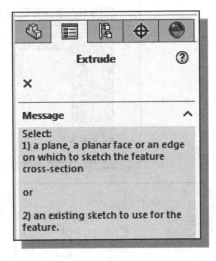

- In the *Extrude Manager* area, SOLIDWORKS indicates the two options to create the new extrusion feature. We will select the back surface of the base feature to align the sketching plane.

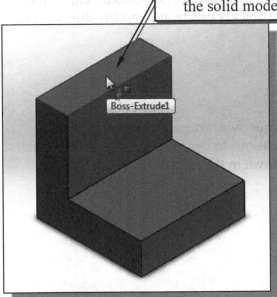

3. Pick the top face of the solid model.

2. On your own, use one of the rotation quick keys/mouse-button and view the back face of the model as shown.

3. Pick the top face of the 3D solid object.

- Note that SOLIDWORKS automatically establishes a *User Coordinate System* (UCS) and records its location with respect to the part on which it was created.

4. Select the **Line** command by clicking once with the **left-mouse-button** on the icon in the *Sketch* toolbar.

➢ To illustrate the usage of dimensions in parametric sketches, we will intentionally create a sketch away from the desired location.

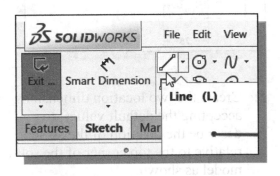

5. Create a sketch with segments perpendicular/parallel to the existing edges of the solid model as shown below. (Hint: Use the **Normal To** command to adjust the display normal to the sketching plane direction.)

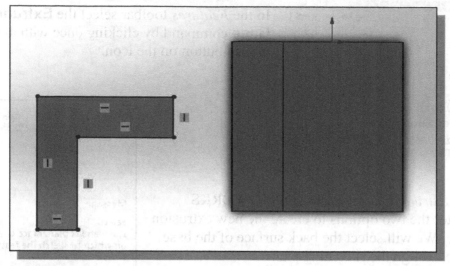

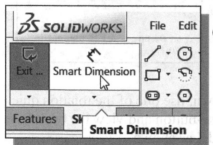

6. Select the **Smart Dimension** command in the *Sketch* toolbar. The Smart Dimension command allows us to quickly create and modify dimensions.

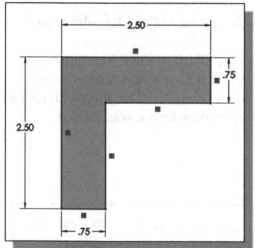

7. On your own, create and modify the size dimensions to describe the size of the sketch as shown in the figure.

8. Create the two location dimensions, accepting the default values, to describe the position of the sketch relative to the top corner of the solid model as shown.

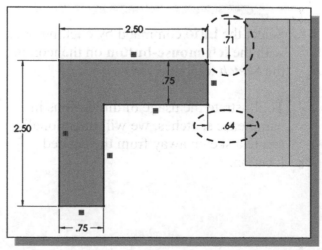

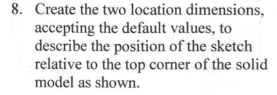

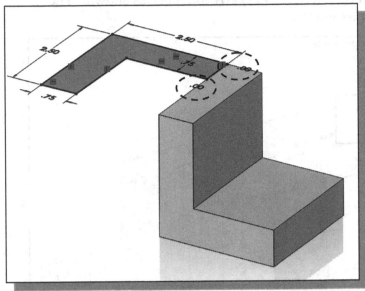

9. On your own, modify the two location dimensions to **0.0** and **0.0** as shown in the figure.

➤ In parametric modeling, the dimensions can be used to quickly control the size and location of the defined geometry.

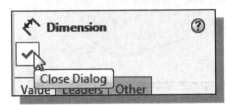

10. Select **Close Dialog** in the *Dimension Property Manager* to end the **Smart Dimension** command.

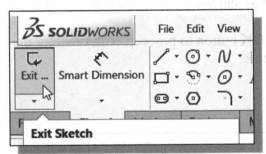

11. Click **Exit Sketch** in the *Sketch* toolbar to end the **Sketch** mode.

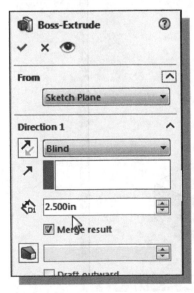

12. In the *Boss-Extrude Property Manager*, enter **2.5** as the extrude *Distance* as shown.

13. Click the **Reverse Direction** button in the *Property Manager* as shown. The extrude preview should appear as shown.

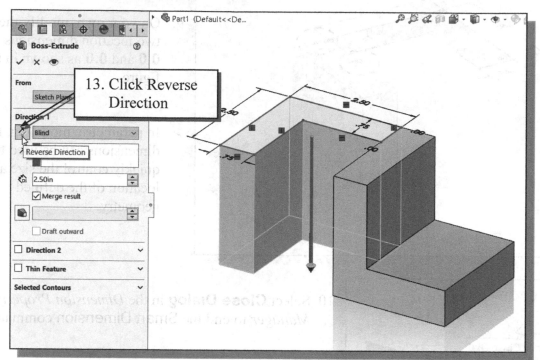

14. Confirm the **Merge result** option is activated as shown.

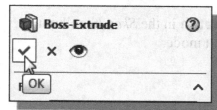

15. Click on the **OK** button to proceed with creating the extruded feature.

Step 4-2: Adding an Extruded Cut Feature

Next, we will create a cut feature that will be added to the existing solid object.

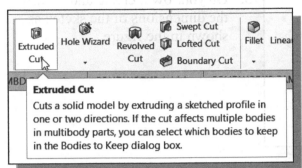

1. In the *Features* toolbar select the **Extruded Cut** command by clicking once with the left-mouse-button on the icon.

2. Pick the vertical face of the last feature we created, as shown.

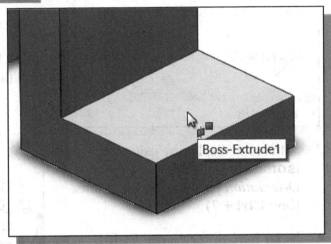

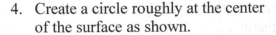

3. Select the **Circle** command by clicking once with the **left-mouse-button** on the icon in the *Sketch* toolbar.

4. Create a circle roughly at the center of the surface as shown.

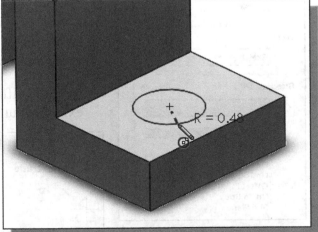

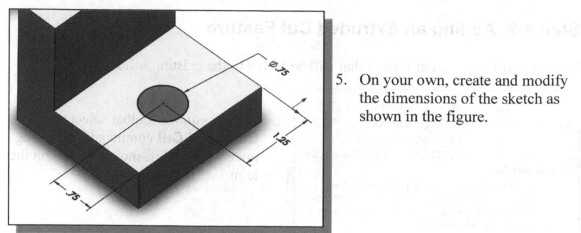

5. On your own, create and modify the dimensions of the sketch as shown in the figure.

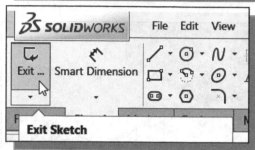

6. Click **Exit Sketch** in the *Sketch* toolbar to end the Sketch mode.

7. Adjust the display by using the **Isometric** option in the *View Orientation* panel as shown. (Quick Key: **Ctrl + 7**)

• Next, we will create and profile another sketch, a rectangle, which will be used to create another extrusion feature that will be added to the existing solid object.

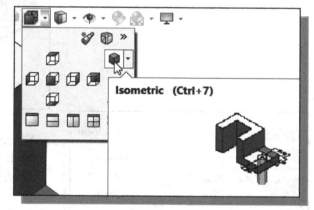

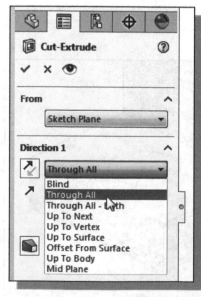

8. Set the *Extents* option to **Through All** as shown. The *Through All* option instructs the software to calculate the extrusion distance and assures the created feature will always cut through the full length of the model.

• Note the different options available to create the extruded cut feature.

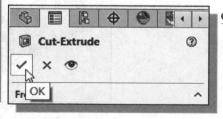

9. Click on the **OK** button to proceed with creating the extruded feature.

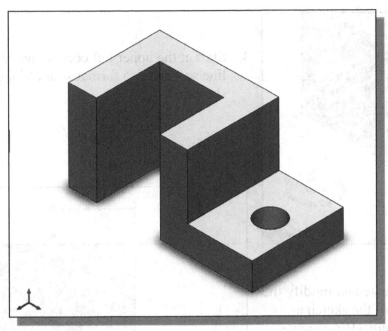

Step 4-3: Adding another Cut Feature

Next, we will create and profile a triangle, which will be used to create a cut feature that will be added to the existing solid object.

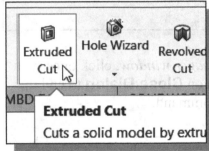

1. In the *Features* toolbar select the **Extruded Cut** command by clicking once with the left-mouse-button on the icon.

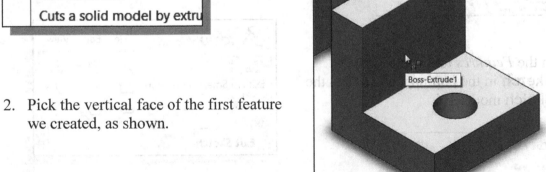

2. Pick the vertical face of the first feature we created, as shown.

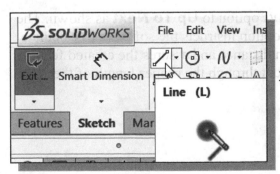

3. Select the **Line** command by clicking once with the **left-mouse-button** on the icon in the *Sketch* toolbar.

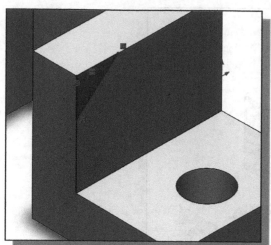

4. Start at the upper left corner and create three line segments to form a small triangle as shown.

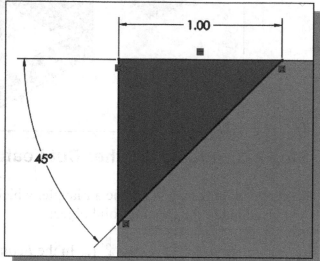

5. On your own, create and modify the two dimensions of the sketch as shown in the figure. (Hint: create the angle dimension by selecting the two adjacent lines as shown and place the angular dimension inside the desired quadrant.)

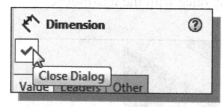

6. In the *Dimension manager window*, click once with the left-mouse-button on **Close Dialog** to end the Smart Dimension command.

7. In the *Features* toolbar, select **Exit Sketch** in the pop-up menu to end the Sketch mode.

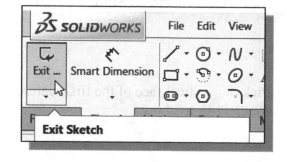

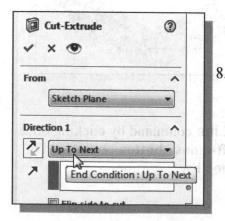

8. Set the *Extents* option to **Up To Next** as shown. The *Up To Next* option instructs the software to calculate the extrusion distance and assures the created feature will always cut through the proper length of the model.

9. Click on the **OK** button to proceed with creating the extruded feature.

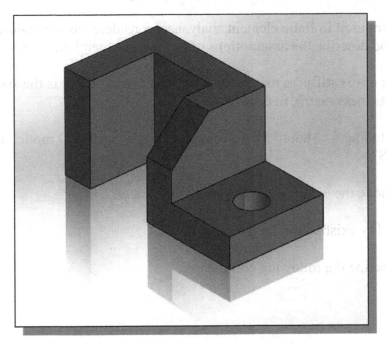

Save the Model

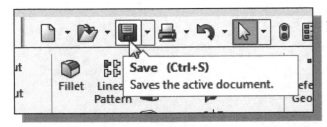

1. Select **Save** in the *Quick Access* toolbar, or you can also use the "**Ctrl-S**" combination (hold down the "Ctrl" key and hit the "S" key once) to save the part.

2. In the pop-up window, select the directory to store the model in and enter **Adjuster** as the name of the file.

3. Click on the **Save** button to save the file.

❖   You should form a habit of saving your work periodically, just in case something might go wrong while you are working on it. In general, one should save one's work at an interval of every 15 to 20 minutes. One should also save before making any major modifications to the model.

Questions:

1. The truss element used in finite element analysis is considered as a two-force member element. List and describe the assumptions of a two-force member.

2. What is the size of the stiffness matrix for a single element? What is the size of the overall global stiffness matrix in example 1.2?

3. What is the first thing we should set up when building a new CAD model in SOLIDWORKS?

4. How do we remove the dimensions created by the *system*?

5. How do we modify existing dimensions?

6. Identify and describe the following commands:

    (a)

    (b)

    (c)

Exercises:

1. For the one-dimensional 3 truss-element system shown, determine the nodal displacements and reaction forces using the direct stiffness method.

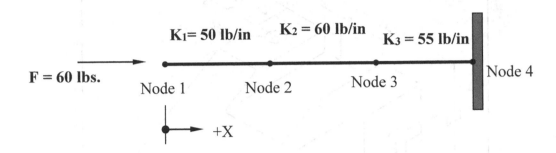

2. For the one-dimensional 4 truss-element system shown, determine the nodal displacements and reaction forces using the direct stiffness method.

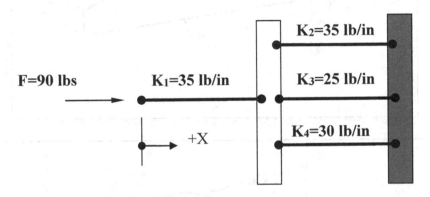

3. For the one-dimensional 4 truss-element system shown, determine the nodal displacements and reaction forces using the direct stiffness method.

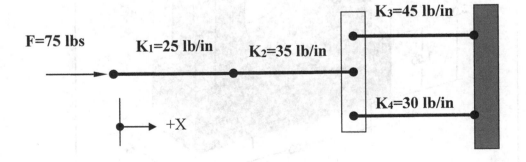

4. Using SOLIDWORKS, create the 3D solid models as shown in the figures;
    (a) Dimensions are in mm. (b) Dimensions are in inches.

a)

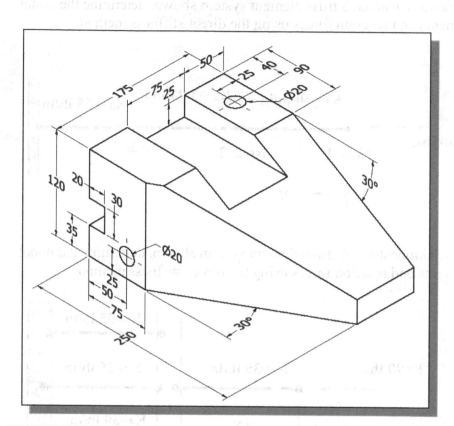

b)

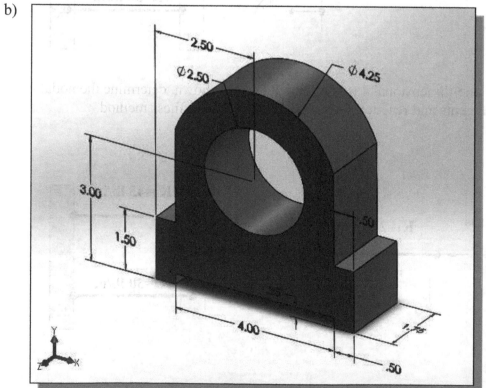

Chapter 2
Truss Elements in Two-Dimensional Spaces

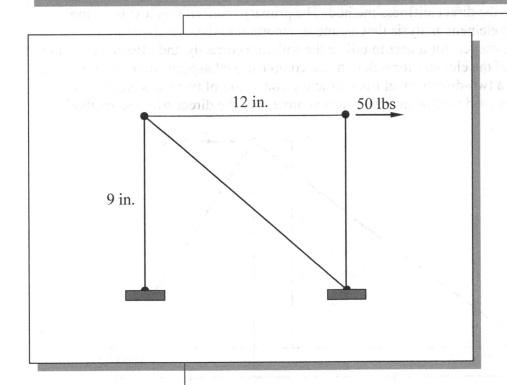

12 in.

50 lbs

9 in.

Learning Objectives

♦ **Perform 2D Coordinate Transformation.**
♦ **Expand the Direct Stiffness Method to 2D Trusses.**
♦ **Derive the general 2D element Stiffness Matrix.**
♦ **Assemble the Global Stiffness Matrix for 2D Trusses.**
♦ **Solve 2D trusses using the Direct Stiffness Method.**

Introduction

This chapter presents the formulation of the direct stiffness method of truss elements in a two-dimensional space and the general procedure for solving two-dimensional truss structures using the direct stiffness method. The primary focus of this text is on the aspects of finite element analysis that are more important to the user than the programmer. However, for a user to utilize the software correctly and effectively, some understanding of the element formulation and computational aspects are also important. In this chapter, a two-dimensional truss structure consisting of two truss elements (as shown below) is used to illustrate the solution process of the direct stiffness method.

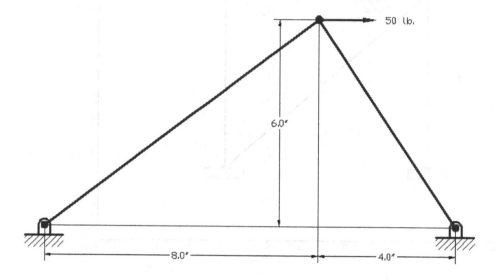

Truss Elements in Two-Dimensional Spaces

As introduced in the previous chapter, the system equations (stiffness matrix) of a truss element can be represented using the system equations of a linear spring in one-dimensional space.

$$K = EA/L$$

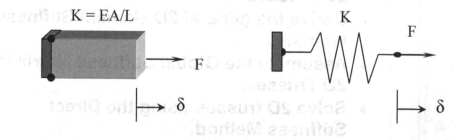

Free Body Diagram:

The general force-displacement equations in matrix form:

$$\left\{ \begin{matrix} F_1 \\ F_2 \end{matrix} \right\} = \left[\begin{matrix} +K & -K \\ -K & +K \end{matrix} \right] \left\{ \begin{matrix} X_1 \\ X_2 \end{matrix} \right\}$$

For a truss element, $\mathbf{K = EA/L}$:

$$\left\{ \begin{matrix} F_1 \\ F_2 \end{matrix} \right\} = \frac{\mathbf{EA}}{\mathbf{L}} \left[\begin{matrix} +1 & -1 \\ -1 & +1 \end{matrix} \right] \left\{ \begin{matrix} X_1 \\ X_2 \end{matrix} \right\}$$

For truss members positioned in two-dimensional space, two coordinate systems are established:

1. The global coordinate system (**X** and **Y** axes) chosen to represent the entire structure.
2. The local coordinate system (**X** and **Y** axes) selected to align the **X** axis along the length of the element.

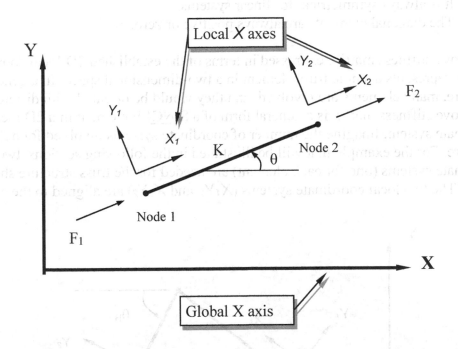

The force-displacement equations expressed in terms of components in the local XY coordinate system:

$$\left\{ \begin{matrix} F_{1X} \\ F_{2X} \end{matrix} \right\} = \frac{\mathbf{EA}}{\mathbf{L}} \left[\begin{matrix} +1 & -1 \\ -1 & +1 \end{matrix} \right] \left\{ \begin{matrix} X_1 \\ X_2 \end{matrix} \right\}$$

The above stiffness matrix (system equations in matrix form) can be expanded to incorporate the two force components at each node and the two displacement components at each node.

Force Components
(Local Coordinate System)

$$\begin{Bmatrix} F_{1X} \\ F_{1Y} \\ F_{2X} \\ F_{2Y} \end{Bmatrix} = \frac{EA}{L} \begin{bmatrix} +1 & 0 & -1 & 0 \\ 0 & 0 & 0 & 0 \\ -1 & 0 & +1 & 0 \\ 0 & 0 & 0 & 0 \end{bmatrix} \begin{Bmatrix} X_1 \\ Y_1 \\ X_2 \\ Y_2 \end{Bmatrix}$$

Nodal Displacements
(Local Coordinate System)

In regard to the expanded local stiffness matrix (system equations in matrix form):

1. It is always a square matrix.
2. It is always symmetrical for linear systems.
3. The diagonal elements are always positive or zero.

The above stiffness matrix, expressed in terms of the established 2D local coordinate system, represents a single truss element in a two-dimensional space. In a general structure, many elements are involved, and they would be oriented with different angles. The above stiffness matrix is a general form of a <u>SINGLE</u> element in a 2D local coordinate system. Imagine the number of coordinate systems involved for a 20-member structure. For the example that will be illustrated in the following sections, two local coordinate systems (one for each element) are needed for the truss structure shown below. The two local coordinate systems (X_1Y_1 and X_2Y_2) are aligned to the elements.

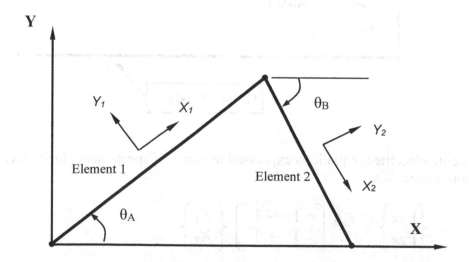

In order to solve the system equations of two-dimensional truss structures, it is necessary to assemble all elements' stiffness matrices into a **global stiffness matrix**, with all the equations of the individual elements referring to a common global coordinate system. This requires the use of *coordinate transformation equations* applied to system equations for all elements in the structure. For a one-dimensional truss structure (illustrated in chapter 2), the local coordinate system coincides with the global coordinate system; therefore, no coordinate transformation is needed to assemble the global stiffness matrix (the stiffness matrix in the global coordinate system). In the next section, the coordinate transformation equations are derived for truss elements in two-dimensional spaces.

Coordinate Transformation

A vector, in a two-dimensional space, can be expressed in terms of any coordinate system set of unit vectors.

For example,

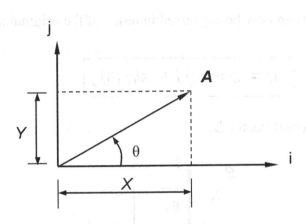

Vector **A** can be expressed as:

$$A = X\,i + Y\,j$$

Where *i* and *j* are unit vectors along the *X* and *Y* axes.

Magnitudes of *X* and *Y* can also be expressed as:

$$X = A\,cos\,(\theta)$$
$$Y = A\,sin\,(\theta)$$

Where *X, Y and A* are scalar quantities.

Therefore,

$$A = X\,i + Y\,j = A\,cos\,(\theta)\,i + A\,sin\,(\theta)\,j \;\text{--------- (1)}$$

Next, establish a new unit vector (u) in the same direction as vector **A**.

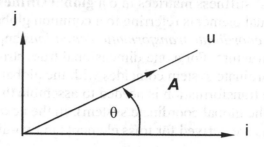

Vector **A** can now be expressed as: $\mathbf{A} = A\,u$ ------------ (2)

Both equations, the above (1) and (2), represent vector **A**:

$$\mathbf{A} = A\,u = A\cos(\theta)\,i + A\sin(\theta)\,j$$

The unit vector u can now be expressed in terms of the original set of unit vectors i and j:

$$\boxed{u = \cos(\theta)\,i + \sin(\theta)\,j}$$

Now consider another vector **B**:

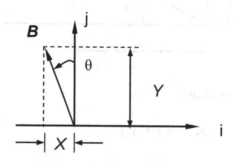

Vector **B** can be expressed as:

$$\mathbf{B} = -X\,i + Y\,j$$

Where i and j are unit vectors along the X- and Y-axes.

Magnitudes of X and Y can also be expressed as components of the magnitude of the vector:

$$X = B\sin(\theta)$$
$$Y = B\cos(\theta)$$

Where X, Y and B are scalar quantities.

Therefore,

$$\mathbf{B} = -X\,i + Y\,j = -B\sin(\theta)\,i + B\cos(\theta)\,j \text{ ---------- (3)}$$

Next, establish a new unit vector (v) along vector **B**.

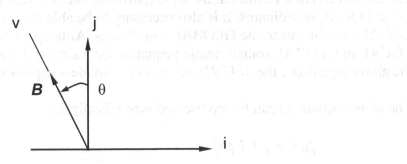

Vector **B** can now be expressed as: $\boldsymbol{B} = B\,v$ ------------ (4)

Equations (3) and (4) represent vector **B**:

$$\boldsymbol{B} = B\,v = -\,B\,\sin{(\theta)}\,i + B\,\cos{(\theta)}\,j$$

The unit vector v can now be expressed in terms of the original set of unit vectors i and j:

$$\boxed{v = -\,\sin{(\theta)}\,i + \cos{(\theta)}\,j}$$

We have established the coordinate transformation equations that can be used to transform vectors from ij coordinates to the rotated uv coordinates.

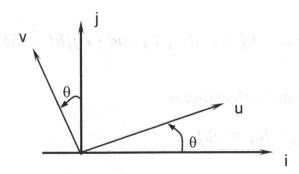

Coordinate Transformation Equations:

$$\boxed{\begin{array}{l} u = \ \ \cos{(\theta)}\,i + \ \sin{(\theta)}\,j \\ v = -\,\sin{(\theta)}\,i + \ \cos{(\theta)}\,j \end{array}}$$

Direction cosines

In matrix form,

$$\left\{ \begin{array}{c} u \\ v \end{array} \right\} = \left[\begin{array}{cc} \cos{(\theta)} & \sin{(\theta)} \\ -\sin{(\theta)} & \cos{(\theta)} \end{array} \right] \left\{ \begin{array}{c} i \\ j \end{array} \right\}$$

The above *direction cosines* allow us to transform vectors from the GLOBAL coordinates to the LOCAL coordinates. It is also necessary to be able to transform vectors from the LOCAL coordinates to the GLOBAL coordinates. Although it is possible to derive the LOCAL to GLOBAL transformation equations in a similar manner as demonstrated for the above equations, the *MATRIX operations* provide a slightly more elegant approach.

The above equations can be represented symbolically as:

$$\{a\} = [\,l\,]\,\{b\}$$

Where {a} and {b} are direction vectors, [l] is the direction cosine.

Perform the matrix operations to derive the reverse transformation equations in terms of the above direction cosines:

$$\{b\} = [\,?\,]\,\{a\}$$

First, multiply by $[\,l\,]^{-1}$ to remove the *direction cosines* from the right hand side of the original equation.

$$\{a\} = [\,l\,]\,\{b\}$$

$$[\,l\,]^{-1}\{a\} = [\,l\,]^{-1}\,[\,l\,]\,\{b\}$$

From matrix algebra, $[\,l\,]^{-1}\,[\,l\,] = [\,I\,]$ and $[\,I\,]\{b\} = \{b\}$

The equation can now be simplified as

$$[\,l\,]^{-1}\{a\} = \{b\}$$

For *linear statics analyses*, the *direction cosine* is an *orthogonal matrix* and the *inverse of the matrix* is equal to the transpose of the matrix.

$$[\,l\,]^{-1} = [\,l\,]^{T}$$

Therefore, the transformation equation can be expressed as:

$$[\,l\,]^{T}\{a\} = \{b\}$$

The transformation equations that enable us to transform any vector from a *LOCAL coordinate system* to the *GLOBAL coordinate system* become:

LOCAL coordinates to the GLOBAL coordinates:

$$\left\{ \begin{array}{c} i \\ j \end{array} \right\} = \left[\begin{array}{cc} \cos(\theta) & -\sin(\theta) \\ \sin(\theta) & \cos(\theta) \end{array} \right] \left\{ \begin{array}{c} u \\ v \end{array} \right\}$$

The reverse transformation can also be established by applying the transformation equations that transform any vector from the *GLOBAL coordinate system* to the *LOCAL coordinate system*:

GLOBAL coordinates to the LOCAL coordinates:

$$\left\{ \begin{array}{c} u \\ v \end{array} \right\} = \left[\begin{array}{cc} \cos(\theta) & \sin(\theta) \\ -\sin(\theta) & \cos(\theta) \end{array} \right] \left\{ \begin{array}{c} i \\ j \end{array} \right\}$$

As is the case with many mathematical equations, derivation of the equations usually appears to be much more complex than the actual application and utilization of the equations. The following example illustrates the application of the two-dimensional *coordinate transformation equations* on a point in between two coordinate systems.

Example 2.1

Given:

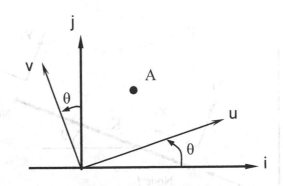

The coordinates of point A: (20 i, 40 j).

Find: The coordinates of point A if the local coordinate system is rotated 15 degrees relative to the global coordinate system.

Solution:

Using the coordinate transformation equations (GLOBAL coordinates to the LOCAL coordinates):

$$\left\{ \begin{array}{c} u \\ v \end{array} \right\} = \left[\begin{array}{cc} \cos(\theta) & \sin(\theta) \\ -\sin(\theta) & \cos(\theta) \end{array} \right] \left\{ \begin{array}{c} i \\ j \end{array} \right\}$$

$$= \left[\begin{array}{cc} \cos(15°) & \sin(15°) \\ -\sin(15°) & \cos(15°) \end{array} \right] \left\{ \begin{array}{c} 20 \\ 40 \end{array} \right\}$$

$$= \left\{ \begin{array}{c} 29.7 \\ 32.5 \end{array} \right\}$$

➢ On your own, perform a coordinate transformation to determine the global coordinates of point A using the *LOCAL* coordinates of (29.7,32.5) with the 15 degrees angle in between the two coordinate systems.

Global Stiffness Matrix

For a single truss element, using the coordinate transformation equations, we can proceed to transform the local stiffness matrix to the global stiffness matrix.

For a single truss element arbitrarily positioned in a two-dimensional space:

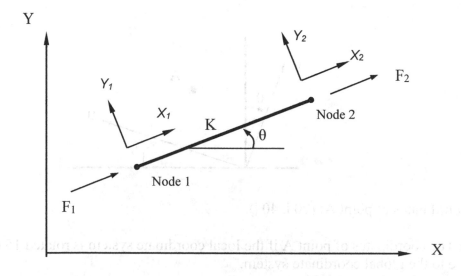

The force-displacement equations (in the local coordinate system) can be expressed as:

$$\begin{Bmatrix} F_{1X} \\ F_{1Y} \\ F_{2X} \\ F_{2Y} \end{Bmatrix} = \frac{EA}{L} \begin{bmatrix} +1 & 0 & -1 & 0 \\ 0 & 0 & 0 & 0 \\ -1 & 0 & +1 & 0 \\ 0 & 0 & 0 & 0 \end{bmatrix} \begin{Bmatrix} X_1 \\ Y_1 \\ X_2 \\ Y_2 \end{Bmatrix}$$

Local Stiffness Matrix

Next, apply the coordinate transformation equations to establish the general GLOBAL STIFFNESS MATRIX of a single truss element in a two-dimensional space.

First, the displacement transformation equations (GLOBAL to LOCAL):

$$\begin{Bmatrix} X_1 \\ Y_1 \\ X_2 \\ Y_2 \end{Bmatrix} = \begin{bmatrix} \cos(\theta) & \sin(\theta) & 0 & 0 \\ -\sin(\theta) & \cos(\theta) & 0 & 0 \\ 0 & 0 & \cos(\theta) & \sin(\theta) \\ 0 & 0 & -\sin(\theta) & \cos(\theta) \end{bmatrix} \begin{Bmatrix} X_1 \\ Y_1 \\ X_2 \\ Y_2 \end{Bmatrix}$$

Local   Global

The force transformation equations (GLOBAL to LOCAL):

$$\begin{Bmatrix} F_{1x} \\ F_{1y} \\ F_{2x} \\ F_{2y} \end{Bmatrix} = \begin{bmatrix} \cos(\theta) & \sin(\theta) & 0 & 0 \\ -\sin(\theta) & \cos(\theta) & 0 & 0 \\ 0 & 0 & \cos(\theta) & \sin(\theta) \\ 0 & 0 & -\sin(\theta) & \cos(\theta) \end{bmatrix} \begin{Bmatrix} F_{1x} \\ F_{1y} \\ F_{2x} \\ F_{2y} \end{Bmatrix}$$

Local   Global

The above three sets of equations can be represented as:

$\{F\} = [K]\{X\}$ ------- *Local force-displacement equation*

$\{X\} = [l]\{X\}$ ------- *Displacement transformation equation*

$\{F\} = [l]\{F\}$ ------- *Force transformation equation*

We will next perform *matrix operations* to obtain the GLOBAL stiffness matrix:

Starting with the local force-displacement equation

$$\{F\} = [K]\{X\}$$

Local   Local

Next, substituting the transformation equations for $\{F\}$ and $\{X\}$,

$$[l]\{F\} = [K][l]\{X\}$$

Multiply both sides of the equation with $[l]^{-1}$,

$$[l]^{-1}[l]\{F\} = [l]^{-1}[K][l]\{X\}$$

The equation can be simplified as:

$$\{F\} = [l]^{-1}[K][l]\{X\}$$

or

Global   Global

$$\{F\} = [l]^{T}[K][l]\{X\}$$

The GLOBAL force-displacement equation is then expressed as:

$$\{F\} = [l]^{T}[K][l]\{X\}$$

or

$$\{F\} = [K]\{X\}$$

The global stiffness matrix [**K**] can now be expressed in terms of the local stiffness matrix.

$$[K] = [l]^{T}[K][l]$$

For a single truss element in a two-dimensional space, the global stiffness matrix is

$$[K] = \frac{EA}{L}\begin{bmatrix} cos^2(\theta) & cos(\theta)sin(\theta) & -cos^2(\theta) & -cos(\theta)sin(\theta) \\ cos(\theta)sin(\theta) & sin^2(\theta) & -cos(\theta)sin(\theta) & -sin^2(\theta) \\ -cos^2(\theta) & -cos(\theta)sin(\theta) & cos^2(\theta) & cos(\theta)sin(\theta) \\ -cos(\theta)sin(\theta) & -sin^2(\theta) & sin(\theta)cos(\theta) & sin^2(\theta) \end{bmatrix}$$

➢ The above matrix can be applied to any truss element positioned in a two-dimensional space. We can now assemble the global stiffness matrix and analyze any two-dimensional multiple-elements truss structures. The following example illustrates, using the general global stiffness matrix derived above, the formulation and solution process of a 2D truss structure.

Example 2.2

Given: A two-dimensional truss structure as shown. (All joints are **Pin Joints**.)

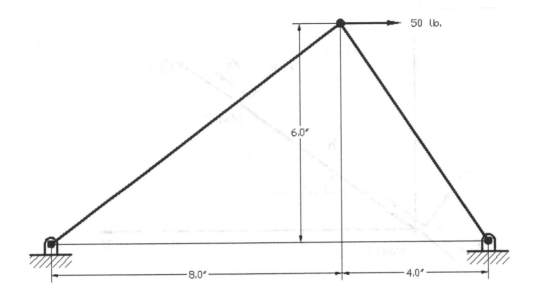

Material: Steel rod, diameter ¼ in.

Find: Displacements of each node and stresses in each member.

Solution:

The system contains two elements and three nodes. The nodes and elements are labeled as shown below.

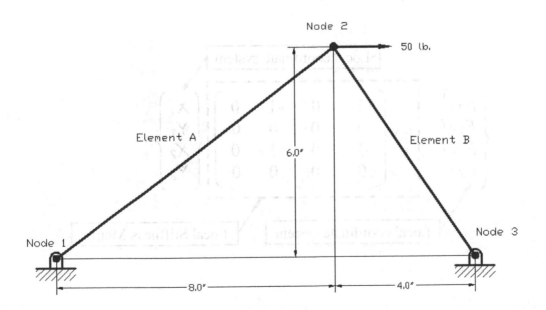

First, establish the GLOBAL stiffness matrix (system equations in matrix form) for each element.

Element A   (Node 1 to Node 2)

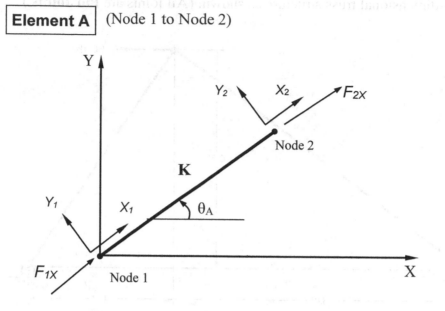

$\theta_A = \tan^{-1}(6/8) = 36.87°$,
E (Young's modulus) $= 30 \times 10^6$ psi
A (Cross sectional area) $= \pi r^2 = 0.049$ in^2,
L (Length of element) $= (6^2 + 8^2)^{1/2} = 10$ in.

Therefore, $\dfrac{EA}{L} = 147262$ lb/in.

The LOCAL force-displacement equations:

Local coordinate system

$$
\begin{Bmatrix} F_{1X} \\ F_{1Y} \\ F_{2X} \\ F_{2Y} \end{Bmatrix} = \frac{EA}{L} \begin{bmatrix} +1 & 0 & -1 & 0 \\ 0 & 0 & 0 & 0 \\ -1 & 0 & +1 & 0 \\ 0 & 0 & 0 & 0 \end{bmatrix} \begin{Bmatrix} X_1 \\ Y_1 \\ X_2 \\ Y_2 \end{Bmatrix}
$$

Local coordinate system   Local Stiffness Matrix

Using the equations we have derived, the GLOBAL system equations for *Element A* can be expressed as:

$$\{ F \} = [K] \{ X \}$$

$$[K] = \frac{EA}{L} \begin{bmatrix} cos^2(\theta) & cos(\theta)sin(\theta) & -cos^2(\theta) & -cos(\theta)sin(\theta) \\ cos(\theta)sin(\theta) & sin^2(\theta) & -cos(\theta)sin(\theta) & -sin^2(\theta) \\ -cos^2(\theta) & -cos(\theta)sin(\theta) & cos^2(\theta) & cos(\theta)sin(\theta) \\ -cos(\theta)sin(\theta) & -sin^2(\theta) & sin(\theta)cos(\theta) & sin^2(\theta) \end{bmatrix}$$

Therefore,

$$\begin{Bmatrix} F_{1X} \\ F_{1Y} \\ F_{2XA} \\ F_{2YA} \end{Bmatrix} = 147262 \begin{bmatrix} .64 & .48 & -.64 & -.48 \\ .48 & .36 & -.48 & -.36 \\ -.64 & -.48 & .64 & .48 \\ -.48 & -.36 & .48 & .36 \end{bmatrix} \begin{Bmatrix} X_1 \\ Y_1 \\ X_2 \\ Y_2 \end{Bmatrix}$$

Global

Global

Global Stiffness Matrix

Element B (Node 2 to Node 3)

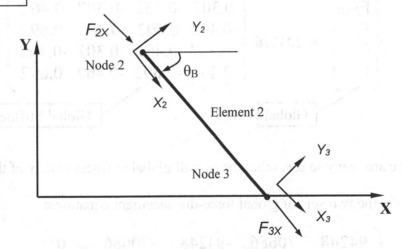

$\theta_B = -\tan^{-1}(6/4) = -56.31°$ (negative angle as shown)

E (Young's modulus) $= 30 \times 10^6$ psi

A (Cross sectional area) $= \pi r^2 = 0.049$ in^2

L (Length of element) $= (4^2 + 6^2)^{1/2} = 7.21$ in.

$\dfrac{EA}{L} = 204216$ lb/in.

$$\begin{Bmatrix} F_{2X} \\ F_{2Y} \\ F_{3X} \\ F_{3Y} \end{Bmatrix} = \frac{EA}{L} \begin{bmatrix} +1 & 0 & -1 & 0 \\ 0 & 0 & 0 & 0 \\ -1 & 0 & +1 & 0 \\ 0 & 0 & 0 & 0 \end{bmatrix} \begin{Bmatrix} X_2 \\ Y_2 \\ X_3 \\ Y_3 \end{Bmatrix}$$

Local system   Local system

Local Stiffness Matrix

Using the equations we derived in the previous sections, the GLOBAL system equation for *Element B* is:

$$\{ F \} = [K] \{ X \}$$

$$[K] = \frac{EA}{L} \begin{bmatrix} cos^2(\theta) & cos(\theta)sin(\theta) & -cos^2(\theta) & -cos(\theta)sin(\theta) \\ cos(\theta)sin(\theta) & sin^2(\theta) & -cos(\theta)sin(\theta) & -sin^2(\theta) \\ -cos^2(\theta) & -cos(\theta)sin(\theta) & cos^2(\theta) & cos(\theta)sin(\theta) \\ -cos(\theta)sin(\theta) & -sin^2(\theta) & sin(\theta)cos(\theta) & sin^2(\theta) \end{bmatrix}$$

Therefore,

Global

$$\begin{Bmatrix} F_{2XB} \\ F_{2YB} \\ F_{3X} \\ F_{3Y} \end{Bmatrix} = 204216 \begin{bmatrix} 0.307 & -0.462 & -0.307 & 0.462 \\ -0.462 & 0.692 & 0.462 & -0.692 \\ -0.307 & 0.462 & 0.307 & -0.462 \\ 0.462 & -0.692 & -0.462 & 0.692 \end{bmatrix} \begin{Bmatrix} X_2 \\ Y_2 \\ X_3 \\ Y_3 \end{Bmatrix}$$

Global   Global Stiffness Matrix

Now we are ready to assemble the overall global stiffness matrix of the structure.

Summing the two sets of global force-displacement equations:

$$\begin{Bmatrix} F_{1X} \\ F_{1Y} \\ F_{2X} \\ F_{2Y} \\ F_{3X} \\ F_{3Y} \end{Bmatrix} = \begin{bmatrix} 94248 & 70686 & -94248 & -70686 & 0 & 0 \\ 70686 & 53014 & -70686 & -53014 & 0 & 0 \\ -94248 & -70686 & 157083 & -23568 & -62836 & 94253 \\ -70686 & -53014 & -23568 & 194395 & 94253 & -141380 \\ 0 & 0 & -62836 & 94253 & 62836 & -94253 \\ 0 & 0 & 94253 & -141380 & -94253 & 141380 \end{bmatrix} \begin{Bmatrix} X_1 \\ Y_1 \\ X_2 \\ Y_2 \\ X_3 \\ Y_3 \end{Bmatrix}$$

Next, apply the following known boundary conditions into the system equations:

(a) Node 1 and Node 3 are fixed-points; therefore, any displacement components of these two node-points are zero (X_1, Y_1 and X_3, Y_3).

(b) The only external load is at Node 2: $F_{2x} = 50$ lbs.

Therefore,

$$\begin{Bmatrix} F_{1X} \\ F_{1Y} \\ 50 \\ 0 \\ F_{3X} \\ F_{3Y} \end{Bmatrix} = \begin{bmatrix} 94248 & 70686 & -94248 & -70686 & 0 & 0 \\ 70686 & 53014 & -70686 & -53014 & 0 & 0 \\ -94248 & -70686 & 157083 & -23568 & -628360 & 94253 \\ -70686 & -53014 & -23568 & 194395 & 94253 & -141380 \\ 0 & 0 & -62836 & 94253 & 62836 & -94253 \\ 0 & 0 & 94253 & -141380 & -94253 & 141380 \end{bmatrix} \begin{Bmatrix} 0 \\ 0 \\ X_2 \\ Y_2 \\ 0 \\ 0 \end{Bmatrix}$$

The two displacements we need to solve are X_2 and Y_2. Let's simplify the above matrix by removing the unaffected/unnecessary columns in the matrix.

$$\begin{Bmatrix} F_{1X} \\ F_{1Y} \\ 50 \\ 0 \\ F_{3X} \\ F_{3Y} \end{Bmatrix} = \begin{bmatrix} -94248 & -70686 \\ -70686 & -53014 \\ 157083 & -23568 \\ -23568 & 194395 \\ -62836 & 94253 \\ 94253 & -141380 \end{bmatrix} \begin{Bmatrix} X_2 \\ Y_2 \end{Bmatrix}$$

Solve for nodal displacements X_2 and Y_2:

$$\begin{Bmatrix} 50 \\ 0 \end{Bmatrix} = \begin{bmatrix} 157083 & -23568 \\ -23568 & 194395 \end{bmatrix} \begin{Bmatrix} X_2 \\ Y_2 \end{Bmatrix}$$

$X_2 = 3.24$ e $^{-4}$ in.
$Y_2 = 3.93$ e $^{-5}$ in.

Substitute the known X_2 and Y_2 values into the matrix and solve for the reaction forces:

$$\begin{Bmatrix} F_{1X} \\ F_{1Y} \\ F_{3X} \\ F_{3Y} \end{Bmatrix} = \begin{bmatrix} -94248 & -70686 \\ -70686 & -53014 \\ -62836 & 94253 \\ 94253 & -141380 \end{bmatrix} \begin{Bmatrix} \mathbf{3.24\ e}^{-4} \\ \mathbf{3.93\ e}^{-5} \end{Bmatrix}$$

Therefore,

$$F_{1X} = -33.33 \text{ lbs.,} \quad F_{1Y} = -25 \text{ lbs.}$$
$$F_{3X} = -16.67 \text{ lbs.,} \quad F_{3Y} = 25 \text{ lbs.}$$

To determine the normal stress in each truss member, one option is to use the displacement transformation equations to transform the results from the global coordinate system back to the local coordinate system.

Element A

$\{X\} = [l]\{X\}$ -------- *Displacement transformation equation*

$$\begin{Bmatrix} X_1 \\ Y_1 \\ X_2 \\ Y_2 \end{Bmatrix} = \begin{bmatrix} .8 & .6 & 0 & 0 \\ -.6 & .8 & 0 & 0 \\ 0 & 0 & .8 & .6 \\ 0 & 0 & -.6 & .8 \end{bmatrix} \begin{Bmatrix} 0 \\ 0 \\ X_2 \\ Y_2 \end{Bmatrix}$$

Local   Global

$$X_2 = 2.83 \text{ e}^{-4}$$
$$Y_2 = -1.63 \text{ e}^{-4}$$

The LOCAL force-displacement equations:

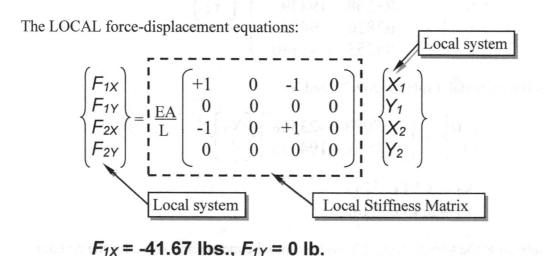

Local system

$$\begin{Bmatrix} F_{1X} \\ F_{1Y} \\ F_{2X} \\ F_{2Y} \end{Bmatrix} = \frac{EA}{L} \begin{bmatrix} +1 & 0 & -1 & 0 \\ 0 & 0 & 0 & 0 \\ -1 & 0 & +1 & 0 \\ 0 & 0 & 0 & 0 \end{bmatrix} \begin{Bmatrix} X_1 \\ Y_1 \\ X_2 \\ Y_2 \end{Bmatrix}$$

Local system   Local Stiffness Matrix

$$F_{1X} = -41.67 \text{ lbs.,} \quad F_{1Y} = 0 \text{ lb.}$$
$$F_{2X} = 41.67 \text{ lbs.,} \quad F_{2Y} = 0 \text{ lb.}$$

Therefore, the normal stress developed in *Element A* can be calculated as (41.67/0.049)=**850 psi**.

➢ On your own, calculate the normal stress developed in *Element B*.

Questions:

1. Determine the coordinates of point A if the local coordinate system is rotated 15 degrees relative to the global coordinate system. The global coordinates of point A: (30,50).

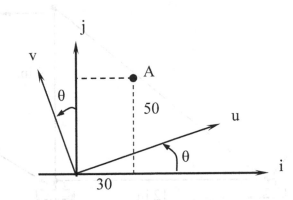

2. Determine the global coordinates of point B if the local coordinate system is rotated 30 degrees relative to the global coordinate system. The local coordinates of point B: (30,15).

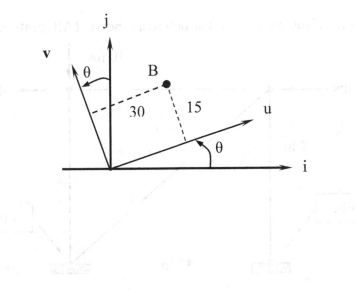

Exercises:

1.   Given: two-dimensional truss structure as shown. (All joints are pin joints.)

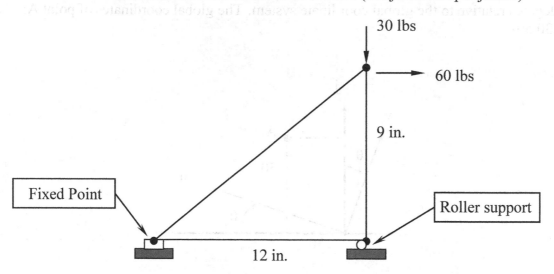

Material: Steel, diameter ¼ in.

Find: (a) Displacements of the nodes.
         (b) Normal stresses developed in the members.

2.   Given: Two-dimensional truss structure as shown. (All joints are pin joints.)

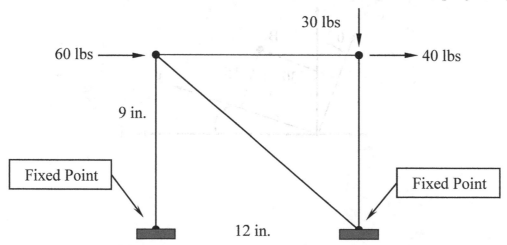

Material: Steel, diameter ¼ in.

Find: (a) Displacements of the nodes.
         (b) Normal stresses developed in the members.

Chapter 3
2D Trusses in MS Excel and Truss Solver

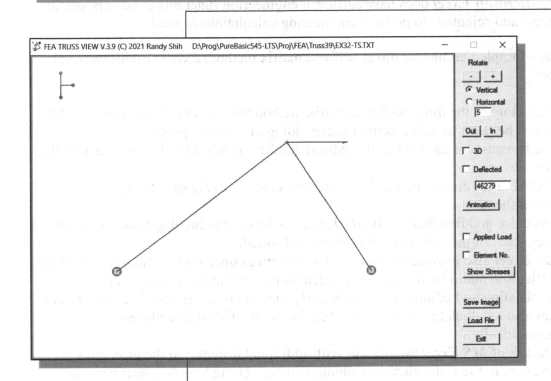

Learning Objectives

♦ **Perform the Direct Stiffness Matrix Method using MS Excel.**

♦ **Use MS Excel built-in commands to solve simultaneous equations.**

♦ **Solve 2D trusses using the Truss Solver program.**

♦ **Understand the general FEA procedure concepts.**

Direct Stiffness Matrix Method using Excel

Microsoft Excel is a very popular general-purpose spreadsheet program, which was originally designed to do accounting related calculations and simple database operations. However, *Microsoft Excel* does have sufficient engineering functions to be very useful for engineers and scientists to perform engineering calculations as well.

In the case of implementing the direct stiffness matrix method, *Excel* provides several advantages:

- The set up of the direct stiffness matrix method using a spreadsheet, such as *MS Excel*, helps us to gain a better understanding of the FEA procedure.
- The spreadsheet takes away the tedious number crunching task, which is usually prone to errors.
- Checking for errors and making corrections are relatively easy to do in a spreadsheet.
- The color and line features in *MS Excel* can be used to highlight and therefore keep track of the series of calculations performed.
- *MS Excel* also provides commonly used matrices operations, which are otherwise difficult to attain in the traditional calculator, pencil and paper approach.
- Duplicating and editing formulas is fairly easy to do in a spreadsheet. *MS Excel* can also recalculate a new set of values in an established spreadsheet automatically.
- The use of *MS Excel* provides us with additional insight into the principles involved in the finite element method, without getting bogged down with the programming aspects of the problem solution.

Example 3.1

Use *MS Excel* to perform the calculations illustrated in Example 2.2.
Material: Steel rod, diameter ¼ in.

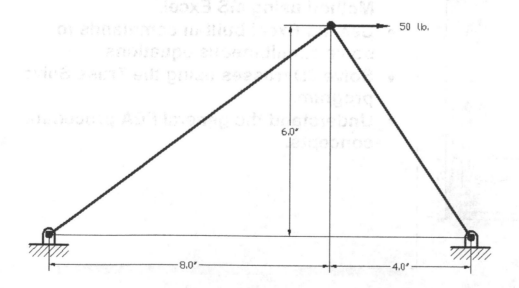

The system contains two elements and three nodes. The nodes and elements are labeled as shown below.

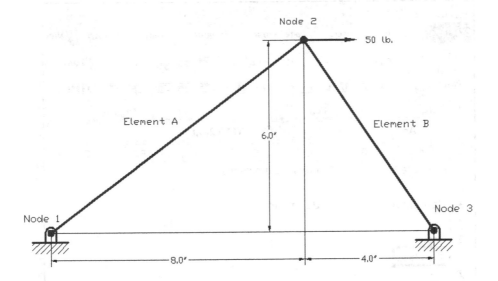

Establish the Global K matrix for each member

1. Start a new excel spreadsheet in *Microsoft Excel* through the desktop icon or the *Explorer* toolbar.

2. First enter the general problem information and element labels near the top of the spreadsheet.

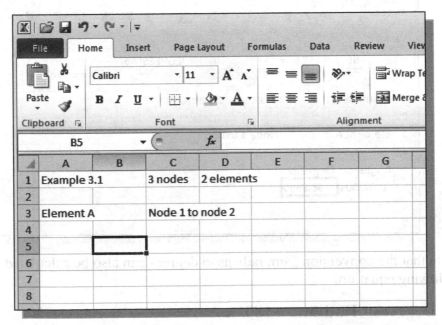

❖ Note that we will be following the same convention as outlined in Example 2.2.

3. To calculate the angle of *Element A*, enter the formula: **=ATAN2(8,6)**; the resulting angle is in **radians**.

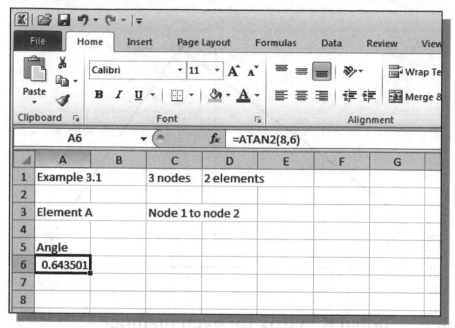

❖ Note the equal sign is used to identify the formula; the **ATAN2** function will calculate the arctangent function in all four quadrants.

4. To convert the radians to degrees, use the **Degrees** function as shown.

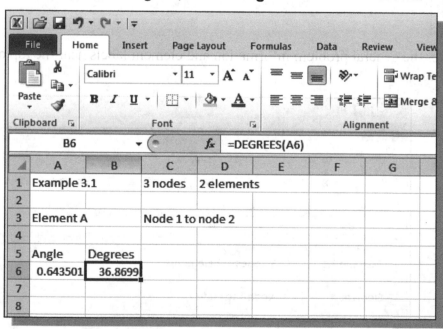

❖ Note that the conversion from radians to degrees can also be calculated using the following equation:

Degrees= Radians × 180/ π

5. Next enter the modulus of elasticity (E) and cross-section area based on the diameter of the truss member.

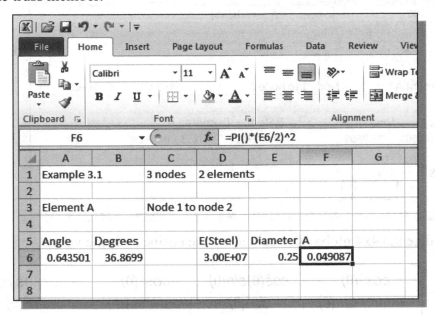

❖ Note the PI() function gives the π value in *Excel*.

6. To calculate the length of the member, enter the formula as shown.

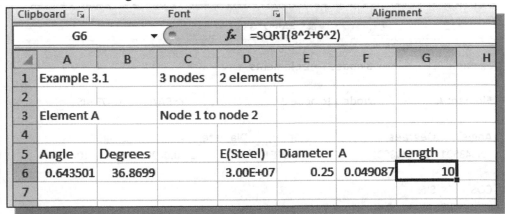

❖ The **SQRT** function can be used to calculate the square root value.

7. Now, we can calculate the k value. Note the equation is **k=EA/L**.

8. Next calculate the cosine and sine of the radian angle as shown.

		B9			f_x	=SIN(A6)	
	A	B	C	D	E	F	G	H
1	Example 3.1		3 nodes	2 elements				
2								
3	Element A		Node 1 to node 2			K=EA/L	1.47E+05	
4								
5	Angle	Degrees		E(Steel)	Diameter	A	Length	
6	0.643501	36.8699		3.00E+07	0.25	0.049087	10	
7								
8	COS	SIN						
9	0.8	0.6						
10								

9. Next calculate the 4x4 global K matrix using the cosine and sine values.

$$[K] = \frac{EA}{L} \begin{bmatrix} \cos^2(\theta) & \cos(\theta)\sin(\theta) & -\cos^2(\theta) & -\cos(\theta)\sin(\theta) \\ \cos(\theta)\sin(\theta) & \sin^2(\theta) & -\cos(\theta)\sin(\theta) & -\sin^2(\theta) \\ -\cos^2(\theta) & -\cos(\theta)\sin(\theta) & \cos^2(\theta) & \cos(\theta)\sin(\theta) \\ -\cos(\theta)\sin(\theta) & -\sin^2(\theta) & \sin(\theta)\cos(\theta) & \sin^2(\theta) \end{bmatrix}$$

		B12			f_x	=A9^2*G3	
	A	B	C	D	E	F	G	H
1	Example 3.1		3 nodes	2 elements				
2								
3	Element A		Node 1 to node 2			K=EA/L	1.47E+05	
4								
5	Angle	Degrees		E(Steel)	Diameter	A	Length	
6	0.643501	36.869898		3.00E+07	0.25	0.049087	10	
7								
8	COS	SIN						
9	0.8	0.6						
10								
11		X1	Y1	X2	Y2			
12		9.425E+04	7.069E+04	-9.425E+04	-7.069E+04			
13		7.069E+04	5.301E+04	-7.069E+04	-5.301E+04			
14		-9.425E+04	-7.069E+04	9.425E+04	7.069E+04			
15		-7.069E+04	-5.301E+04	7.069E+04	5.301E+04			
16								

$K_{11}=k*(\cos(\theta))^2$, $K_{12}=k*\cos(\theta)\sin(\theta)$,
$K_{13}=-k*(\cos(\theta))^2$, $K_{14}=-k*\cos(\theta)\sin(\theta)$,
$K_{21}=k*\cos(\theta)\sin(\theta)$, $K_{22}=k*(\sin(\theta))^2$........

❖ On your own, compare the calculated values to the numbers shown on page 2-15.

10. Once we have confirmed the calculations are correct for *Element A*, we can simply *copy and paste* the established formulas for any other elements. Select the established formulas as shown.

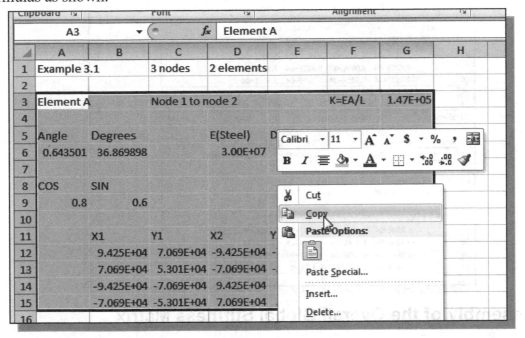

11. Click the **Copy** icon or right-mouse-click to bring up the option menu and choose Copy as shown in the above figure.

12. Move the cursor a few cells below the selected cells as shown.

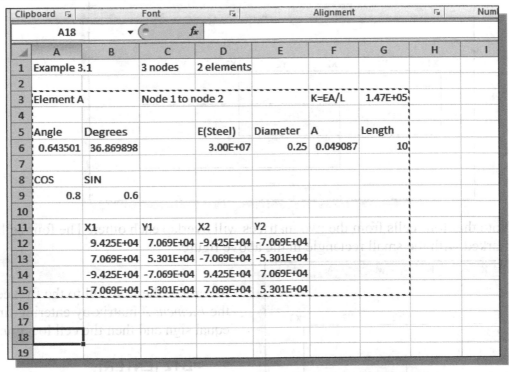

13. Press the [**ENTER**] key once to create a copy of the selected cells.

14. On your own, update the second set for *Element B* as shown in the figure below. Hint: Update the values by changing the values in the following two cells: **Radian Angle** and **Length**.

	A	B	C	D	E	F	G	H	I
	A21			f_x	=ATAN2(4,-6)				
13		7.069E+04	5.301E+04	-7.069E+04	-5.301E+04				
14		-9.425E+04	-7.069E+04	9.425E+04	7.069E+04				
15		-7.069E+04	-5.301E+04	7.069E+04	5.301E+04				
16									
17									
18	Element B		Node 2 to node 3				K=EA/L	2.04E+05	
19									
20	Angle	Degrees		E(Steel)	Diameter	A	Length		
21	-0.98279	-56.30993		3.00E+07	0.25	0.049087	7.211103		
22									
23	COS	SIN							
24	0.5547	-0.83205							
25									
26		X2	Y2	X3	Y3				
27		6.284E+04	-9.425E+04	-6.284E+04	9.425E+04				
28		-9.425E+04	1.414E+05	9.425E+04	-1.414E+05				
29		-6.284E+04	9.425E+04	6.284E+04	-9.425E+04				
30		9.425E+04	-1.414E+05	-9.425E+04	1.414E+05				
31									

Assembly of the Overall Global Stiffness Matrix

1. On your own, set up the labels for the overall global stiffness matrix as shown. The two large rectangles indicate the locations for where the two element matrices will fit. You can also use different colors to highlight the cells.

J	K	L	M	N	O	P	Q
	X1	Y1	X2	Y2	X3	Y3	

❖ Note that four cells from the two matrices will overlap each other. The four cells are marked with the small rectangle in the above figure.

	K	L	M	N	O
	X1	Y1	X2	Y2	X3
	=B12				

2. Reference the first cell to the first cell of the *Element A* matrix by entering first the equal sign and then the cell location:

=B12 [ENTER]

3. Repeat the last step to fill in the first rectangle, which is simply referencing the cells in the *Element A* matrix. (For the cell locations, refer to the image shown on page 3-6.)

	K	L	M	N	O	P
X1		Y1	X2	Y2	X3	Y3
	9.425E+04	7.069E+04	-9.425E+04	-7.069E+04		
	7.069E+04	5.301E+04	-7.069E+04	-5.301E+04		
	-9.425E+04	-7.069E+04	9.425E+04	7.069E+04		
	-7.069E+04	-5.301E+04	7.069E+04	5.301E+04		

4. Edit the third cell in the X2 column; this is where the first overlap occurs. Edit the cell so that it includes the first cell of the *Element B* matrix: **=D14+B27**

f_x	=E15+C28					
	K	L	M	N	O	P
X1		Y1	X2	Y2	X3	Y3
	9.425E+04	7.069E+04	-9.425E+04	-7.069E+04		
	7.069E+04	5.301E+04	-7.069E+04	-5.301E+04		
	-9.425E+04	-7.069E+04	1.571E+05	-2.357E+04		
	-7.069E+04	-5.301E+04	-2.357E+04	1.944E+05		

5. Repeat the above step and complete the four overlapped cells as shown.

f_x	=D14+B27					
	K	L	M	N	O	P
X1		Y1	X2	Y2	X3	Y3
	9.425E+04	7.069E+04	-9.425E+04	-7.069E+04		
	7.069E+04	5.301E+04	-7.069E+04	-5.301E+04		
	-9.425E+04	-7.069E+04	1.571E+05	7.069E+04		
	-7.069E+04	-5.301E+04	7.069E+04	5.301E+04		

6. Fill in the rest of the second rectangle by referencing to the *Element B* matrix; the results should appear as shown.

f_x	=E27				
K	L	M	N	O	P
X1	Y1	X2	Y2	X3	Y3
9.425E+04	7.069E+04	-9.425E+04	-7.069E+04		
7.069E+04	5.301E+04	-7.069E+04	-5.301E+04		
-9.425E+04	-7.069E+04	1.571E+05	-2.357E+04	-6.284E+04	9.425E+04
-7.069E+04	-5.301E+04	-2.357E+04	1.944E+05	9.425E+04	-1.414E+05
		-6.284E+04	9.425E+04	6.284E+04	-9.425E+04
		9.425E+04	-1.414E+05	-9.425E+04	1.414E+05

7. Enter eight **zeros** in the remaining blank cells. Note that these zeros are necessary for the calculations of the reaction forces, which we will do once the global displacements are calculated.

f_x	0				
K	L	M	N	O	P
X1	Y1	X2	Y2	X3	Y3
9.425E+04	7.069E+04	-9.425E+04	-7.069E+04	0	0
7.069E+04	5.301E+04	-7.069E+04	-5.301E+04	0	0
-9.425E+04	-7.069E+04	1.571E+05	-2.357E+04	-6.284E+04	9.425E+04
-7.069E+04	-5.301E+04	-2.357E+04	1.944E+05	9.425E+04	-1.414E+05
0	0	-6.284E+04	9.425E+04	6.284E+04	-9.425E+04
0	0	9.425E+04	-1.414E+05	-9.425E+04	1.414E+05

Solving the Global Displacements

1. Label the global displacements to the right of the overall global matrix; enter the four zeros as shown. (Node 1 and Node 3 are fixed points.)

	X3	Y3			
04	0	0		X1	0
04	0	0		Y1	0
04	-6.284E+04	9.425E+04		X2=?	
05	9.425E+04	-1.414E+05		Y2=?	
04	6.284E+04	-9.425E+04		X3	0
05	-9.425E+04	1.414E+05		Y3	0

2. Label the global forces to the left of the overall global matrix; enter **50** and **0** to represent the loads that are applied at Node 2.

		X1	Y1	X2	Y2
	FX1=?	9.425E+04	7.069E+04	-9.425E+04	-7.0
	FY1=?	7.069E+04	5.301E+04	-7.069E+04	-5.3
50	FX2	-9.425E+04	-7.069E+04	1.571E+05	-2.3
0	FY2	-7.069E+04	-5.301E+04	-2.357E+04	1.9
	FX3=?	0	0	-6.284E+04	9.4
	FY3=?	0	0	9.425E+04	-1.4

❖ We will solve the two unknown displacements, X2 and Y2, first. This can be done using the established matrix equations.

   For $\{ F \} = [K] \{ X \}$, $\{X\}$ can be solved by finding the inverse [K] matrix.
   $\{ X \} = [K]^{-1} \{ F \}$, where the $\{X\}$ vector represents the two unknowns.

3. Place a label for calculating the **Inverse K** as shown and select a 2x2 array as shown in the figure.

		X1	Y1	X2	Y2	X3	Y3
	FX1=?	9.425E+04	7.069E+04	-9.425E+04	-7.069E+04	0	0
	FY1=?	7.069E+04	5.301E+04	-7.069E+04	-5.301E+04	0	0
50	FX2	-9.425E+04	-7.069E+04	1.571E+05	-2.357E+04	-6.284E+04	9.425E+04
0	FY2	-7.069E+04	-5.301E+04	-2.357E+04	1.944E+05	9.425E+04	-1.414E+05
	FX3=?	0	0	-6.284E+04	9.425E+04	6.284E+04	-9.425E+04
	FY3=?	0	0	9.425E+04	-1.414E+05	-9.425E+04	1.414E+05

Inverse K for X2 Y2

4. Click on the **Insert Function** icon that is located in front of the input edit box as shown in the figure.

❖ *Excel* provides a variety of functions, which includes many of the more commonly used financial, math and trigonometric functions.

		X1	Y1	X2	Y2
	FX1=?	9.425E+04	7.069E+04	-9.425E+04	-7.069E+04
	FY1=?	7.069E+04	5.301E+04	-7.069E+04	-5.301E+04
50	FX2	-9.425E+04	-7.069E+04	1.571E+05	-2.357E+04
0	FY2	-7.069E+04	-5.301E+04	-2.357E+04	1.944E+05
	FX3=?	0	0	-6.284E+04	9.425E+04
	FY3=?	0	0	9.425E+04	-1.414E+05

Inverse K for X2 Y2

5. In the **Insert Function** dialog box, select **Math & Trig** as shown.

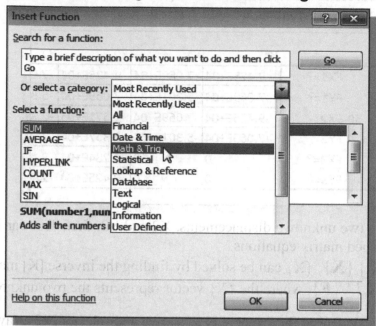

6. Select the **MINVERSE** command as shown.

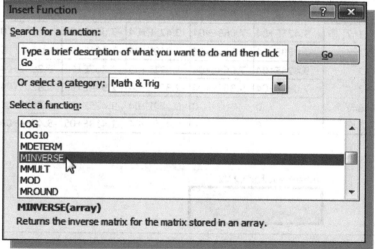

 7. Click **OK** to proceed with the **MINVERSE** command.

8. The **Function Arguments** dialog box appears on the screen with a brief description of the function.

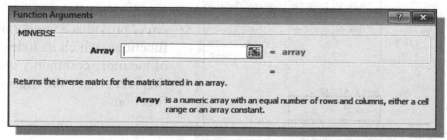

9. Select the 2x2 array as shown in the figure below.

X1	Y1	X2	Y2	X3	Y3
9.425E+04	7.069E+04	-9.425E+04	-7.069E+04	0	0
7.069E+04	5.301E+04	-7.069E+04	-5.301E+04	0	0
-9.425E+04	-7.069E+04	1.571E+05	-2.357E+04	-6.284E+04	9.425E+04
-7.069E+04	-5.301E+04	-2.357E+04	1.944E+05	9.425E+04	-1.414E+05
0	0	-6.284E+04	9.425E+04	6.284E+04	-9.425E+04
0	0	9.425E+04	-1.414E+05	-9.425E+04	1.414E+05

10. It is important to note that when using the matrix commands in *Excel*, we must press **CTRL+SHIFT+ENTER** to perform the calculations. The Inverse K is calculated as shown in the figure below.

		X1	Y1	X2	Y2	X3	Y3
	FX1=?	9.425E+04	7.069E+04	-9.425E+04	-7.069E+04	0	0
	FY1=?	7.069E+04	5.301E+04	-7.069E+04	-5.301E+04	0	0
50	FX2	-9.425E+04	-7.069E+04	1.571E+05	-2.357E+04	-6.284E+04	9.425E+04
0	FY2	-7.069E+04	-5.301E+04	-2.357E+04	1.944E+05	9.425E+04	-1.414E+05
	FX3=?	0	0	-6.284E+04	9.425E+04	6.284E+04	-9.425E+04
	FY3=?	0	0	9.425E+04	-1.414E+05	-9.425E+04	1.414E+05

Inverse K for X2 Y2

6.484E-06	7.861E-07
7.8609E-07	5.239E-06

11. Label the two global displacements, and pre-select the 2x1 array as shown.

B=?		0	0	9.425E+04	-1.414E+05	-9.425E+04	1.414E+05

Inverse K for X2 Y2

6.484E-06	7.861E-07	X2	
7.8609E-07	5.239E-06	Y2	

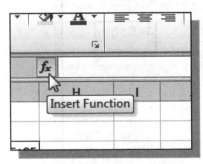

12. Click on the **Insert Function** icon that is located in front of the input edit box as shown in the figure.

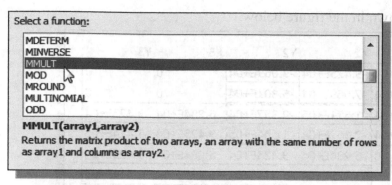

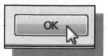

13. Select the **MMULT** command as shown.

14. Click **OK** to proceed with the **MMULT** command.

15. The ***Function Arguments*** dialog box appears on the screen with a brief description of the function. Note that two arrays are required for this function.

16. Select the calculated **Inverse [K] for X2 and Y2**, which is a 2x2 array, as shown.

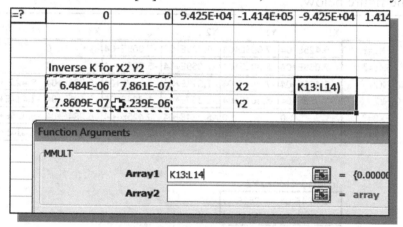

17. Hit the **Tab** key once to move the cursor to the next input box and select the two known forces at node 2 (*FX2* and *FY2*).

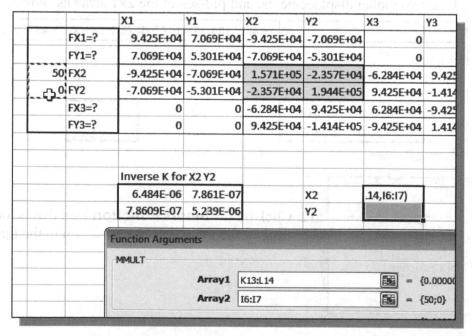

18. It is important to note that when using the matrix commands in *Excel*, we must press **CTRL+SHIFT+ENTER** to perform the calculations. The global displacements X2 and Y2 are calculated as shown in the figure below.

		X1	Y1	X2	Y2	X3	Y3
	FX1=?	9.425E+04	7.069E+04	-9.425E+04	-7.069E+04	0	0
	FY1=?	7.069E+04	5.301E+04	-7.069E+04	-5.301E+04	0	0
50	FX2	-9.425E+04	-7.069E+04	1.571E+05	-2.357E+04	-6.284E+04	9.425E+04
0	FY2	-7.069E+04	-5.301E+04	-2.357E+04	1.944E+05	9.425E+04	-1.414E+05
	FX3=?	0	0	-6.284E+04	9.425E+04	6.284E+04	-9.425E+04
	FY3=?	0	0	9.425E+04	-1.414E+05	-9.425E+04	1.414E+05

	Inverse K for X2 Y2			
	6.484E-06	7.861E-07	X2	0.000324199
	7.8609E-07	5.239E-06	Y2	3.93046E-05

❖ Note the calculation is based on the equation $\{X\} = [K]^{-1}\{F\}$, and in matrix multiplication, the order of the matrices are non-interchangeable. So, the inverse K matrix is selected first, followed by the Force matrix.

19. Set the X2 value, in the overall global displacement matrix, to reference the X2 value just calculated as shown.

	Y2	X3	Y3			
E+04	-7.069E+04	0	0	X1		0
E+04	-5.301E+04	0	0	Y1		0
E+05	-2.357E+04	-6.284E+04	9.425E+04	X2=?	=O13	
E+04	1.944E+05	9.425E+04	-1.414E+05	Y2=?		
E+04	9.425E+04	6.284E+04	-9.425E+04	X3		0
E+04	-1.414E+05	-9.425E+04	1.414E+05	Y3		0

	X2	0.000324199
	Y2	3.93046E-05

E+04	-5.301E+04	0	0	Y1		0
E+05	-2.357E+04	-6.284E+04	9.425E+04	X2=?	0.000324	
E+04	1.944E+05	9.425E+04	-1.414E+05	Y2=?	=O14	
E+04	9.425E+04	6.284E+04	-9.425E+04	X3		0
E+04	-1.414E+05	-9.425E+04	1.414E+05	Y3		0

	X2	0.000324199
	Y2	3.93046E-05

20. Repeat the above step and set the Y2 value as shown as well.

Calculating Reaction Forces

1. Pre-select the **six cells** in front of the global forces labels as shown.

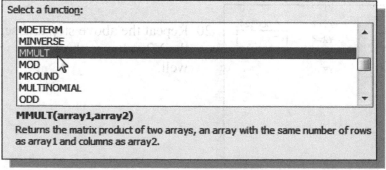

	H	I	J	K	L	M	
05				X1	Y1	X2	Y2
			FX1=?	9.425E+04	7.069E+04	-9.425E+04	-7.0
			FY1=?	7.069E+04	5.301E+04	-7.069E+04	-5.3
10		50	FX2	-9.425E+04	-7.069E+04	1.571E+05	-2.3
		0	FY2	-7.069E+04	-5.301E+04	-2.357E+04	1.9
			FX3=?	0	0	-6.284E+04	9.4
			FY3=?	0	0	9.425E+04	-1.4

Inverse K for X2 Y2

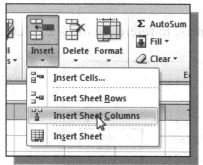

❖ Note the **Insert Sheet Columns** and the **Insert Sheet Rows** commands are also available to add additional columns and rows in between existing cells.

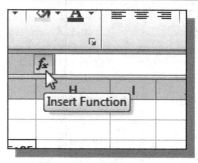

2. Click on the **Insert Function** icon that is located in front of the input edit box as shown in the figure.

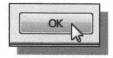

Select a function:

| MDETERM |
| MINVERSE |
| MMULT |
| MOD |
| MROUND |
| MULTINOMIAL |
| ODD |

MMULT(array1,array2)
Returns the matrix product of two arrays, an array with the same number of rows as array1 and columns as array2.

3. Select the **MMULT** command as shown.

4. Click **OK** to proceed with the **MMULT** command.

OK

5. The *Function Arguments* dialog box appears on the screen with a brief description of the function. Note that two arrays are required for this function.

6. Select the **overall global [k] matrix**, which is the 6x6 array, as shown in the first array.

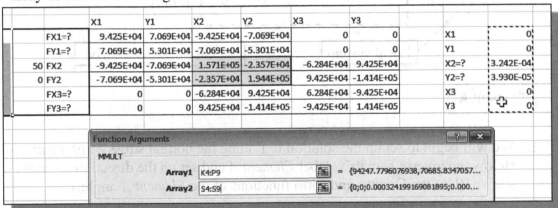

7. Hit the **Tab** key once to move the cursor to the second array input box.

8. Select the **overall global displacement matrix** as the second array. This is the 6x1 array as shown in the figure.

9. It is important to note that when using the matrix commands in *Excel*, we must press **CTRL+SHIFT+ENTER** to perform the calculations. The global forces are calculated as shown in the figure.

❖ Note this calculation also provides a quick check against the initial forces at node 2. The small FY2 represents the rounding error that exists in computer software.

Determining the Stresses in Elements

To determine the normal stress in each truss member, one option is to use the displacement transformation equations to transform the results from the global coordinate system back to the local coordinate system.

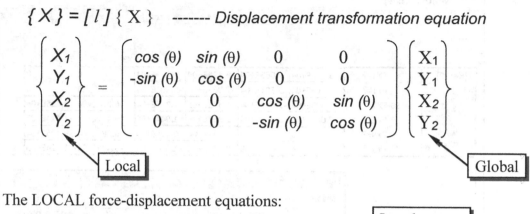

$$\{X\} = [l]\{X\} \quad \text{------- Displacement transformation equation}$$

$$
\begin{Bmatrix} X_1 \\ Y_1 \\ X_2 \\ Y_2 \end{Bmatrix} =
\begin{bmatrix}
\cos(\theta) & \sin(\theta) & 0 & 0 \\
-\sin(\theta) & \cos(\theta) & 0 & 0 \\
0 & 0 & \cos(\theta) & \sin(\theta) \\
0 & 0 & -\sin(\theta) & \cos(\theta)
\end{bmatrix}
\begin{Bmatrix} X_1 \\ Y_1 \\ X_2 \\ Y_2 \end{Bmatrix}
$$

Local   Global

The LOCAL force-displacement equations:

Local system

$$
\begin{Bmatrix} F_{1X} \\ F_{1Y} \\ F_{2X} \\ F_{2Y} \end{Bmatrix} =
\frac{EA}{L}
\begin{bmatrix}
+1 & 0 & -1 & 0 \\
0 & 0 & 0 & 0 \\
-1 & 0 & +1 & 0 \\
0 & 0 & 0 & 0
\end{bmatrix}
\begin{Bmatrix} X_1 \\ Y_1 \\ X_2 \\ Y_2 \end{Bmatrix}
$$

Local system   Local Stiffness Matrix

Element A

1. We will begin to set up the solution in finding the normal stress of *Element A*. Below the inverse K matrix, label *Element A* and set up the direction cosines as shown. (Use the **Cosine** and **Sine** functions of the *Element A* angle.)

		7.8609E-07	5.239E-06		Y2		3.93(
Element A							
	x=lX						
		0.8	0.6	0		0	
		-0.6	0.8	0		0	
		0	0	0.8		0.6	
		0	0	-0.6		0.8	

2. Pre-select the **four cells** in front of the local displacement labels as shown.

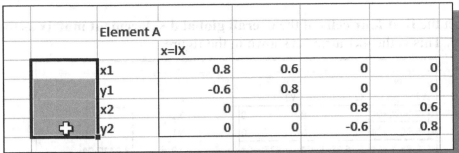

Element A				
	x=lX			
x1	0.8	0.6	0	0
y1	-0.6	0.8	0	0
x2	0	0	0.8	0.6
y2	0	0	-0.6	0.8

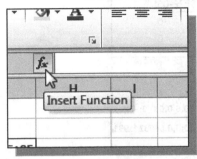

3. Click on the **Insert Function** icon that is located in front of the input edit box as shown in the figure.

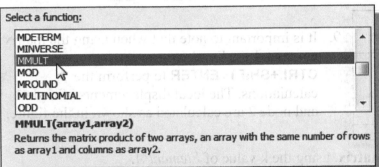

Select a function:

MDETERM
MINVERSE
MMULT
MOD
MROUND
MULTINOMIAL
ODD

MMULT(array1,array2)
Returns the matrix product of two arrays, an array with the same number of rows as array1 and columns as array2.

4. Select the **MMULT** command as shown.

5. Click **OK** to proceed with the **MMULT** command.

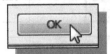
OK

6. Select the **directional cosine matrix**, which is the 4x4 array, which we just set up as the first array.

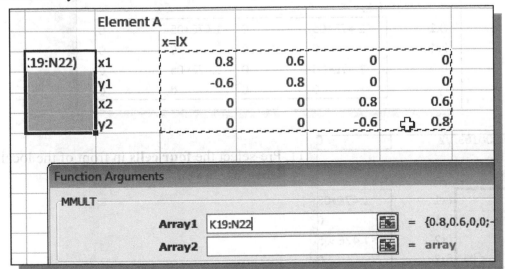

	Element A			
		x=lX		
:19:N22)	x1	0.8	0.6	0	0
	y1	-0.6	0.8	0	0
	x2	0	0	0.8	0.6
	y2	0	0	-0.6	0.8

Function Arguments

MMULT

Array1 K19:N22   = {0.8,0.6,0,0;–

Array2   = array

7. Hit the **Tab** key once to move the cursor to the second array input box.

8. Select the **first four cells** of the **overall global displacement matrix** as the second array. This is the 4x1 array as shown in the figure.

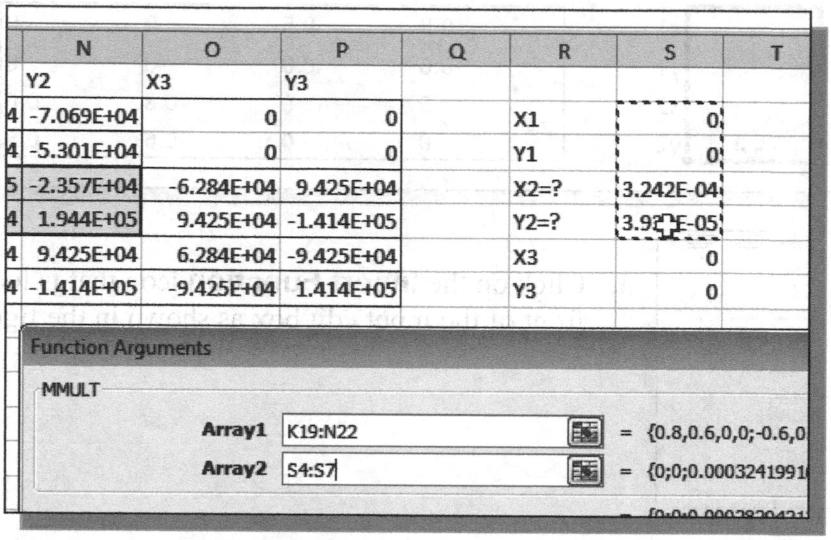

	N	O	P	Q	R	S	T
	Y2	X3	Y3				
4	-7.069E+04	0	0		X1	0	
4	-5.301E+04	0	0		Y1	0	
5	-2.357E+04	-6.284E+04	9.425E+04		X2=?	3.242E-04	
4	1.944E+05	9.425E+04	-1.414E+05		Y2=?	3.92?E-05	
4	9.425E+04	6.284E+04	-9.425E+04		X3	0	
4	-1.414E+05	-9.425E+04	1.414E+05		Y3	0	

Function Arguments

MMULT

Array1	K19:N22	= {0.8,0.6,0,0;-0.6,0
Array2	S4:S7	= {0;0;0.0003241991

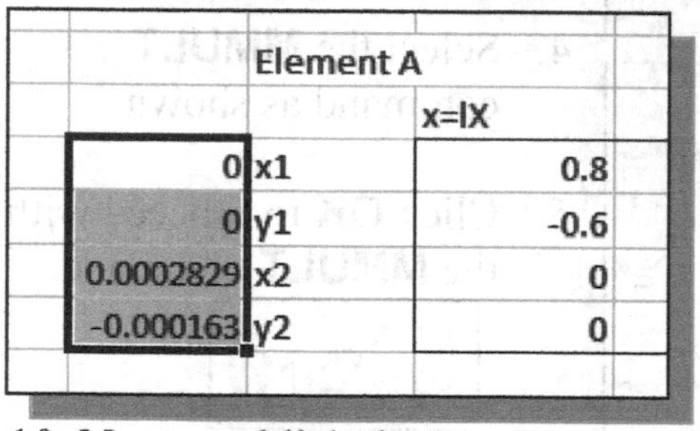

9. It is important to note that when using the matrix commands in *Excel*, we must press **CTRL+SHIFT+ENTER** to perform the calculations. The local displacements of node 1 and node 2 are calculated as shown in the figure.

	Element A	
		x=lX
0	x1	0.8
0	y1	-0.6
0.0002829	x2	0
-0.000163	y2	0

10. Next establish the **local k matrix**, using the k value of *Element A*.

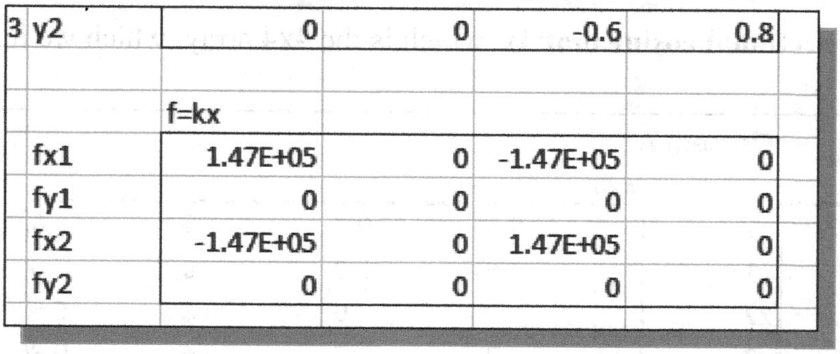

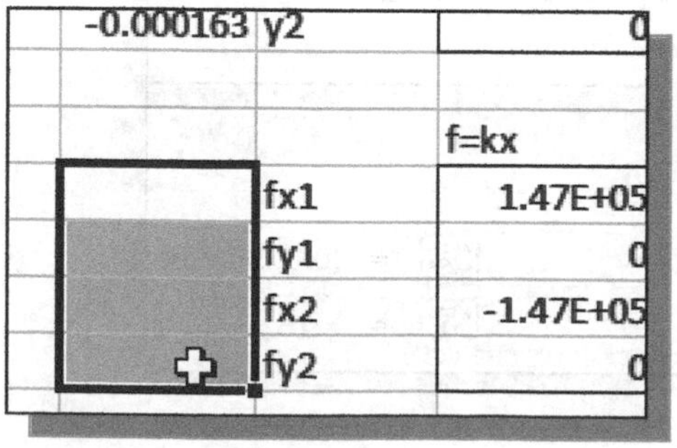

3	y2	0	0	-0.6	0.8
		f=kx			
	fx1	1.47E+05	0	-1.47E+05	0
	fy1	0	0	0	0
	fx2	-1.47E+05	0	1.47E+05	0
	fy2	0	0	0	0

-0.000163	y2		0
		f=kx	
	fx1		1.47E+05
	fy1		0
	fx2		-1.47E+05
	fy2		0

11. Pre-select the **four cells** in front of the local force labels as shown.

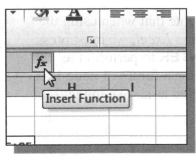

Insert Function

12. Click on the **Insert Function** icon that is located in front of the input edit box as shown in the figure.

13. Select the **MMULT** command as shown.

14. Click **OK** to proceed with the **MMULT** command.

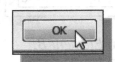

Select a function:

MDETERM
MINVERSE
MMULT
MOD
MROUND
MULTINOMIAL
ODD

MMULT(array1,array2)
Returns the matrix product of two arrays, an array with the same number of rows as array1 and columns as array2.

15. Select the **local k matrix**, which is the 4x4 array, which we just set up as the first array.

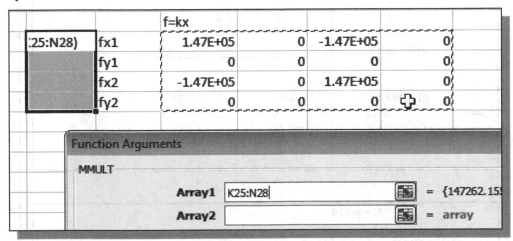

		f=kx			
:25:N28)	fx1	1.47E+05	0	-1.47E+05	0
	fy1	0	0	0	0
	fx2	-1.47E+05	0	1.47E+05	0
	fy2	0	0	0	0

Function Arguments

MMULT

Array1 K25:N28   = {147262.15!

Array2   = array

16. Select the **local displacement matrix** as the second array. This is the 4x1 array which we calculated in step 9.

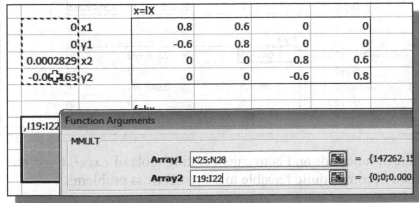

		x=lX			
0	x1	0.8	0.6	0	0
0	y1	-0.6	0.8	0	0
0.0002829	x2	0	0	0.8	0.6
-0.0...163	y2	0	0	-0.6	0.8

Function Arguments

MMULT

Array1 K25:N28   = {147262.15

Array2 I19:I22   = {0;0;0.000

-0.000163	y2		0
		f=kx	
-41.66667	fx1		1.47E+05
0	fy1		0
41.666667	fx2		-1.47E+05
0	fy2		0

17. It is important to note that when using the matrix commands in *Excel*, we must press **CTRL+SHIFT+ENTER** to perform the calculations. The local forces of node 1 and node 2 are calculated as shown in the figure.

18. To calculate the normal stress of *Element A*, use the equation $\sigma = F/A$.

			f=kx			
	-41.66667	fx1	1.47E+05	0	-1.47E+05	0
Stress	0	fy1	0	0	0	0
=I27/F6	41.666667	fx2	-1.47E+05	0	1.47E+05	0
	0	fy2	0	0	0	0

19. On your own, repeat the above steps and find the normal stress in *Element B*.

		Element A				
5			x=lX			
	0	x1	0.8	0.6	0	0
	0	y1	-0.6	0.8	0	0
5	0.0002829	x2	0	0	0.8	0.6
	-0.000163	y2	0	0	-0.6	0.8
			f=kx			
	-41.66667	fx1	1.47E+05	0	-1.47E+05	0
Stress	0	fy1	0	0	0	0
848.82636	41.666667	fx2	-1.47E+05	0	1.47E+05	0
	0	fy2	0	0	0	0

		Element B				
			x=lX			
	0.0001471	x2	0.5547002	-0.83205	0	0
	0.0002916	y2	0.83205029	0.5547002	0	0
	0	x3	0	0	0.5547002	-0.83205
	0	y3	0	0	0.8320503	0.5547002
			f=kx			
	30.046261	fx1	2.04E+05	0	-2.04E+05	0
Stress	0	fy1	0	0	0	0
-612.0974	-30.04626	fx2	-2.04E+05	0	2.04E+05	0
	0	fy2	0	0	0	0

❖ With the built-in functions and convenient editing tools of *Excel*, the direct stiffness matrix method becomes quite feasible to solve 2D truss problems.

The Truss Solver and the Truss View programs

The *Truss Solver* program is a custom-built *Windows* based computer program that can be used to solve 1D, 2D and 3D trusses. The *Truss Solver* program is based on the direct stiffness matrix method with a built-in editor and uses the *Cholesky* decomposition method in solving the simultaneous equations. The program is very compact, 62 Kbytes in size, and it will run in any *Windows* based system.

The *Truss View* program is a simple viewer program that can be used to view the results of the *Truss Solver* program. It simply provides a graphical display of the information that is stored in the output file of the *Truss Solver* program; no FEA analysis is done in the *Truss View* program.

❖ First download the *Truss39.zip* file from the *SDC Publications* website.

1. Launch your internet browser, such as the *MS Internet Explorer* or *Mozilla Firefox* web browsers.

2. In the *URL address* box, enter
   www.SDCPublications.com/downloads/978-1-63057-484-0

3. Click the **Download Now** button to download the *Truss Solver* software (*Truss39.zip*) to your computer.

4. On your own, extract the content of the ZIP file to a folder or on the desktop. Two files are included in the ZIP file: (1) the *Truss Solver* program, *Truss39.exe,* and (2) the *Truss View* program, *TrusVW39.exe.*

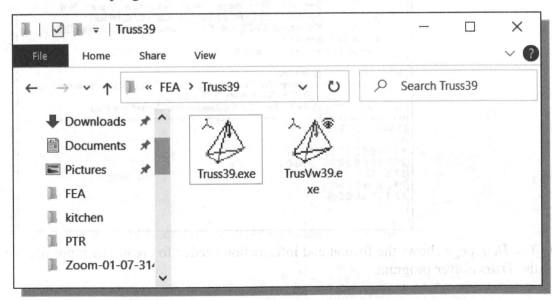

5. Start the *Truss Solver* by double-clicking on the **Truss39.exe** icon.

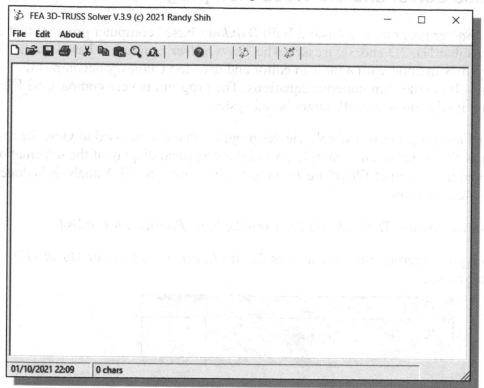

6. Hit the [**F1**] key once or click on the **Help** button to display the *Help* page.

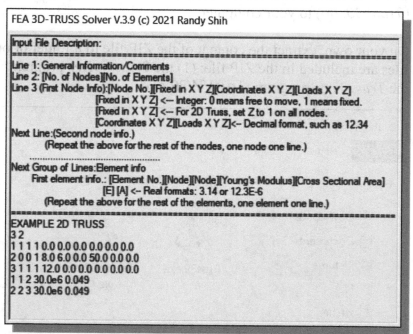

❖ The *Help* page shows the format and information needed to create the input file for the *Truss Solver* program.

7. The first line is a comment line, which can be used to describe any general information about the truss system being solved. For this example, enter **Example 3.2, 2D truss**.

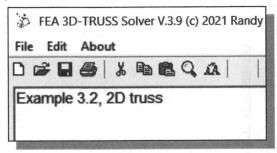

8. The second line contains two numbers: the number of nodes and the number of elements in the system. For our example, we enter **3 2** (the two numbers are separated by a space) to indicate the system has 3 nodes and 2 elements.

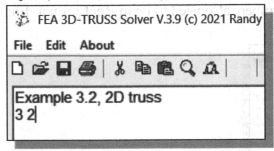

9. Since we have indicated there are three nodes, the next three lines will be describing the three nodes. Enter **1**, **2** and **3** as the first number of the three lines.

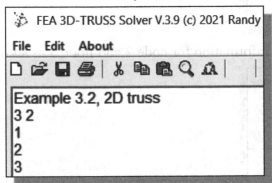

10. The next three numbers, behind the node number, are used to identify the restraints of the node in X, Y and Z directions. Use 1 to indicate it cannot move in that direction; use 0 if it is free to move in that direction. Enter **1 1 1** to indicate node 1 cannot move in all three directions.

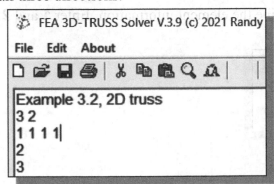

11. The next three numbers are the coordinates of the node; the coordinates need to be expressed in Real format. Enter **0.0 0.0 0.0** to identify the node is aligned to the origin of the coordinate system.

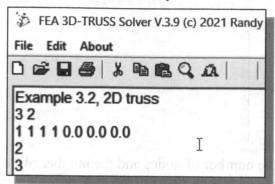

12. The last three numbers describe the external loads applied, components in the X, Y and Z directions, at the node. The numbers also need to be in Real format. Enter **0.0 0.0 0.0** to indicate no loads at node 1.

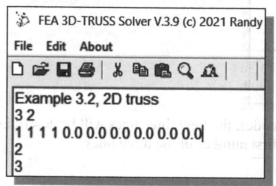

13. Repeat the above steps and enter the information for node 2 and node 3 as shown.

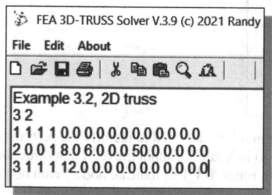

14. For the next two lines, we will describe the element information. Enter **1** and **2** as the element numbers as shown.

```
2 0 0 1 8.0 6.0 0.0 50.0 0.0 0.0
3 1 1 1 12.0 0.0 0.0 0.0 0.0 0.0
1
2
```

15. The two numbers after the element number are used to describe the two nodes connected to the element. Enter **1 2** to indicate the first element is connected to node 1 and node 2.

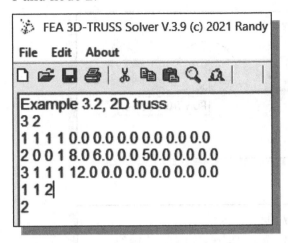

16. The next two numbers describe the **Modulus of elasticity** for the material used and the **cross-sectional area** of the first element.

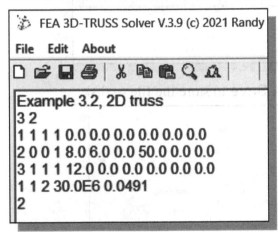

17. Repeat the above step and enter the material and cross-sectional information of the second element.

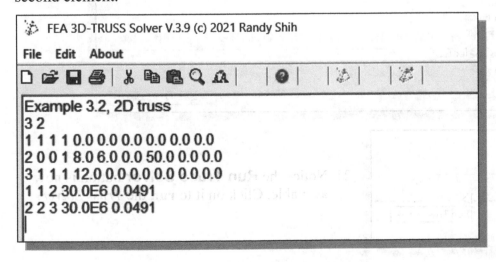

18. Notice the **Truss Solver** button is currently grayed-out, which indicates the command is unavailable. The *Truss Solver* will read the data from a disk file; therefore, it is necessary to save the data to disk.

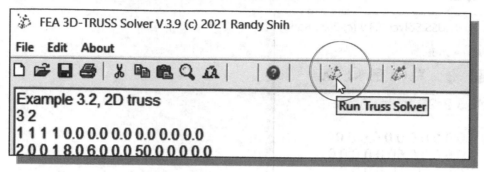

19. Click on the **Save** button as shown.

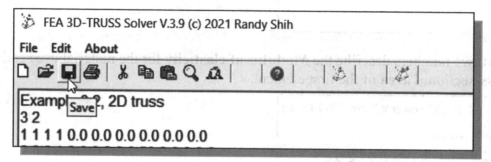

20. Enter **Ex32.txt** as the filename and click **Save** to store the file to disk.

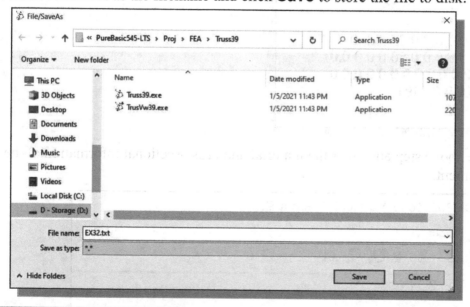

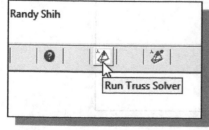

21. Notice the **Run Truss Solver** button is now available. Click on it to **run** the *Truss Solver*.

22. The *Truss Solver* will then process the data and the solution is displayed inside the built-in editor.

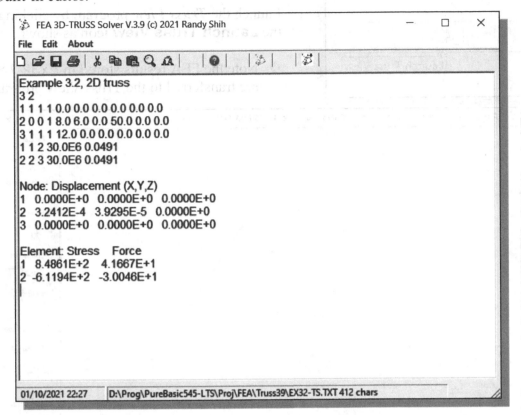

- The solution contains two sections: (1) displacements of the nodes in X, Y and Z directions, and (2) the stresses and internal forces of the elements.

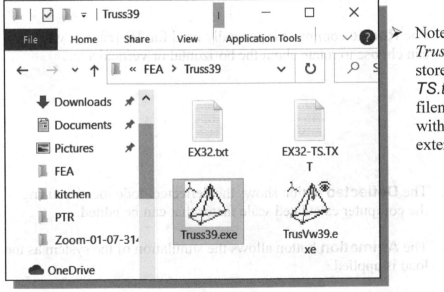

➤ Note the results of the *Truss Solver* are also stored in the *Ex32-TS.txt* file. Same filename as the input file, with a different filename extension.

❖ Also note that both *Ex32.txt* and *Ex32-TS.txt* are plain text files, which can be opened with any text editor, such as *Windows Notepad*.

The Truss View program

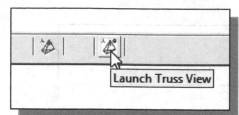

1. Launch the *Truss View* program by clicking on the **Launch Truss View** icon as shown.

   ❖ Note the FEA results stored in *Ex32-TS.txt* are transferred to the Truss View program.

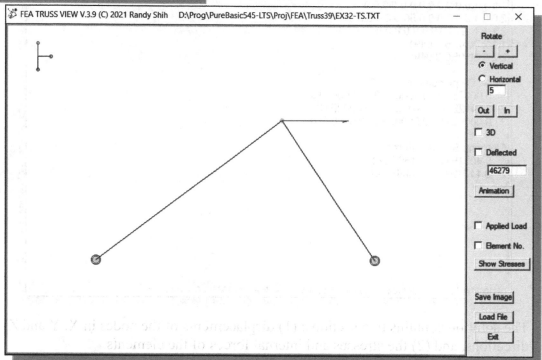

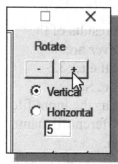

2. The **Rotate** command is generally used for **3D trusses**, which we can choose to rotate about the horizontal or vertical axes.

3. The **Deflected** option shows the deflected node location using the computer calculated scale factor that can be edited.

4. The **Animation** button allows the simulation of the system as the load is applied.

5. Click on the **Show Stresses** button to display the list of the stresses and forces for the elements.

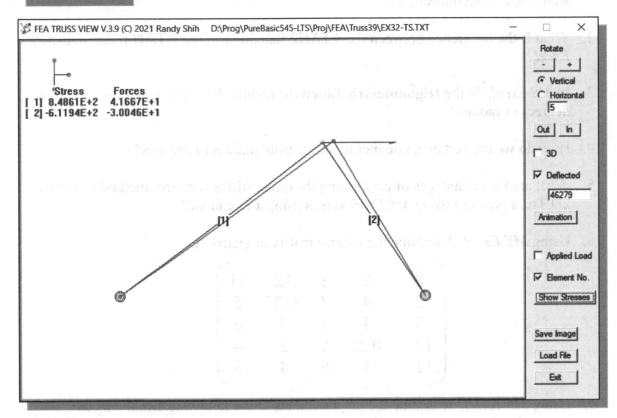

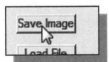

6. Click on the **Save Image** button to save a copy of the screen image as a jpg image file.

> The use of the *Truss View* program provides the user with a visual representation of the FEA results, which makes interpreting the FEA results much easier than reading the lengthy text-based output file. The use of the *Truss Solver* and the *Truss View* programs also illustrates the basic three steps of the FEA procedure: (1) **Pre-processing**, where the user creates an input file that contains all of the system information; (2) **Solver**, the FEA procedure solving of the system; and (3) **Post-processing**, displaying and viewing the FEA results.

Questions:

1. List and describe two of the commands in *MS Excel* to perform matrix operations that were used in the tutorial.

2. What is the difference between the **ATAN2** function and the **ATAN** function in *MS Excel*?

3. In *MS Excel*, do the **trigonometric functions** require the associated angles to be in degrees or radians?

4. How do we convert an angle measurement from radians to degrees?

5. What are the advantages of performing the direct stiffness matrix method to solve a 2D Truss problem using *MS Excel* versus using a calculator?

6. Using *MS Excel*, determine the inverse matrix of matrix A:

$$\begin{pmatrix} 8 & 2 & 3 & 12 & 11 \\ 2 & 4 & 7 & 0.25 & 5 \\ 3 & 7 & 3 & 5 & 6 \\ 12 & 0.25 & 5 & 2 & 4 \\ 11 & 5 & 6 & 4 & 8 \end{pmatrix}$$

7. Using *MS Excel*, determine the displacements of the following simultaneous equations:

$$\begin{Bmatrix} 25 \\ 13.25 \\ 18 \\ 19.25 \end{Bmatrix} = \begin{bmatrix} 8 & 2 & 3 & 12 \\ 2 & 4 & 7 & 0.25 \\ 3 & 7 & 3 & 5 \\ 12 & 0.25 & 5 & 2 \end{bmatrix} \begin{Bmatrix} X_1 \\ Y_1 \\ X_2 \\ Y_2 \end{Bmatrix}$$

Exercises: Solve the following problems using *MS Excel* and the *Truss Solver* program.

1.   Given: two-dimensional truss structure as shown. (All joints are pin joints.)

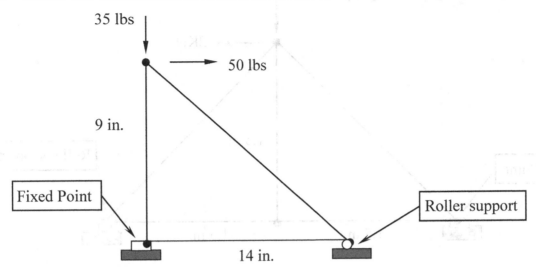

Material: Steel, diameter ¼ in.

Find: (a) Displacements of the nodes.
        (b) Normal stresses developed in the members.

2.   Given: Two-dimensional truss structure as shown. (All joints are pin joints.)

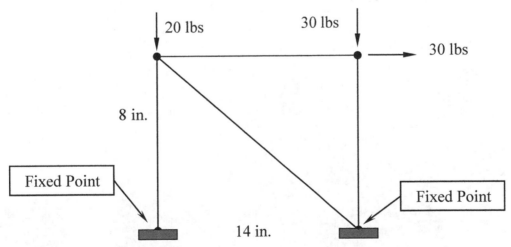

Material: Steel, diameter ¼ in.

Find: (a) Displacements of the nodes.
       (b) Normal stresses developed in the members.

3.   Given: two-dimensional truss structure as shown. (All joints are pin joints.)

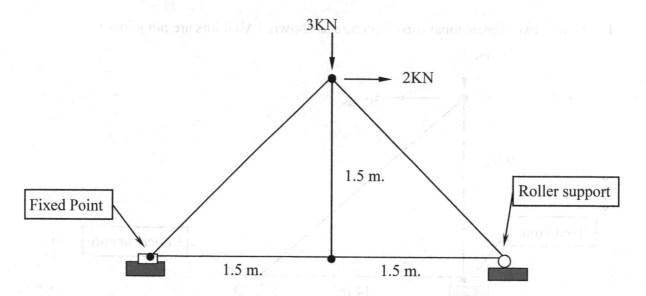

3KN

2KN

1.5 m.

Fixed Point

Roller support

1.5 m.   1.5 m.

Material: Steel, diameter 50mm.

Find: (a) Displacements of the nodes.
      (b) Normal stresses developed in the members.

Chapter 4
Truss Elements in SOLIDWORKS Simulation

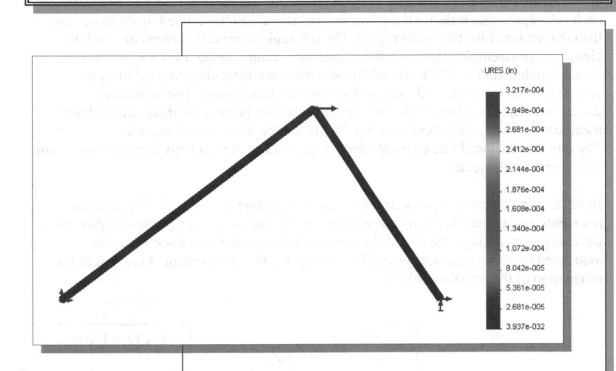

URES (in)

3.217e-004
2.949e-004
2.681e-004
2.412e-004
2.144e-004
1.876e-004
1.608e-004
1.340e-004
1.072e-004
8.042e-005
5.361e-005
2.681e-005
3.937e-032

Learning Objectives

- ◆ **Create SOLIDWORKS Truss Models.**
- ◆ **Create Truss Elements and FEA models in SOLIDWORKS Simulation.**
- ◆ **Apply Loads and Boundary conditions at Nodes.**
- ◆ **Run the SOLIDWORKS Simulation FEA Solver.**
- ◆ **View and Examine the SOLIDWORKS Simulation FEA Results.**
- ◆ **Understand the General Computer FEA Procedure.**

One-dimensional Line Elements

The finite element analysis method is a numerical solution technique that finds an approximate solution by dividing a region into small sub-regions. The solution within each sub-region that satisfies the governing equations can be reached much more simply than that required for the entire region. The sub-regions are called *elements*, and the elements are assembled through interconnecting a finite number of points on each element called *nodes*. While all real-life structures are three-dimensional in nature, solutions of many stress analyses are done on two-dimensional spaces and one-dimensional spaces. Quite often, the approximations represent the three-dimensional members very well, and there is no need to do a three-dimensional analysis.
The **one-dimensional** line elements that are commonly used in FEA include truss, beam and boundary elements.

In FEA, a line element is geometrically a line connecting nodes at which *loads* and *restraints* are applied. We can consider the line represents the axis of the member, where the axis passes through the centroid locations of the member cross sections. The associated cross-sections are generally shown along the line elements to represent the orientation of the actual members.

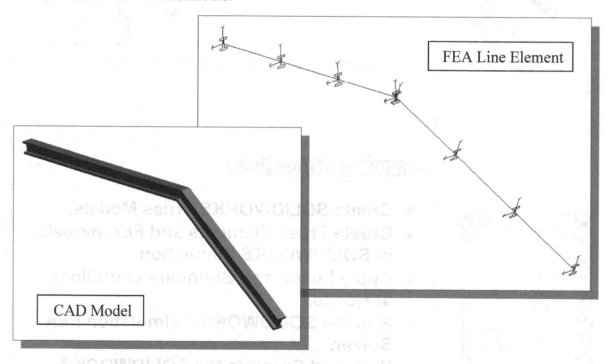

In modern FEA packages such as SOLIDWORKS Simulation, a CAD model is generally created first and the FEA model is then created on top of the CAD model.

In SOLIDWORKS Simulation, to create an FEA model we are required to construct the associated CAD model first. There are several options in creating the truss/beam CAD models in SOLIDWORKS. Once the CAD model is created, it is then transferred into the SOLIDWORKS Simulation module to perform the necessary FEA analysis.

In this chapter, we will examine the process of creating, solving and viewing the results of a finite element analysis on a two-dimensional truss structure using the FEA application software SOLIDWORKS Simulation, which is available in SOLIDWORKS. Note that several options are available in SOLIDWORKS to create truss/beam systems. The same 2D truss system that was analyzed in the previous chapter will be used as the first FEA example.

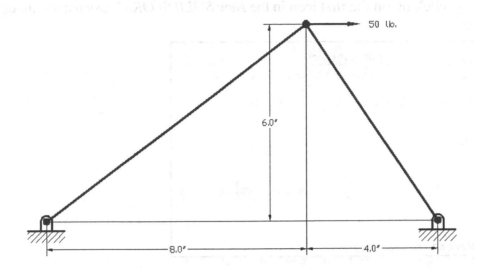

To create trusses and beams models in SOLIDWORKS, two different approaches are available. We can either (1) create the truss and beam members using the standard parametric **solid modeling tools**, an approach which treats the members the same as any other solid model; or (2) use the **Structural Members** tool specifically designed for creating long and slender members. In this chapter, we will create and analyze the truss system using the solid modeling approach. The **Structural Members** tool approach will be demonstrated in the next chapter.

The typical procedure of creating and analyzing an FEA truss or beam element model in SOLIDWORKS involves the following steps:

1. Create the solid models of the truss or beam systems using either the standard parametric modeling tools or the **Structural Members** tool.
2. Transfer the CAD model into SOLIDWORKS Simulation.
3. Define the type of finite element analysis to be performed.
4. Select the proper element type and create the FEA model.
5. Assign material properties to the FEA model.
6. Prescribe how the system is supported.
7. Prescribe how the loads are applied to the system.
8. Run the *FEA Solver* to compute displacements, strains and stresses.
9. View the results of the FEA procedure and confirm the results of the analysis.

Starting SOLIDWORKS

1. Select the **SOLIDWORKS** option on the *Start* menu or select the **SOLIDWORKS** icon on the desktop to start SOLIDWORKS. The SOLIDWORKS main window will appear on the screen.

2. Select **Part** by clicking on the first icon in the *New SOLIDWORKS Document* dialog box as shown.

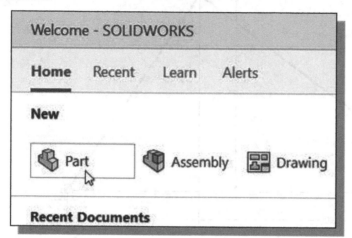

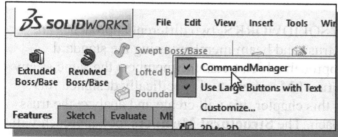

3. In the *Standard* toolbar area, right-click on any icon and activate the **Command Manager** in the option list.

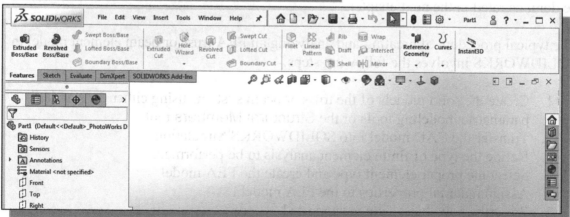

❖ The ***Command Manager*** is a context-sensitive toolbar that dynamically updates based on the user's selection. When you click a tab below the *Command Manager*, it updates to display the corresponding toolbar. By default, the *Command Manager* has toolbars embedded in it based on the document type.

Units Setup

When starting a new CAD file, the first thing we should do is choose the units we would like to use. We will use the *English* setting (inches) for this example.

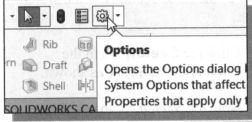

1. Select the **Options** icon from the *Menu Bar* toolbar to open the *Options* dialog box.

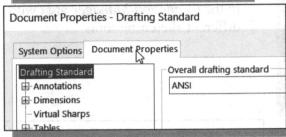

2. Switch to the **Document Properties** tab and reset the *Drafting Standard* to **ANSI** as shown in the figure.

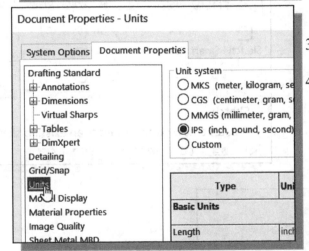

3. Click **Units** as shown in the figure.

4. Select **IPS (inch, pound, second)** under the *Unit system* options.

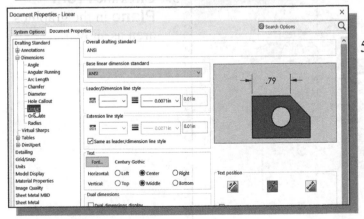

5. On your own, examine the different settings available, such as defining the degree of accuracy with which the units will be displayed.

6. Click **OK** in the *Options* dialog box to accept the selected settings.

Creating the CAD Model – Solid Modeling Approach

To create the two-member truss system, we will first establish the locations of the two center axes of the members by creating a 2D sketch. Two additional datum planes will then be created to allow the placement of the profiles of the cross sections of the system.

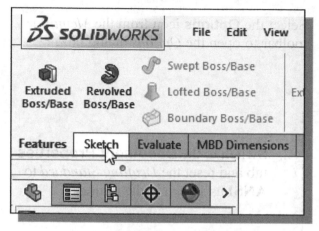

1. Click the **Sketch** tab in the *Command Manager* as shown.

❖ Note the *Sketch* toolbar is displayed in the *Command Manager* area with different sketching tools.

2. Click the **Sketch** icon, in the *Sketch* toolbar, to create a new sketch.

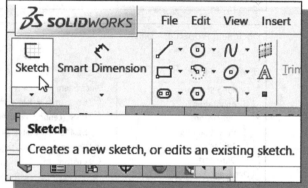

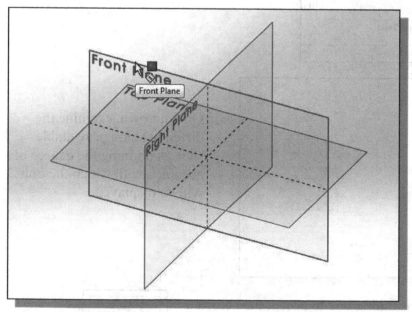

3. Click the **Front Plane** in the graphics area as shown.

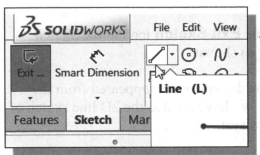

4. Click the **Line** icon in the *Sketch* toolbar as shown.

5. Start the line segment at the **origin** and create the two connected line segments as shown. (Note: **Do not** make the two lines perpendicular to each other.)

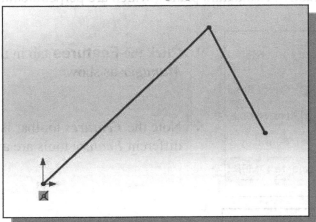

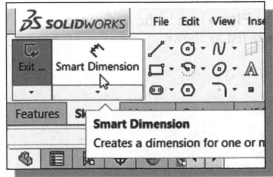

6. Click the **Smart Dimension** icon in the *Sketch* toolbar as shown.

7. On your own, create the four dimensions to adjust the locations of the endpoints as shown. (Note the **0.00** dimension is used to align the right endpoint to the **origin** in the vertical direction.)

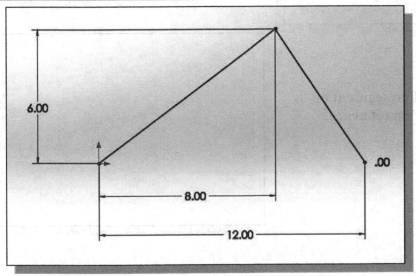

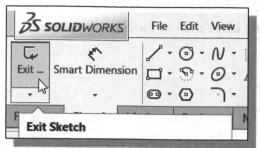

8. Click the **Exit Sketch** icon in the *Sketch* toolbar as shown.

❖ Note the dimensions disappeared from the screen. We have created the 2D line sketch of the truss system.

❖ The two lines we just created represent the center axes of the two truss members. Next, we will establish two datum planes, which are perpendicular to the two established lines.

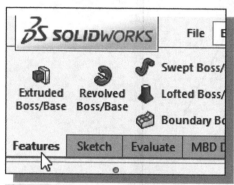

9. Click the **Features** tab in the *Command Manager* as shown.

❖ Note the *Features* toolbar is activated and different *Feature* tools are available.

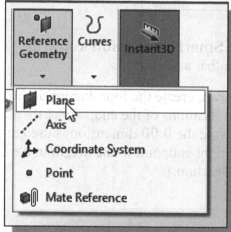

10. Click the **Plane** icon in the *Reference Geometry* list as shown.

11. Select the **line segment** on the left as the first reference.

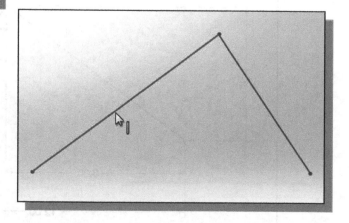

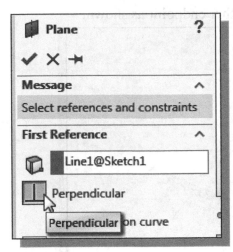

12. In the *Plane* option list, select **Perpendicular**. This option sets the new datum plane to be *perpendicular* to the selected geometry.

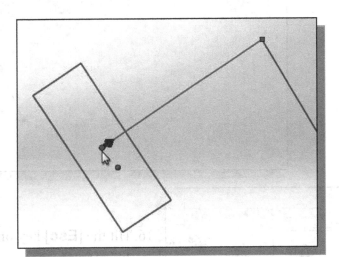

13. Select the **bottom endpoint** on the left as the *second reference*.

❖ Note the selected endpoint sets the location of the new datum plane.

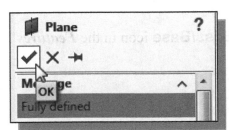

14. Click **OK** to accept the settings and create the new datum plane.

❖ The new datum plane is perpendicular to the line segment and placed exactly at the bottom endpoint of the left line segment. This will allow the construction of the truss element by using the **Extrude** command.

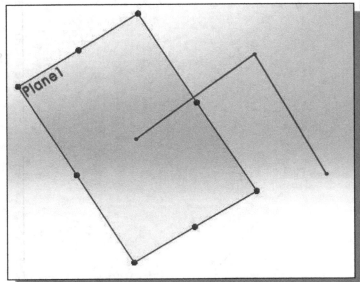

15. On your own, create the other datum plane at the other endpoint as shown.

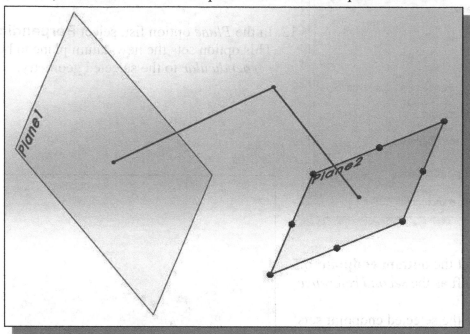

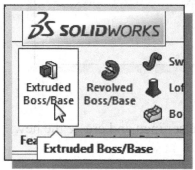

16. Hit the [**Esc**] key once to de-select any pre-selected items.

17. Click the **Extruded Boss/Base** icon in the *Features* toolbar as shown.

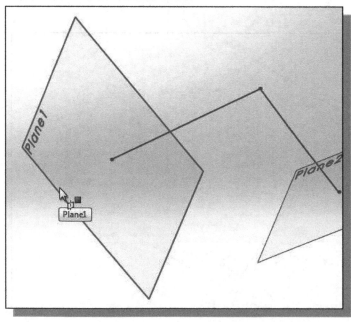

18. Select the **first datum plane**, **Plane 1**, to align the sketching plane.

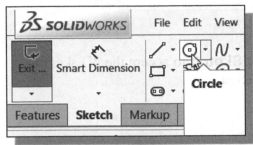

19. Click the **Circle** icon in the *Sketch* toolbar as shown.

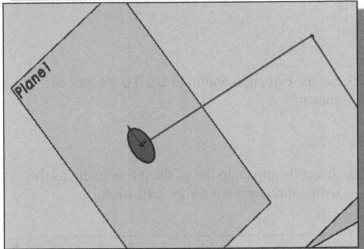

20. Select the bottom endpoint of the left line segment to align the center of the circle.

21. On your own, create a circle of arbitrary size.

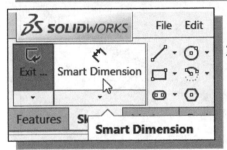

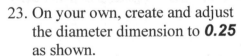

22. Click the **Smart Dimension** icon in the *Sketch* toolbar as shown.

23. On your own, create and adjust the diameter dimension to *0.25* as shown.

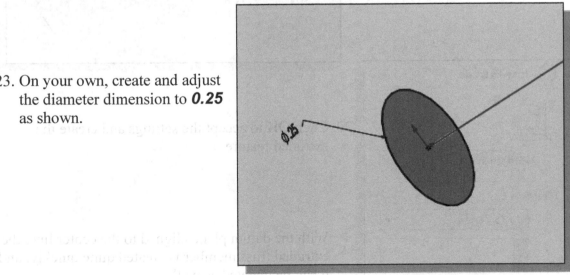

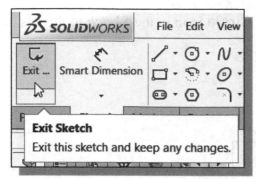

24. Click **Exit Sketch** to accept the 2D sketch we just completed.

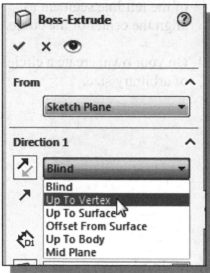

25. Set the extrusion option to **Up To Vertex** as shown.

26. Select the top endpoint of the left segment as the termination location for the extrusion.

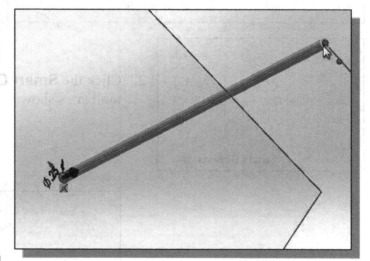

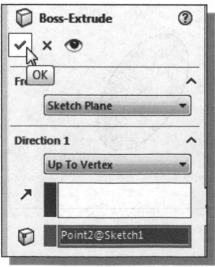

27. Click **OK** to accept the settings and create the extruded feature.

❖ With the datum plane aligned to the center line, the extruded truss member is created quite quickly, and it can be updated as well.

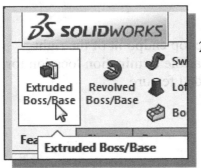

28. Click the **Extruded Boss/Base** icon in the *Features* toolbar as shown.

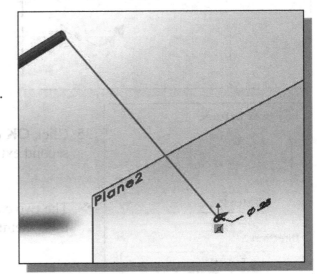

29. Select the **second datum plane**, **Plane 2**, to align the sketching plane.

30. On your own, create a **diameter 0.25 circle** that is aligned to the bottom endpoint of the right line segment as shown.

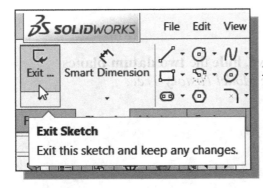

31. Click **Exit Sketch** to accept the 2D sketch we just completed.

32. Confirm the extrusion option is still set to **Up To Vertex** as shown.

33. Turn *off* the **Merge result** option as shown.

❖ This step is necessary as SOLIDWORKS Simulation will treat each extruded feature as a separate truss member.

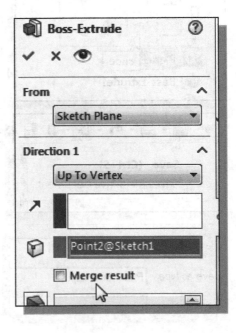

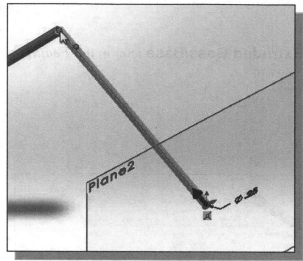

34. Select the top endpoint of the right segment as the termination location for the extruded feature.

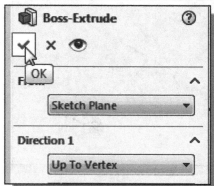

35. Click **OK** to accept the settings and create the second extruded feature.

❖ The two extruded features represent two separate truss members.

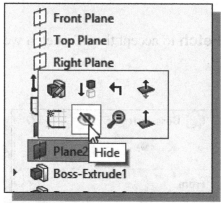

36. On your own, **hide** the **two datum planes** through the *Model History Tree*.

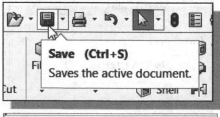

37. Click the **Save** button in the *Standard* toolbar.

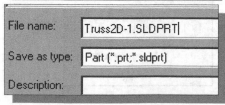

38. Enter **Truss2D-1.SLDPRT** as the *File name* and save the model.

A CAD Model is NOT an FEA Model

The two extruded features we just created represent the solid model of the truss system that we will be analyzing. This model is created under SOLIDWORKS, and it is a CAD model that contains geometric information about the system. The CAD model can be used to provide geometric information needed in the finite element model. However, the CAD model is not an FEA model. In the previous chapters, we have examined the FEA procedure, which concentrated on the nodes and elements. So far, none of the nodes or elements exist in our CAD model. The geometric information of the system is necessary for the FEA analysis, and what we have established in the CAD model is just the geometric information.

The CAD model is typically developed to provide geometric information necessary for manufacturing. All details must be specified, and all dimensions are required. The manufactured part and the CAD model are identical in terms of geometric information. The FEA model uses the geometric information of the CAD model as the starting point, but the FEA model usually will adjust some of the basic geometric information of the CAD model. The FEA model usually contains additional nodes and elements. Idealized boundary conditions and external loads are also required in the model. The goal of finite element analysis is to gain sufficient reliable insights into the behaviors of the real-life system. Many assumptions are made in the finite element analysis procedure to simplify the analysis, since it is not possible or practical to simulate all factors involved in real-life systems. The finite element analysis procedure provides an idealized approximation of the real-life system. It is therefore not practical to include all details of the system in the FEA model; the associated computational cost cannot be justified in doing so. It is a common practice to begin with a more simplified FEA model. Once the model has been solved accurately and the result has been interpreted, it is feasible to consider a more refined model in order to increase the accuracy of the prediction of the actual system. In SOLIDWORKS Simulation, creating an FEA model by using the geometric information of the CAD model is known as *idealization*.

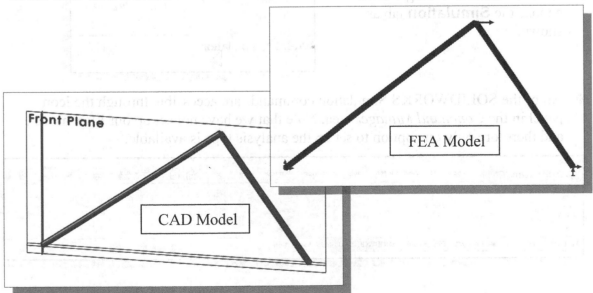

The SOLIDWORKS Simulation Module

SOLIDWORKS Simulation is a multi-discipline Computer Aided Engineering (CAE) tool that enables the user to simulate the physical behavior of a model, and therefore enables the user to improve the design. SOLIDWORKS Simulation can be used to predict how a design will behave in the real world by calculating stresses, deflections, frequencies, etc. All of the SOLIDWORKS Simulation functions are integrated as part of SOLIDWORKS, which offers the convenience and power of SOLIDWORKS' parametric modeling capabilities.

1. Click the **SOLIDWORKS Add-Ins** tab in the *Command Manager* as shown.

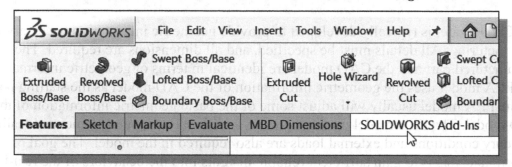

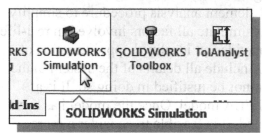

2. Start the SOLIDWORKS Simulation module by selecting the **SOLIDWORKS Simulation** option in the *Command Manager* as shown.

3. In the *Command Manager* area, choose the **Simulation** tab as shown.

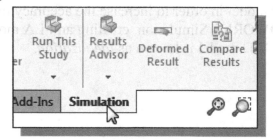

❖ All of the SOLIDWORKS Simulation commands are accessible through the icon panel in the *Command Manager* area. Note that we have not set up our FEA model, and therefore only one option to set up the analysis type is available.

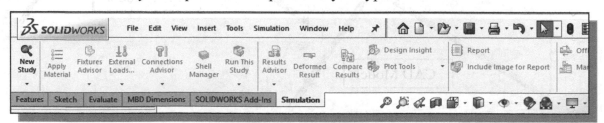

Creating an FEA Model

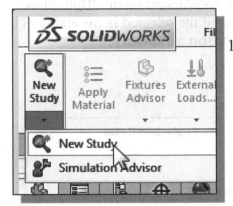

1. Choose **New Study**, the second option in the *Study Advisor* icon list, in the *Command Manager* area.

2. Select the analysis type to **Static** as shown in the figure.

❖ The **Static** option allows us to perform the basic linear elastic FEA analysis of the system.

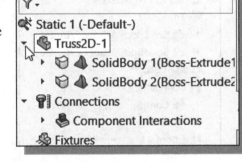

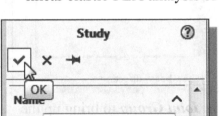

3. Click **OK** to accept the selection.

4. Click on the [triangle] in front of the model name to expand the list as shown.

❖ Note that, by default, the two extruded features are treated as two separate *SolidBodies*.

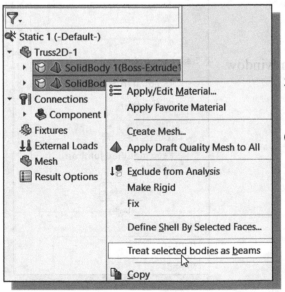

5. Select both *SolidBodies* while holding down the [**Ctrl**] key.

6. Right-click once on one of the selected items and select **Treat selected bodies as beams** as shown.

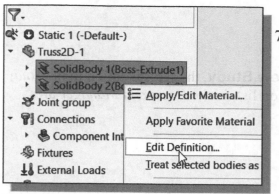

7. Select both *SolidBodies* again. Right-click once on one of the selected items and select **Edit Definition** as shown.

8. Select **Truss** in the *Apply/Edit Beam* option window as shown.

9. Click **OK** to accept the setting.

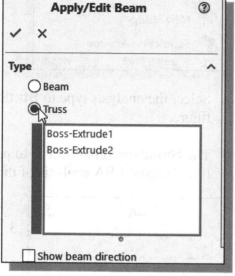

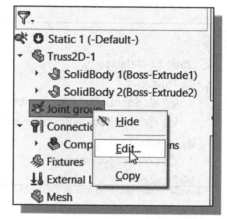

10. Right-click once on ***Joint Group*** to bring up the option list and select **Edit** as shown.

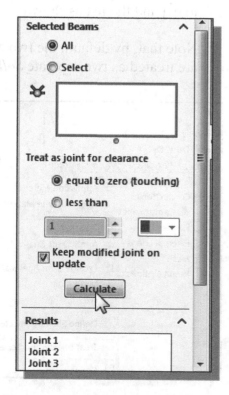

11. Select **Calculate** in the ***Edit Joints*** option window to update the joints as shown.

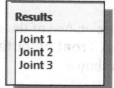

12. Click **OK** to accept the settings.

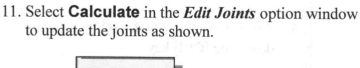

Assign the Element Material Property

Next, we will set up the *Material Property* for the elements. The *Material Property* contains the general material information, such as *Modulus of Elasticity*, *Poisson's Ratio*, etc. that are necessary for the FEA analysis.

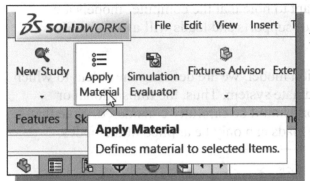

1. Choose the **Apply Material** option from the pull-down menu as shown.

❖ Note the default list of materials, which is available in the pre-defined SOLIDWORKS Simulation material library, is displayed.

2. Select **ALLOY STEEL** in the *Material* list as shown.

3. Set the **Units** option to display **English (IPS)** to make the selected material available for use in the current FEA model.

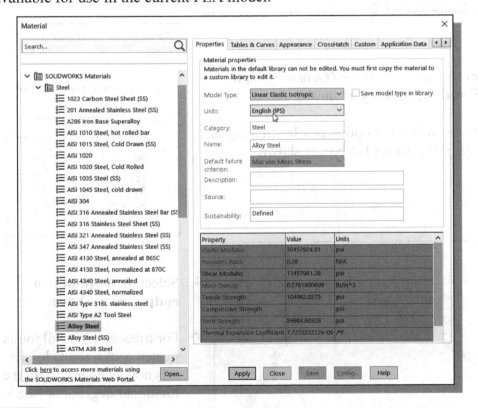

4. Click **Apply** to assign the material property, then click **Close** to exit the Material Assignment command.

Apply Boundary Conditions – Constraints

In SOLIDWORKS Simulation FEA analysis, the real-world conditions are simulated through the use of **constraints** and **loads**. In defining constraints for a SOLIDWORKS Simulation model, the goal is to model how the real-life system is supported. Loads are also generalized and idealized. It is important to note that the computer models are idealized; our selections of different types of supports and loads will affect the FEA results.

In constraining a SOLIDWORKS Simulation model, we are defining the extent to which the model can move in reference to a coordinate system. Thus, the translational or rotational movements of a particular support should be carefully considered. Note that *truss members* are two-force members; the loads can only be applied at the nodes.

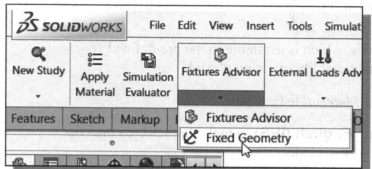

1. Choose **Fixed Geometry** by clicking the icon in the *Fixtures Advisor* tool as shown.

2. Confirm the *Standard (Fixed Geometry)* option is set to **Fixed Geometry** as shown.

3. Activate the *Joints* list option box by clicking on the inside of the *Joints* list box as shown.

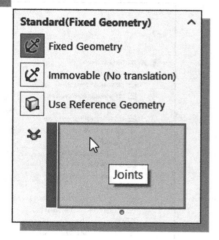

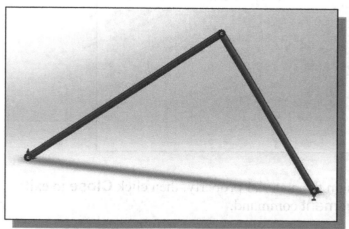

4. Select the **bottom two endpoints** as shown.

❖ For truss systems, all joints are treated as pin joints by default; it is not necessary to fix the rotational degrees of freedom.

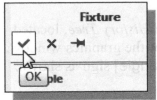

5. Click on the **OK** button to accept the first **Fixture** constraint settings.

❖ Note that for 2D truss systems, any translation motion in the Z direction is not allowed. We will need to add another constraint set to restrict the motion of the top node in the Z direction. Not restricting the movement in the Z direction will cause errors in the FEA results.

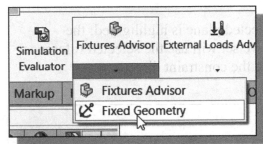

6. Choose **Fixed Geometry** by clicking the icon in the *Fixtures Advisor* toolbar as shown.

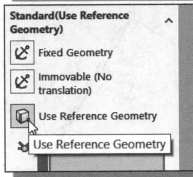

7. Change the *Standard (Fixed Geometry)* option to **Use Reference Geometry** as shown.

8. Select the **top node point** as shown.

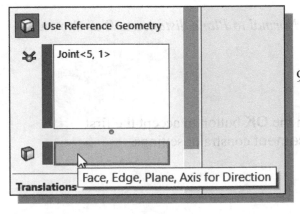

9. Activate the ***Direction Reference*** list option box by clicking on the inside of the ***Reference*** list box as shown.

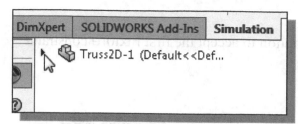

10. Expand the *Model History Tree*, located at the top left corner of the graphics window, by clicking on the [triangle] sign as shown.

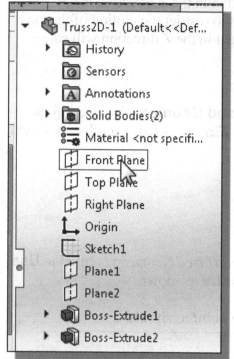

11. Select **Front Plane** as the *direction reference* as shown.

❖ Note the selected plane is highlighted; the constraints we set will use the selected reference to determine the constraint direction.

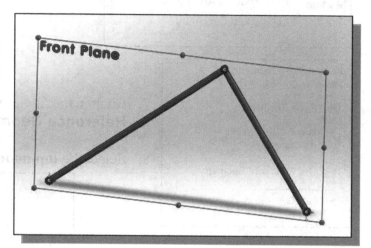

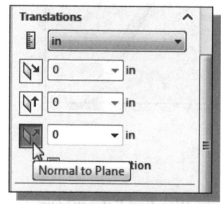

12. Set the distance measurement to **inches** to match with the system's units we are using, as shown.

13. Click on the **Normal to Plane** icon to activate the constraint.

14. Set the *Normal to Plane distance* to **0** as shown.

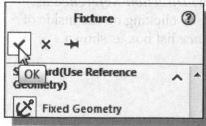

15. Click on the **OK** button to accept the first Displacement constraint settings.

Apply External Loads

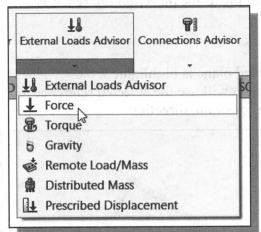

1. Choose **External Loads → Force** by clicking the icon in the toolbar as shown.

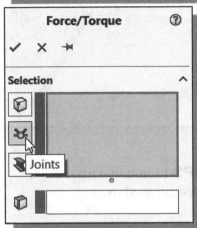

2. Change the *Force/Torque* option to **Joints** as shown.

3. Select the **top node point** as shown.

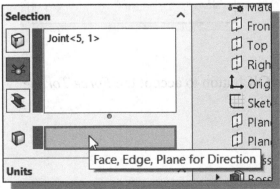

4. Activate the *Direction Reference* list option box by clicking on the inside of the *Reference* list box as shown.

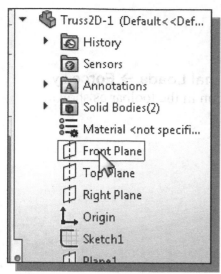

5. Select **Front Plane** as the *direction reference* as shown.

❖ Note the selected plane is highlighted; the constraints we set will use the selected reference to determine the constraint direction.

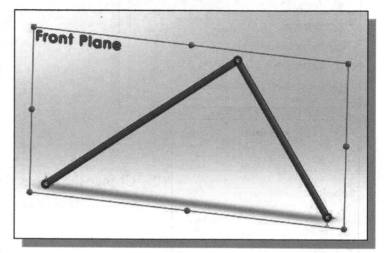

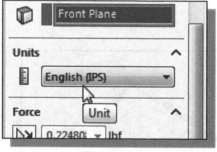

6. Set the *Unit Option* to **IPS** to match with the system's units we are using, as shown.

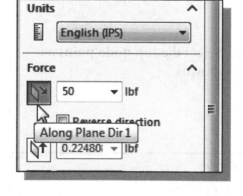

7. Click on the **Along Plane Dir 1** icon to activate the force direction.

8. Set the *Force* to **50** as shown.

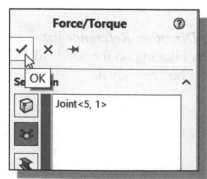

9. Click on the **OK** button to accept the *Force/Torque* settings.

Create the FEA Mesh and Run the Solver

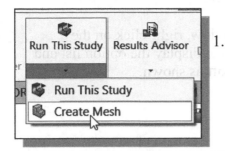

1. Choose **Create Mesh** by clicking the icon under *Run This Study* in the toolbar as shown.

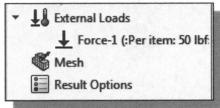

❖ Note the Mesh icon has been changed, which indicates the FEA Mesh has been created.

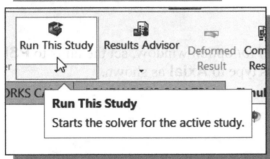

2. Click on the **Run This Study** button to start the *FEA Solver* to calculate the results.

❖ Note the stress results are displayed when the *Solver* has completed the FEA calculations.

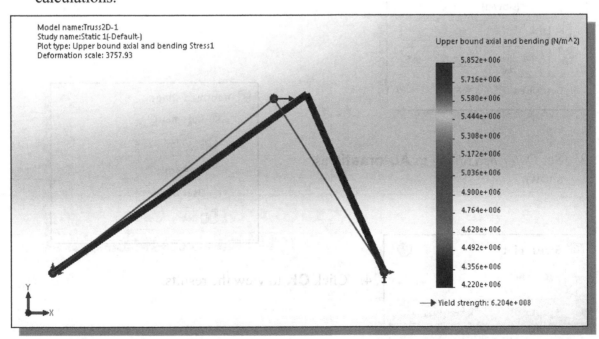

Viewing the Stress Results

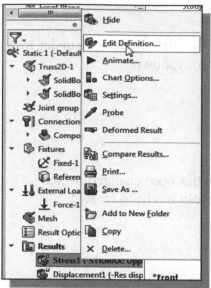

1. In the *FEA Study* window, right-click on the **Result/Stress** item to display the option list and select **Edit Definition** as shown.

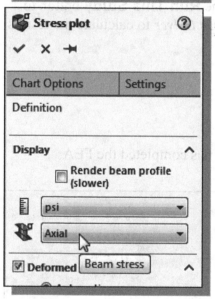

2. In the *Stress Plot* option window, set the units to **PSI** and the stress type to **Axial** as shown.

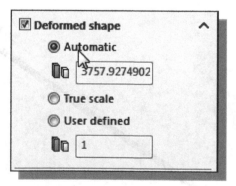

3. Set *Deformed Shape* to **Automatic** as shown.

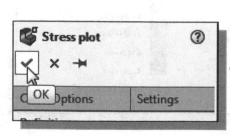

4. Click **OK** to view the results.

❖ SOLIDWORKS Simulation calculated the stresses for the two members as (1) left member: **848.8** psi in tension and (2) right member: **612.1** psi in compression. These values match with the results of the *Excel* analysis on page 3-22.

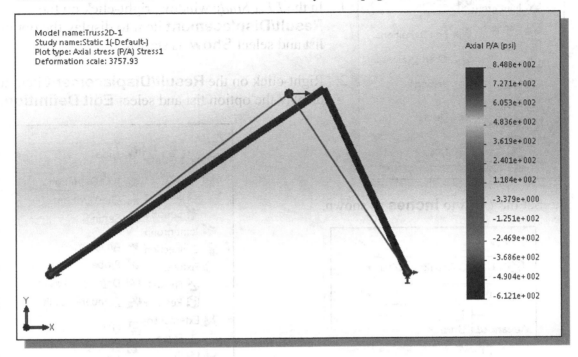

Model name:Truss2D-1
Study name:Static 1(-Default-)
Plot type: Axial stress (P/A) Stress1
Deformation scale: 3757.93

Axial P/A (psi)

8.488e+002
7.271e+002
6.053e+002
4.836e+002
3.619e+002
2.401e+002
1.184e+002
-3.379e+000
-1.251e+002
-2.469e+002
-3.686e+002
-4.904e+002
-6.121e+002

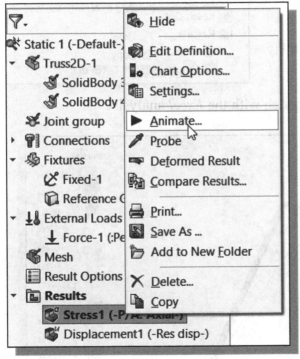

5. In the *FEA Study* window, right-click on the **Result/Stress** item to display the option list and select **Animate** as shown.

6. On your own, adjust the speed by dragging the slide bar as shown.

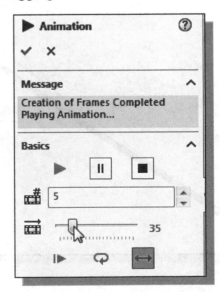

Viewing the Displacement Results

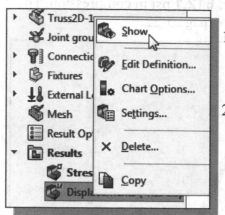

1. In the *FEA Study* window, right-click on the **Result/Displacement** item to display the option list and select **Show** as shown.

2. Right-click on the **Result/Displacement** item to display the option list and select **Edit Definition**.

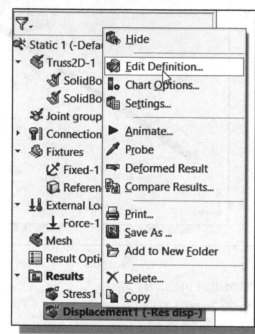

3. Set the *Units* to **inches** as shown.

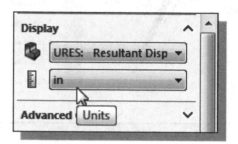
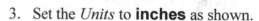

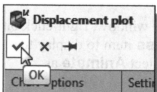

4. Click **OK** to display the results.

❖ Do the results match with the *Excel* analysis?

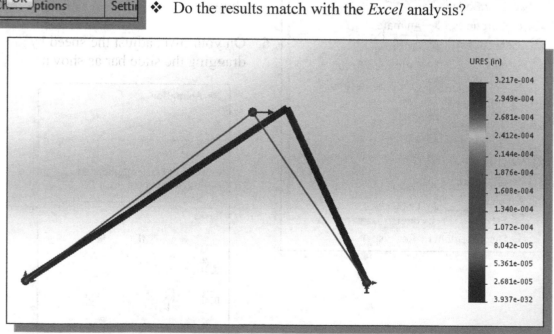

Questions:

1. Describe the typical finite element analysis steps.

2. In the tutorial, how were the nodes and elements created in SOLIDWORKS Simulation?

3. Can loads be applied to the midpoint of a truss element?

4. How do you identify the ZERO-FORCE members in a truss structure?

5. What are the basic assumptions, as defined in Statics, for truss members?

6. How do we assign the material properties for truss elements in SOLIDWORKS Simulation?

7. In the tutorial, how did we define the cross sections for truss elements in SOLIDWORKS Simulation?

8. How do we apply constraints on an existing system in SOLIDWORKS Simulation?

9. How do we obtain the stress values of truss members in SOLIDWORKS Simulation?

10. What is the type of FEA analysis performed in the tutorial?

Exercises:

Determine the normal stress in each member of the truss structures shown. (All joints are pin-joint.)

1. Material: Steel,
   Diameter: ¼ in.

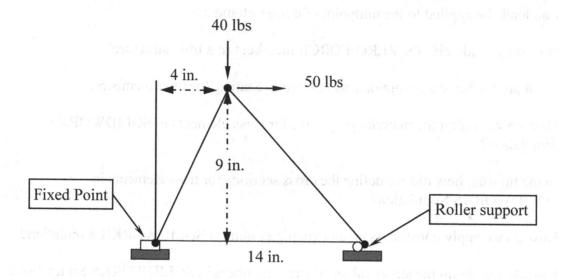

40 lbs

4 in.

50 lbs

9 in.

Fixed Point

Roller support

14 in.

2. Material: Steel,
   Diameter: 3 cm.

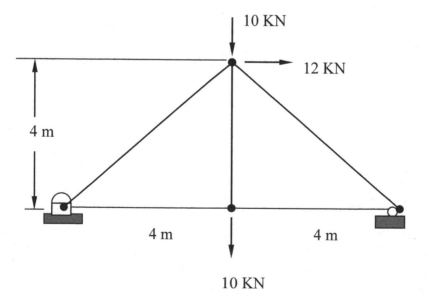

10 KN

12 KN

4 m

4 m   4 m

10 KN

3.   Material: Steel,
                Diameter: 3.0 in.

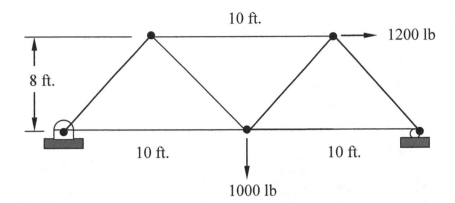

Notes:

Chapter 5
SOLIDWORKS Simulation Two-Dimensional Truss Analysis

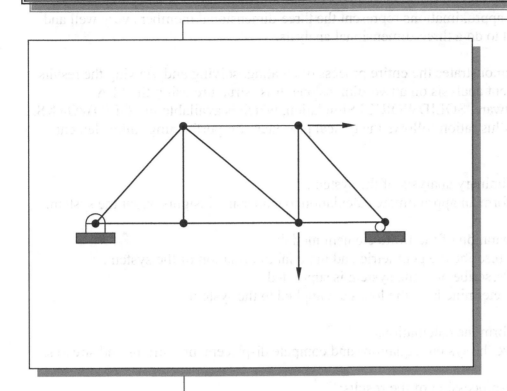

Learning Objectives

- ♦ **Understand the General Computer FEA Procedure.**
- ♦ **Create Structural Members in SOLIDWORKS.**
- ♦ **Understand the use of Weldment Profiles in SOLIDWORKS.**
- ♦ **Create Truss Elements using the Structural Members.**
- ♦ **View the Internal Loads of All members.**

Finite Element Analysis Procedure

While all real-life structures are three-dimensional in nature, solutions of many stress analyses are done on two-dimensional spaces and sometimes one-dimensional spaces. Quite often the approximations represent the three-dimensional members very well and there is no need to do a three-dimensional analysis.

This chapter demonstrates the entire process of creating, solving and viewing the results of a finite element analysis on a two-dimensional truss structure using the FEA application software, SOLIDWORKS Simulation, which is available in SOLIDWORKS. The following illustration follows the typical procedure for performing finite element analysis:

    a. Preliminary analysis of the system:
        Perform an approximate calculation to gain some insights about the system.

    b. Preparation of the finite element model:
        1. Prescribe the geometric and material information of the system.
        2. Prescribe how the system is supported.
        3. Determine how the loads are applied to the system.

    c. Perform the calculations:
        Solve the system equations and compute displacements, strains and stresses.

    d. Post-processing of the results:
        View the results of the FEA procedure and confirm the results of the analysis by comparing it to the preliminary analysis.

➢ Before going through the tutorial, perform a preliminary analysis of the two-dimensional truss structure as shown below. First, create a free body diagram of the entire structure to find the reactions at the supports. Then use either the classical *joint method* or the *section method* to find the forces in each of the members. Which member would you expect to have the highest stress? On your own, calculate the stress for the member you identified as the highest-stress member and compare to the computer solution at the end of this chapter.

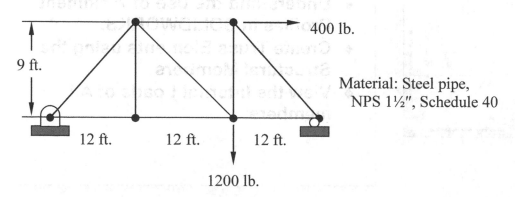

Preliminary Analysis

Determine the normal stress in each member of the truss structure shown.

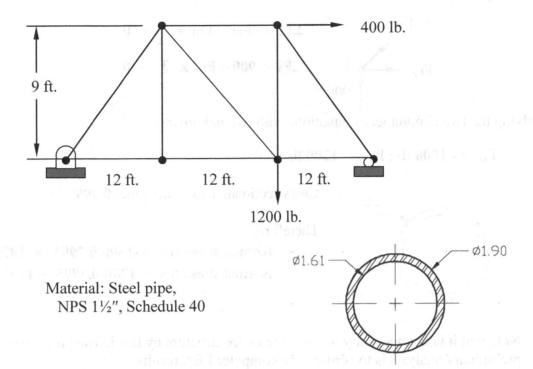

Material: Steel pipe,
NPS 1½″, Schedule 40

Prior to carrying out the finite element analysis, it is important to do an approximate preliminary analysis to gain some insights into the problem and as a means to check the finite element analysis results.

Free Body Diagram of the structure:

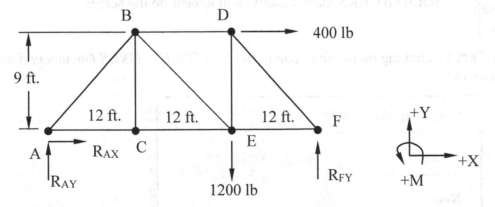

By inspection, member **BC** can be identified as a ZERO-FORCE member. Therefore, the stress in **BC** will be zero.

$\Sigma M_A = 36 \times R_{FY} - 24 \times 1200 - 9 \times 400 = 0$
Solving for R_{FY}: $R_{FY} = 900$ lb.

Next, using the *joint method* (conventional statics analysis technique), solve for internal forces in members **DF** and **EF**.

FBD of point **F**:

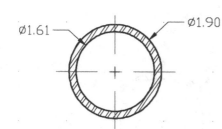

$$\Sigma F_x = - F_{EF} - F_{DF} \times \frac{4}{5} = 0$$

$$\Sigma F_Y = 900 + F_{DF} \times \frac{3}{5} = 0$$

Solving the two simultaneous equations with two unknowns:

$F_{DF} = - 1500$ lb.; $F_{EF} = 1200$ lb

Cross sectional area of the pipe: **0.7995 in²**

Therefore,

Normal stress $\sigma_{DF} = -1500/0.7995 = -1876$ psi

Normal stress $\sigma_{EF} = 1200/0.7995 = 1501$ psi

➤ Note that it is not necessary to solve the entire structure by hand. The purpose of the preliminary analysis is to confirm the computer FEA results.

Starting SOLIDWORKS

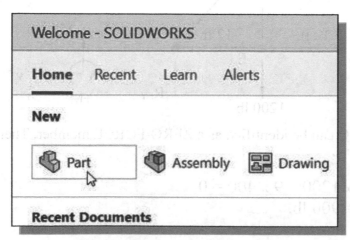

1. Select the **SOLIDWORKS** option on the *Start* menu or select the **SOLIDWORKS** icon on the desktop to start SOLIDWORKS. The SOLIDWORKS main window will appear on the screen.

2. Select **Part** by clicking on the first icon in the *New SOLIDWORKS Document* dialog box as shown.

Units Setup

When starting a new CAD file, the first thing we should do is choose the units we would like to use. We will use the English setting (inches) for this example.

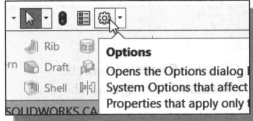

1. Select the **Options** icon from the *Menu Bar* toolbar to open the *Options* dialog box.

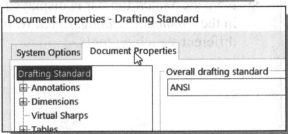

2. Switch to the **Document Properties** tab and reset the *Drafting Standard* to **ANSI** as shown in the figure.

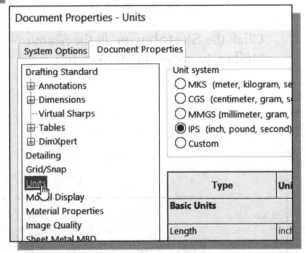

3. Click **Units** as shown in the figure.

4. Select **IPS (inch, pound, second)** under the *Unit system* options.

5. On your own, examine the different settings available, such as defining the degree of accuracy with which the units will be displayed.

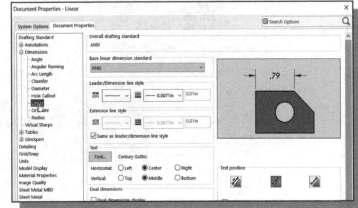

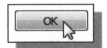

6. Click **OK** in the *Options* dialog box to accept the selected settings.

Creating the CAD Model – Structural Member approach

To create the nine member truss system, we will first establish the locations of the nine center axes of the members by creating a 2D sketch. We will then use the SOLIDWORKS **Structural Member tool** to select the predefined profiles of the cross sections of the system.

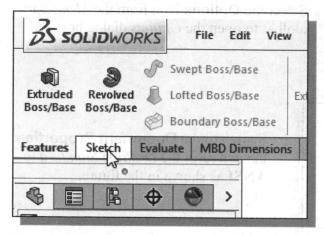

1. Click the **Sketch** tab in the *Command Manager* as shown.

❖ Note the *Sketch* toolbar is displayed in the *Command Manager* area with different sketching tools.

2. Click the **Sketch** icon, in the *Sketch* toolbar, to create a new sketch.

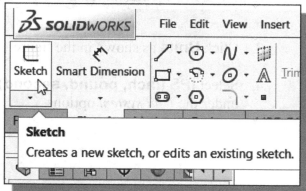

Sketch

Creates a new sketch, or edits an existing sketch.

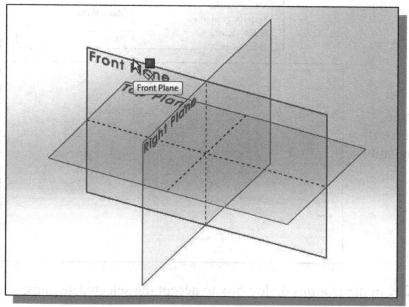

3. Click the **Front Plane**, in the graphics area, to set the sketching plane of our sketch as shown.

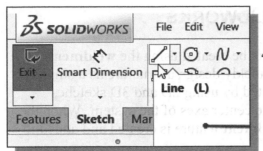

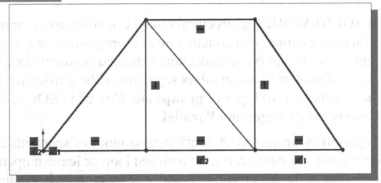

4. Click the **Line** icon in the *Sketch* toolbar as shown.

5. Start the line segment at the origin and create the **nine** connected line segments as shown.

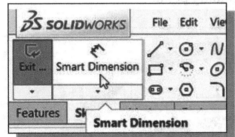

6. On your own, apply the **Equal** constraint to the bottom three members and use **Smart Dimension** to adjust the sketch as shown.

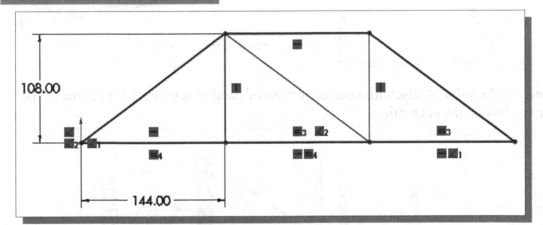

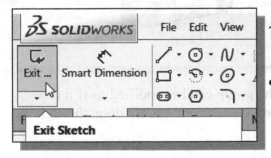

7. Click the **Exit sketch** icon in the *Sketch* toolbar as shown.

-   Note the dimensions disappeared from the screen. The nine lines we just created represent the center axes of the nine truss members.

Creating Structural Members in SOLIDWORKS

In SOLIDWORKS, both 2D and 3D structures can be created using the **weldments** options. The *Weldments* options can be used to quickly design a structure as a single multi-body part. The structural members are created by using 2D and 3D sketches defining the basic framework, which represent the center axes of the system. When the first structural member in a part is created, a **weldment** feature is created and added to the *Feature Manager Design Tree*.

In SOLIDWORKS, a structural model can contain one or multiple structural members. A structural member can contain one or more groups, and a group can contain one or more segments. A **group** is a collection of related segments in a structural system. Forming a group allows us to affect all its segments without affecting other segments or groups in the structure. Two types of **groups** are allowed in SOLIDWORKS: **Segments Connected** or **Segments Parallel**.

Segments Connected: A continuous contour of segments connected end-to-end. The connected segments can form a closed loop or leave it open-ended. Also, note the last endpoint of the group can optionally connect to its beginning point.

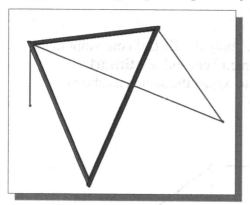
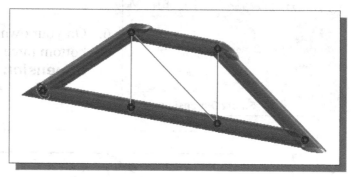

Segments Parallel: A discontinuous collection of parallel segments. Segments in the group cannot touch each other.

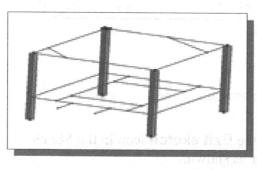
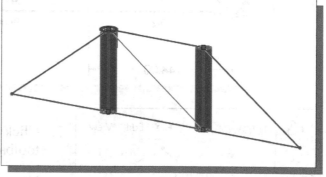

Once a **group** has been defined, all segments of the group can be modified as if it is a single object. Note that most of the properties of structural members are done through the *Structural Member Property Manager*.

Weldment Profiles

In SOLIDWORKS, all structural members use **weldment profiles**, which provide the definitions of the properties of the associated cross sections. All groups in a single structural member will use the same profile. *Weldment profiles* are identified by **standard**, **type** and **size**. *Weldment profiles* are *library files* of 2D sketches; note that we can create our own profiles and add them to the existing library. We can also download additional library files of weldment profiles from the SOLIDWORKS website; the available packages contain many standard profiles commonly used in structural members.

Note: The current list of weldment profiles were installed during the installation of SOLIDWORKS. To download additional weldment profiles, first go to the **Design Library** tab, and under *SOLIDWORKS Content*, hold the [**Ctrl**] key and click on one of the *Weldments* items to begin downloading the compressed **zip** file for the selected standard. Once the download is complete, unzip the compressed weldment profiles to the following folder: **<install_ dir>\lang\english\weldment profiles**.

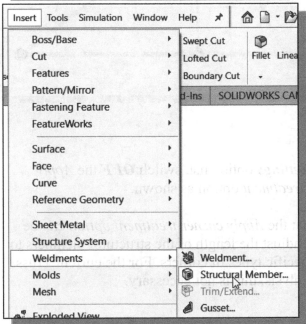

1. Select **Insert → Weldments → Structural Member** in the pull-down menu as shown.

❖ Note a *Weldment* feature is automatically added in the *Model History Tree* when the **Structural Member** command is activated.

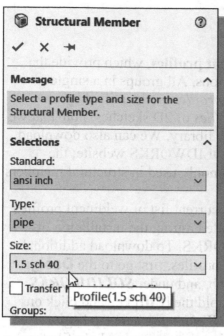

2. In the *Structural Member Property Manager* dialog box, set the *Standard* option to **ansi inch** as shown.

3. In the *Structural Member Property Manager* dialog box, set the *Type* option to **pipe** as shown.

4. In the *Structural Member Property Manager* dialog box, set the *Size* option to **1.5 sch 40** as shown.

❖ The three options we just set are used to define the **profile** of the cross section, which will be used on all segments of the same group.

5. Select the **six line segments** that form the outside loop of the structure as shown.

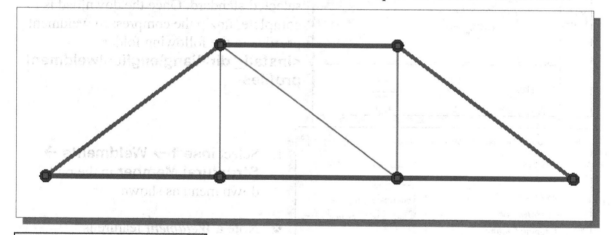

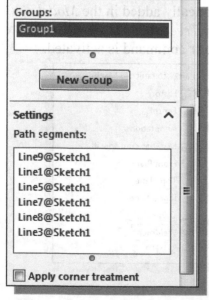

6. In the **Settings** option list, switch **OFF** the *Apply corner treatment* option as shown.

❖ Note that the *Apply corner treatment option* can be used to adjust the length of the structural members to form specific types of corners. For the current Truss FEA analysis, this is not necessary.

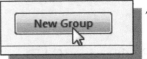

7. Click on the **New Group** button to create a new group under the same structural member.

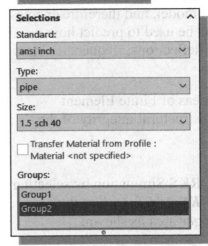

8. Confirm the new group is highlighted and also all of the settings for the weldment profile remain the same as *Group1* as shown.

❖ All groups in a single structural member will use the same profile.

9. Select the **three line segments**, forming the letter N, that are on the inside loop of the structure as shown.

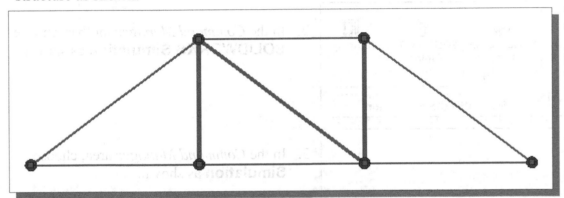

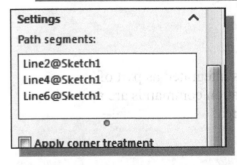

10. In the *Settings* option list, switch *OFF* the *Apply corner treatment* option as shown.

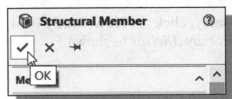

11. Click **OK** to create the structural member, which contains two groups with the same profile.

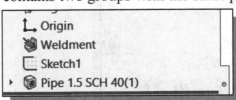

Activate the SOLIDWORKS Simulation Module

SOLIDWORKS Simulation is a multi-discipline Computer Aided Engineering (CAE) tool that enables users to simulate the physical behavior of a model, and therefore enables users to improve the design. SOLIDWORKS Simulation can be used to predict how a design will behave in the real world by calculating stresses, deflections, frequencies, heat transfer paths, etc.

The SOLIDWORKS Simulation product line features two areas of Finite Element Analysis: **Structure** and **Thermal**. *Structure* focuses on the structural integrity of the design, and *thermal* evaluates heat-transfer characteristics.

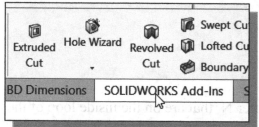

1. Start SOLIDWORKS Simulation by selecting the ***SOLIDWORKS Add-Ins*** tab in the *Command Manager* area as shown.

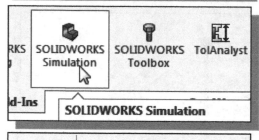

2. In the *Command Manager* toolbar, choose **SOLIDWORKS Simulation** as shown.

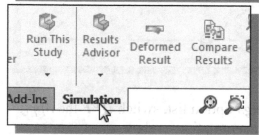

3. In the *Command Manager* area, choose **Simulation** as shown.

❖ Note that the SOLIDWORKS Simulation module is integrated as part of SOLIDWORKS. All of the SOLIDWORKS Simulation commands are accessible through the icon panel in the *Command Manager* area.

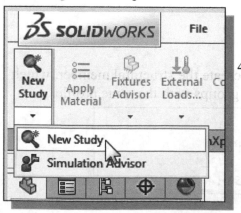

4. To start a new study, click the **New Study** item listed under the *Study Advisor* as shown.

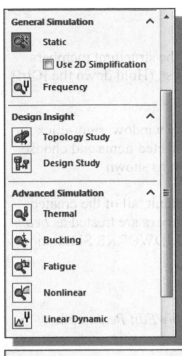

5. Select **Static** as the type of analysis to be performed with SOLIDWORKS Simulation.

❖ Note the different types of analyses available, which include both structural static and dynamic analyses, as well as the thermal analysis.

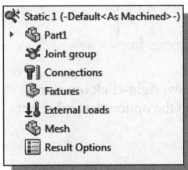

6. Click **OK** to start the definition of a structural static analysis.

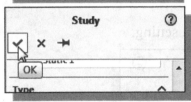

❖ In the *Feature Manager* area, note that a new panel, the *FEA Study* window, is displayed with all the key items listed.

❖ Also note that **Static 1** tab is activated, which indicates the use of the FEA model.

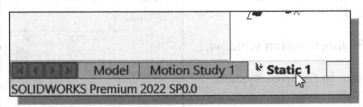

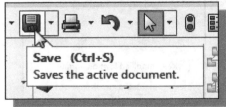

7. On your own, save a copy of the current model as **Truss2D-2**.

Setting Up Truss Elements

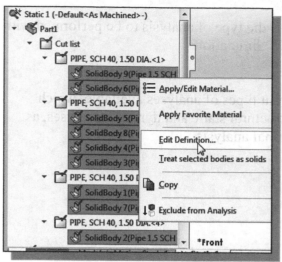

1. Preselect all of the structural members under the *Part* list. (Hold down the [**Ctrl**] key to select.)

2. In the *FEA Study* window, right-click on one of the preselected items and choose **Edit Definition** as shown.

❖   Note that, by default, all of the created **structural members** are treated as *beam elements* in SOLIDWORKS Simulation.

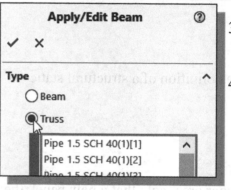

3. Select **Truss** in the *Apply/Edit Beam* option window as shown.

4. Click **OK** to accept the setting.

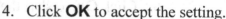
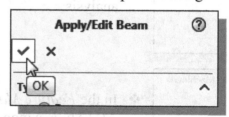

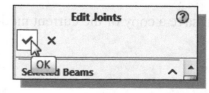

5. In the *FEA Study* window, right-click once on *Joint Group* to bring up the option list and select **Edit** as shown.

6. Click **Calculate** in the *Edit Joints* option window.

❖   Note that six nodes are present in the truss system.

7. Click **OK** to accept the settings.

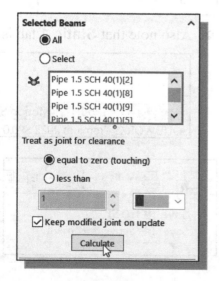

Assign the Element Material Property

Next, we will set up the *Material Property* for the elements. The *Material Property* contains the general material information, such as *Modulus of Elasticity, Poisson's Ratio,* etc. that is necessary for the FEA analysis.

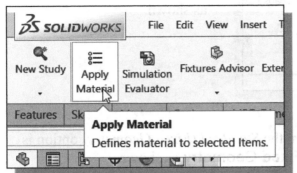

1. Choose the **Apply Material** option from the pull-down menu as shown.

❖ Note the default list of materials, which are available in the predefined SOLIDWORKS Simulation material library, is displayed.

2. Select **Alloy Steel** in the *Material* list as shown.

3. Set the **Units** option to display **English (IPS)** to make the selected material available for use in the current FEA model.

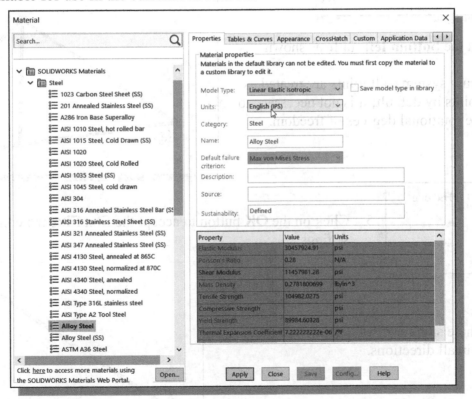

4. Click **Apply** to assign the material property then click **Close** to exit the Material Assignment command.

Applying Boundary Conditions – Constraints

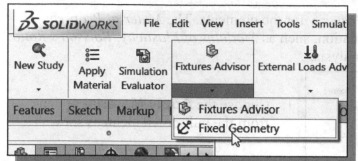

1. Choose **Fixed Geometry** by clicking the icon in the toolbar as shown.

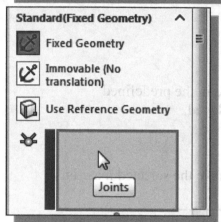

2. Confirm the *Standard (Fixed Geometry)* option is set to **Fixed Geometry** as shown.

3. Activate the *Joints* list option box by clicking on the inside of the **Joints** list box as shown.

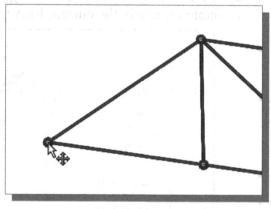

4. Select the **bottom left node** as shown.

❖ For truss systems, all joints are treated as pin-joints by default; it is not necessary to fix the rotational degrees of freedom.

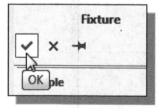

5. Click on the **OK** button to accept the first **Fixture** constraint settings.

❖ The three small arrows indicate the node is fixed in all directions.

❖ Next, we will add constraint to the bottom right node, only one translation DOF is allowed.

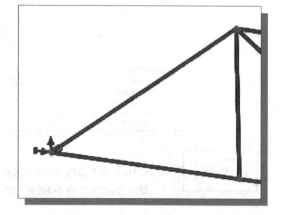

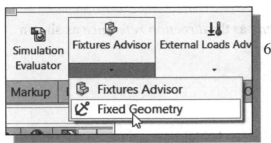

6. Choose **Fixed Geometry** by clicking the icon in the toolbar as shown.

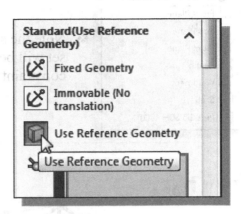

7. Change the *Standard (Fixed Geometry)* option to **Use Reference Geometry** as shown.

8. Select the **bottom right node** as shown.

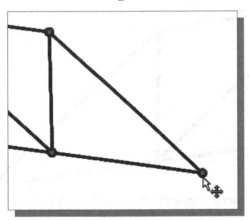

9. Activate the *Direction Reference* list option box by clicking on the inside of the **Reference** list box as shown.

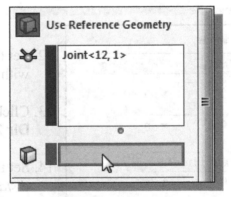

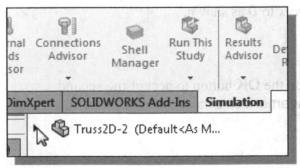

10. Expand the *Model History Tree*, located at the top left corner of the graphics window, by clicking on the **[Triangle]** as shown.

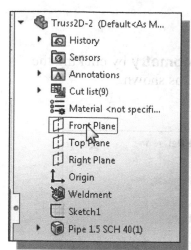

11. Select *Front Plane* as the *direction reference* as shown.

❖ Note the selected plane is highlighted; the constraints we set will be using the selected reference to determine the constraint direction.

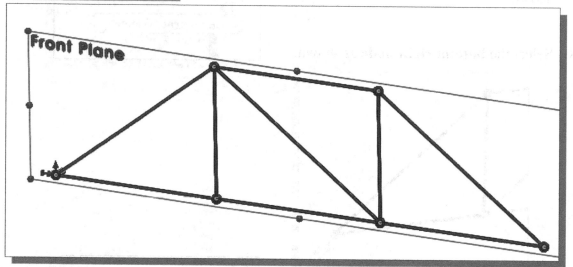

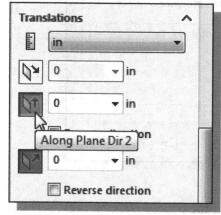

12. Set the distance measurement to **inches**, to match with the system units we are using.

13. Click on the **Normal to Plane** and **Along Plane Dir 2** icons to activate the constraint.

14. Set the *Normal to Plane* and *Along Plane Dir 2 distances* to **0** as shown.

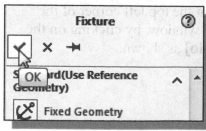

15. Click on the **OK** button to accept the second Displacement constraint settings.

❖ Note that for 2D truss systems, any translation motion in the Z direction is not allowed. We will need to add another constraint set to restrict the motion of the unconstrained nodes in the Z direction. Not restricting the movement in the Z direction will result in an erroneous FEA analysis.

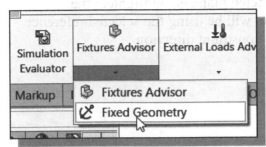

16. Choose **Fixed Geometry** by clicking the icon in the toolbar as shown.

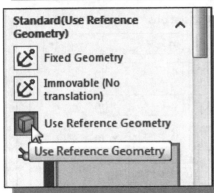

17. Change the *Standard (Fixed Geometry)* option to **Use Reference Geometry** as shown.

18. Select the **four nodes in the middle** as shown.

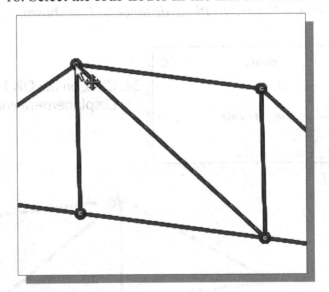

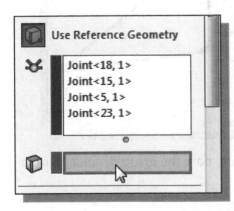

19. Activate the *Direction Reference* list option box by clicking on the inside of the **Reference** list box as shown.

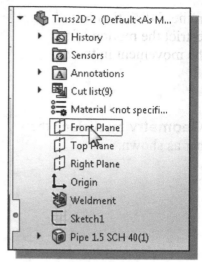

20. Select **Front Plane** as the *direction reference* as shown.

❖ Note that the selected plane is highlighted; the constraints we set will be using the selected reference to determine the constraint direction.

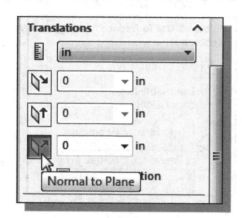

21. Set the distance measurement to **inches**, to match with the systems units we are using, as shown.

22. Click on the **Normal to Plane** icon to activate the constraint.

23. Set the *Normal to Plane distance* to **0** as shown.

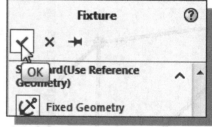

24. Click on the **OK** button to accept the first Displacement constraint settings.

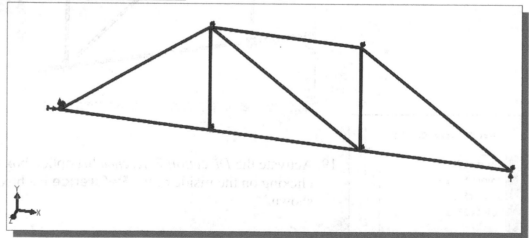

❖ On your own, examine the applied support constraints; note the arrows indicating the directions of constraints.

Applying External Loads

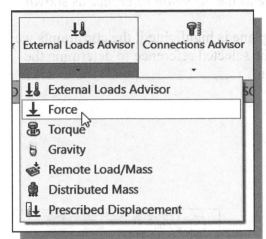

1. Choose **External Loads → Force** by clicking the icon in the toolbar as shown.

2. Change the *Force/Torque* option to **Joints** as shown.

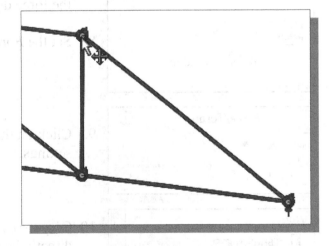

3. Select the **top right node point** as shown.

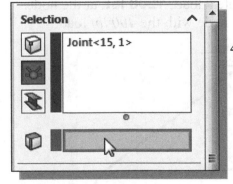

4. Activate the *Direction Reference* list option box by clicking on the inside of the **Reference** list box as shown.

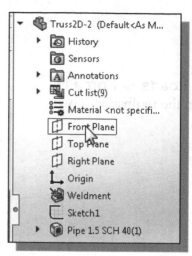

5. Select **Front Plane** as the *direction reference* as shown.

❖ Note the selected plane is highlighted; the constraints we set will be using the selected reference to determine the constraint direction.

6. Set the *Units* option to **English (IPS)**, to match with the system units we are using, as shown.

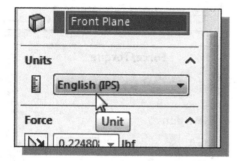

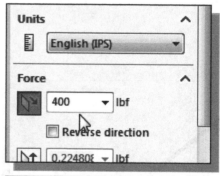

7. Click on the **Along Plane Dir 1** icon to activate the force direction.

8. Set the *Force* to **400** as shown.

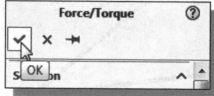

9. Click on the **OK** button to accept the *Force/Torque* settings.

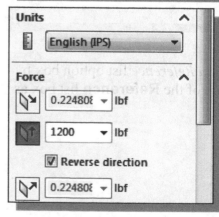

10. On your own, create a load, **1200 lbf**, at the node directly below the node with the *400 lbf* load.

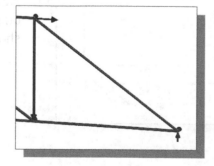

Create the FEA Mesh and Run the Solver

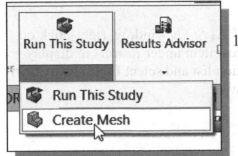

1. Choose **Create Mesh** by clicking the icon under *Run This Study* in the toolbar as shown.

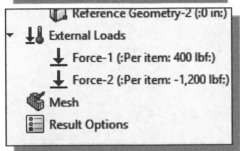

❖ Note the **Mesh** icon has changed, which indicates the FEA Mesh has been created.

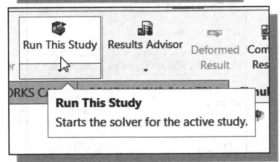

2. Click on the **Run This Study** button to start the *FEA Solver* to calculate the results.

❖ Note the stress results are displayed when the *Solver* has completed the FEA calculations.

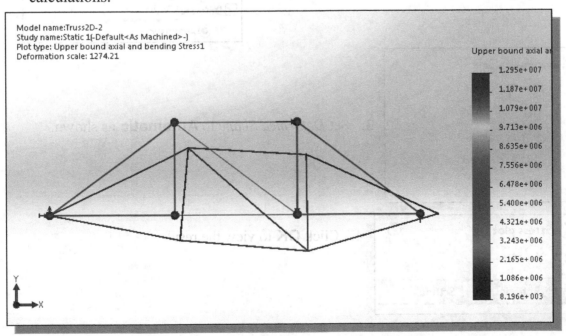

Viewing the Stress results

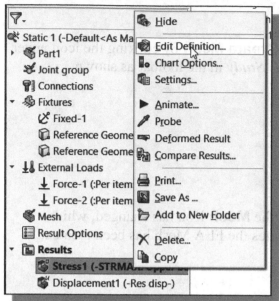

1. In the *FEA Study* window, right-click on the **Stress** item under *Results* to display the option list and select **Edit Definition** as shown.

2. In the *Stress Plot* option window, set the units to **psi** and the stress type to **Axial** as shown.

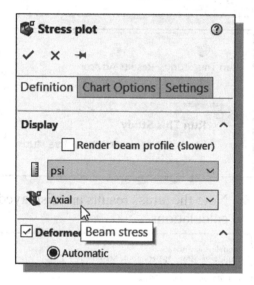

3. Set *Deformed Shape* to **Automatic** as shown.

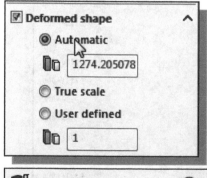

4. Click **OK** to view the results.

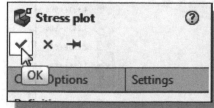

❖ SOLIDWORKS Simulation calculated the Max stresses as **1502** and **-1878 psi** which match with the results of the preliminary analysis on page 5-4.

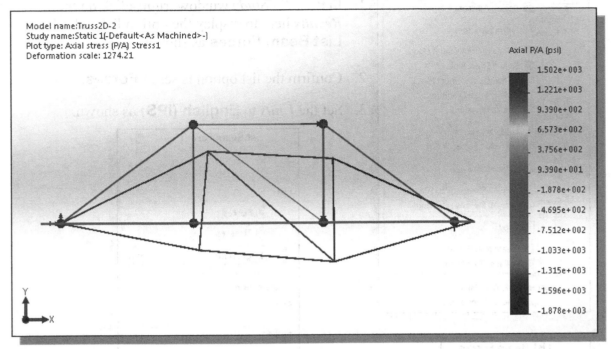

Model name:Truss2D-2
Study name:Static 1(-Default<As Machined>-)
Plot type: Axial stress (P/A) Stress1
Deformation scale: 1274.21

Axial P/A (psi)

1.502e+003
1.221e+003
9.390e+002
6.573e+002
3.756e+002
9.390e+001
-1.878e+002
-4.695e+002
-7.512e+002
-1.033e+003
-1.315e+003
-1.596e+003
-1.878e+003

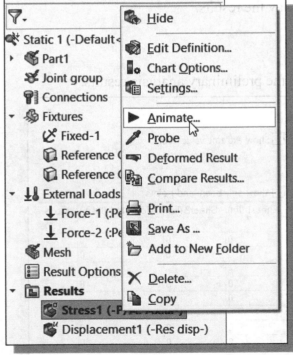

5. In the *FEA Study* window, right-click on the ***Results/Stress*** item to display the option list and select **Animate** as shown.

6. On your own, adjust the speed by dragging the slide bar as shown.

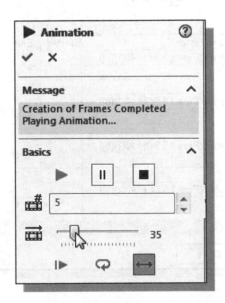

Viewing the Internal Loads of All members

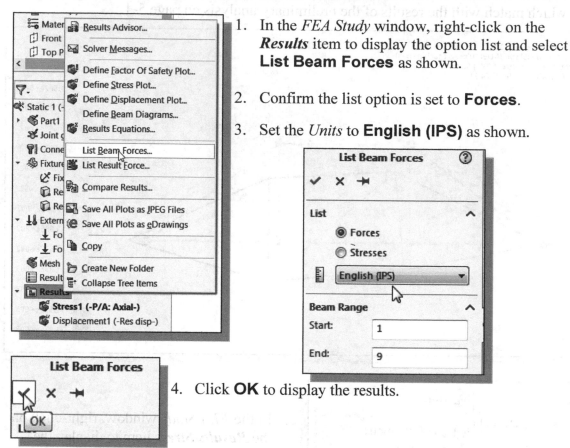

1. In the *FEA Study* window, right-click on the *Results* item to display the option list and select **List Beam Forces** as shown.

2. Confirm the list option is set to **Forces**.

3. Set the *Units* to **English (IPS)** as shown.

4. Click **OK** to display the results.

❖ On your own, compare the FEA results to the preliminary analysis results.

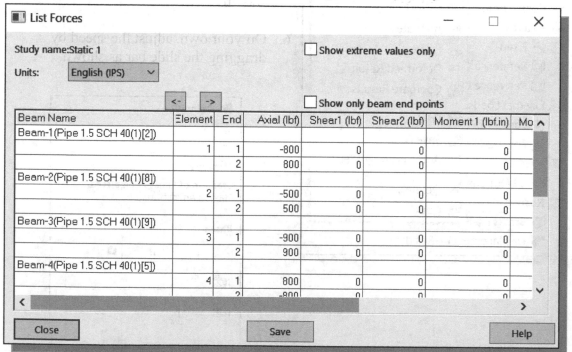

Beam Name	Element	End	Axial (lbf)	Shear1 (lbf)	Shear2 (lbf)	Moment 1 (lbf.in)	Mo
Beam-1(Pipe 1.5 SCH 40(1)[2])							
	1	1	-800	0	0	0	
		2	800	0	0	0	
Beam-2(Pipe 1.5 SCH 40(1)[8])							
	2	1	-500	0	0	0	
		2	500	0	0	0	
Beam-3(Pipe 1.5 SCH 40(1)[9])							
	3	1	-900	0	0	0	
		2	900	0	0	0	
Beam-4(Pipe 1.5 SCH 40(1)[5])							
	4	1	800	0	0	0	
		2	-800	0	0	0	

Viewing the Reaction Forces at the supports

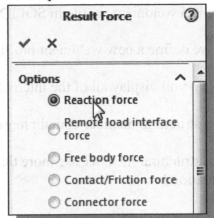

1. In the *FEA Study* window, right-click on the **Results** item to display the option list and select **List Result Forces** as shown.

2. Confirm the Option is set to **Reaction Force**.

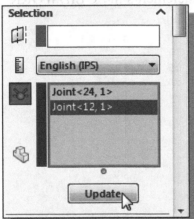

3. Set the *Units* to **English (IPS)** as shown.

4. Select the **two bottom nodes** with fixed support and the slider support.

5. Click **Update** to display the results on the screen.

❖ On your own, compare the FEA results to the preliminary analysis results.

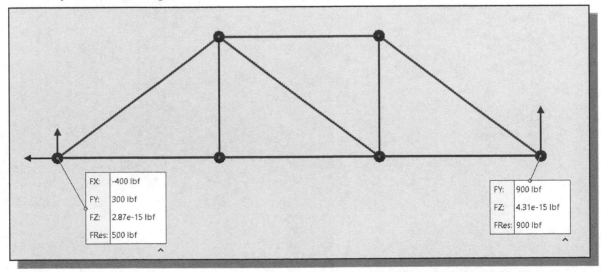

FX:	-400 lbf
FY:	300 lbf
FZ:	2.87e-15 lbf
FRes:	500 lbf

FY:	900 lbf
FZ:	4.31e-15 lbf
FRes:	900 lbf

Questions:

1. What is a structural member in SOLIDWORKS?

2. In the tutorial, how were the nodes and elements created in SOLIDWORKS Simulation?

3. What is a weldment profile in SOLIDWORKS?

4. Can we define a new weldment profile in SOLIDWORKS?

5. How do you display all of the internal loads for members in a truss structure?

6. Can a structural member contain more than one group?

7. Can a structural member use more than one weldment profile in SOLIDWORKS Simulation?

8. How do we define the cross sections orientation for truss elements in SOLIDWORKS Simulation?

Exercises:
Determine the normal stress in each member of the truss structures shown.

1. Material: Steel pipe

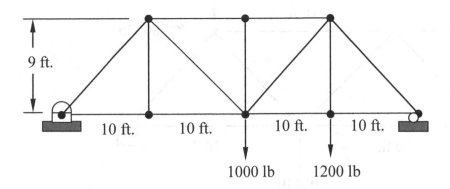

Material: Steel pipe,
   NPS 1½″, Schedule 40

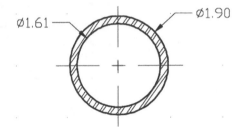

2. Material: Steel pipe,
         O.D. 33.7 mm, pipe thickness 4.0 mm.

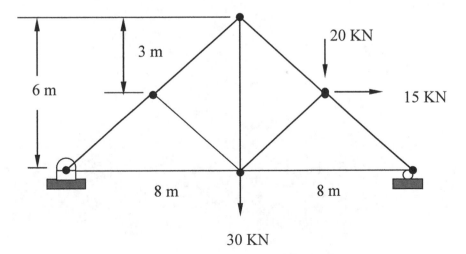

3. Material: Steel pipe,
    NPS 1½″, Schedule 40

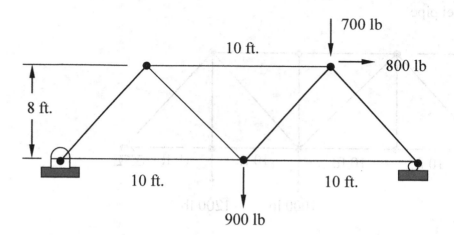

Chapter 6
Three-Dimensional Truss Analysis

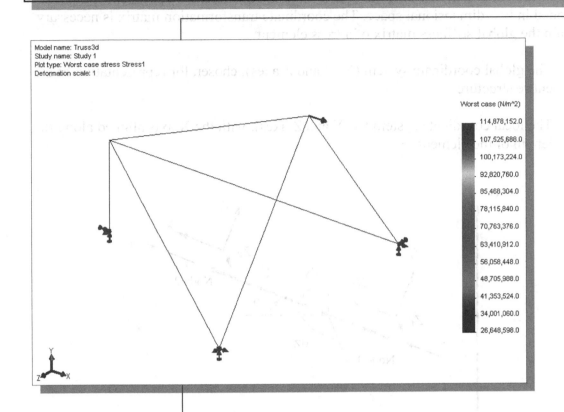

Model name: Truss3d
Study name: Study 1
Plot type: Worst case stress Stress1
Deformation scale: 1

Worst case (N/m^2)

114,878,152.0
107,525,688.0
100,173,224.0
92,820,760.0
85,468,304.0
78,115,840.0
70,763,376.0
63,410,912.0
56,058,448.0
48,705,988.0
41,353,524.0
34,001,060.0
26,648,598.0

Learning Objectives

- ◆ **Determine the Number of Degrees of Freedom in Elements.**
- ◆ **Create 3D FEA Truss Models.**
- ◆ **Apply proper boundary conditions to FEA Models.**
- ◆ **Use SOLIDWORKS Simulation Solver for 3D Trusses.**
- ◆ **Use SOLIDWORKS Simulation to determine Axial Loads.**

Three-Dimensional Coordinate Transformation Matrix

For truss members positioned in a three-dimensional space, the coordinate transformation equations are more complex than the transformation equations for truss members positioned in two-dimensional space. The coordinate transformation matrix is necessary to obtain the global stiffness matrix of a truss element.

1. The global coordinate system (X, Y and Z axes), chosen for representation of the entire structure.

2. The local coordinate system (X, Y and Z axes), with the X axis aligned along the length of the element.

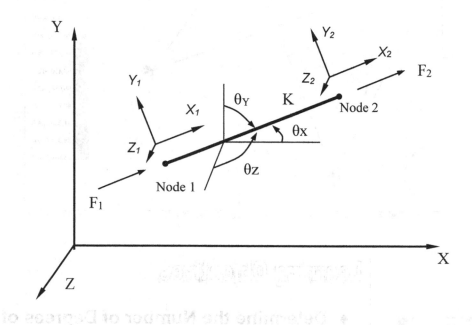

Since truss elements are two-force members, the displacements occur only along the local X axis. The GLOBAL to LOCAL transformation matrix can be written as:

$$
\left\{ \begin{array}{c} X_1 \\ X_2 \end{array} \right\} = \left[\begin{array}{cccccc} \cos(\theta_X) & \cos(\theta_Y) & \cos(\theta_Z) & 0 & 0 & 0 \\ 0 & 0 & 0 & \cos(\theta_X) & \cos(\theta_Y) & \cos(\theta_Z) \end{array} \right] \left\{ \begin{array}{c} X_1 \\ Y_1 \\ Z_1 \\ X_2 \\ Y_2 \\ Z_2 \end{array} \right\}
$$

Local

Global

Stiffness Matrix

The displacement and force transformations can be expressed as:

$$\{X\} = [l]\{X\}$$ ------- *Displacement transformation equation*

$$\{F\} = [l]\{F\}$$ ------- *Force transformation equation*

Combined with the local stiffness matrix, $\{F\} = [K]\{X\}$, we can then derive the global stiffness matrix for an element:

$$[K] = \frac{EA}{L}\begin{bmatrix} C_X{}^2 & C_XC_Y & C_XC_Z & -C_X{}^2 & C_XC_Y & C_XC_Z \\ & C_Y{}^2 & C_YC_Z & -C_XC_Y & -C_Y{}^2 & -C_YC_Z \\ & & C_Z{}^2 & -C_XC_Z & -C_YC_Z & -C_Z{}^2 \\ & & & C_X{}^2 & C_XC_Y & C_XC_Z \\ & \text{Symmetry} & & & C_Y{}^2 & C_YC_Z \\ & & & & & C_Z{}^2 \end{bmatrix}$$

where $C_X = cos(\theta_X)$, $C_Y = cos(\theta_Y)$, $C_Z = cos(\theta_Z)$.

The resulting matrix is a 6×6 matrix. The size of the stiffness matrix is related to the number of nodal displacements. The nodal displacements are also used to determine the number of **degrees of freedom** at each node.

Degrees of Freedom

◆ For a truss element in a one-dimensional space:

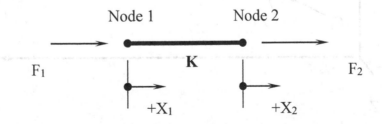

One nodal displacement at each node: one degree of freedom at each node. Each element possesses two degrees of freedom, which forms a 2×2 stiffness matrix for the element. The global coordinate system coincides with the local coordinate system.

♦ For a truss element in a two-dimensional space:

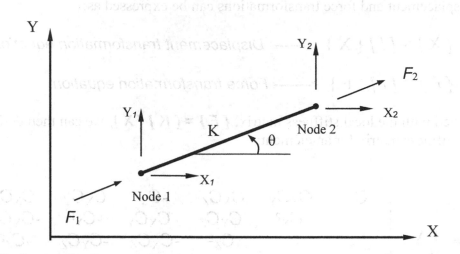

Two nodal displacements at each node: two degrees of freedom at each node. Each element possesses four degrees of freedom, which forms a 4×4 stiffness matrix for the element.

♦ For a truss element in a three-dimensional space:

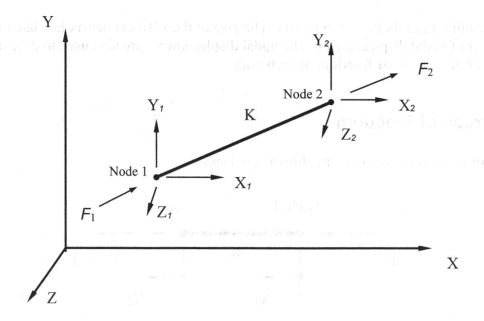

Three nodal displacements at each node: three degrees of freedom at each node. Each element possesses six degrees of freedom, which forms a 6×6 stiffness matrix for the element.

Problem Statement

➢ Determine the normal stress in each member of the truss structure shown. All joints are ball-joints.

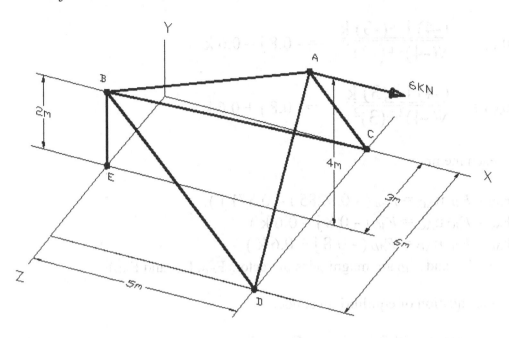

Material: Steel   Cross Section: ▨ a   a = b = 7.5mm
                                           b

Preliminary Analysis

❖ This section demonstrates using conventional *vector algebra* to solve the three-dimensional truss problem.

The coordinates of the nodes:
      A(5,4,3), B(0,2,3), C(5,0,0), D(5,0,6), E(0,0,3)

Free Body Diagram of Node A:

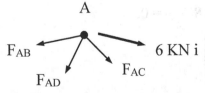

Position vectors AB, AC and AD:

$$AB = (0\text{-}5)\,i + (2\text{-}4)\,j + (3\text{-}3)\,k = (\text{-}5)\,i + (\text{-}2)\,j$$
$$AC = (5\text{-}5)\,i + (0\text{-}4)\,j + (0\text{-}3)\,k = (\text{-}4)\,j + (\text{-}3)\,k$$
$$AD = (5\text{-}5)\,i + (0\text{-}4)\,j + (6\text{-}3)\,k = (\text{-}4)\,j + (3)\,k$$

Unit vectors along AB, AC and AD:

$$u_{AB} = \frac{(-5)\,i + (-2)\,j}{\sqrt{(-5)^2+(-2)^2}} = -\,0.9285\,i - 0.371\,j$$

$$u_{AC} = \frac{(-4)\,j + (-3)\,k}{\sqrt{(-4)^2+(-3)^2}} = -\,0.8\,j - 0.6\,k$$

$$u_{AD} = \frac{(-4)\,j + (3)\,k}{\sqrt{(-4)^2+(3)^2}} = -\,0.8\,j + 0.6\,k$$

Forces in each member:

$$F_{AB} = F_{AB}\,u_{AB} = F_{AB}\,(-\,0.9285\,i - 0.371\,j\,)$$
$$F_{AC} = F_{AC}\,u_{AC} = F_{AC}\,(-\,0.8\,j - 0.6\,k\,)$$
$$F_{AD} = F_{AD}\,u_{AD} = F_{AD}\,(-\,0.8\,j + 0.6\,k\,)$$
$(F_{AB}, F_{AC}$ and F_{AD} are magnitudes of vectors F_{AB}, F_{AC} and F_{AD})

Applying the equation of equilibrium at node A:

$$\sum F_{@A} = 0 = 6000\,i + F_{AB} + F_{AC} + F_{AD}$$
$$= 6000\,i - 0.9285\,F_{AB}\,i - 0.371\,F_{AB}\,j - 0.8\,F_{AC}\,j - 0.6\,F_{AC}\,k$$
$$-\,0.8\,F_{AD}\,j + 0.6\,F_{AD}\,k$$
$$= (6000 - 0.9285\,F_{AB}\,)i + (-\,0.371\,F_{AB} - 0.8\,F_{AC} - 0.8\,F_{AD}\,)j$$
$$+ (-\,0.6\,F_{AC} + 0.6\,F_{AD}\,)k$$

Also, since the structure is symmetrical, $F_{AC} = F_{AD}$

Therefore,

$6000 - 0.9285\,F_{AB} = 0$, $\boxed{F_{AB} = \textbf{6462 N}}$

$-0.371\,F_{AB} - \textbf{0.8}\,F_{AC} - \textbf{0.8}\,F_{AD} = 0$,

$$\boxed{F_{AC} = F_{AD} = -\,\textbf{1500 N}}$$

The stresses:

$$\boxed{\begin{array}{l}\sigma_{AB} = 6460 / (5.63 \times 10^{-5}) = \textbf{115 MPa}\\[4pt]\sigma_{AC} = \sigma_{AD} = -\,1500 / (5.63 \times 10^{-5}) = \textbf{-26.7 MPa}\end{array}}$$

Start SOLIDWORKS

1. Select the **SOLIDWORKS** option on the *Start* menu or select the **SOLIDWORKS** icon on the desktop to start SOLIDWORKS. The SOLIDWORKS main window will appear on the screen.

2. Select **Part** by clicking on the first icon in the *New SOLIDWORKS Document* dialog box as shown.

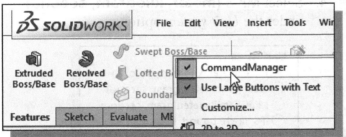

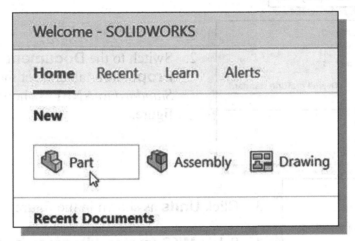

3. In the *Standard* toolbar area, right-click on any icon and activate the **Command Manager** in the option list.

❖ The ***Command Manager*** is a context-sensitive toolbar that dynamically updates based on the user's selection. When you click a tab below the *Command Manager*, it updates to display the corresponding toolbar. By default, the *Command Manager* has toolbars embedded in it based on the document type.

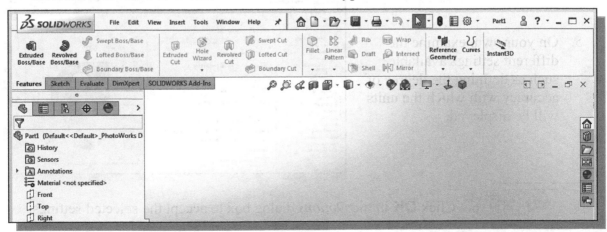

Units Setup

For this example, we will set up the model to using the *Metric* units (meters).

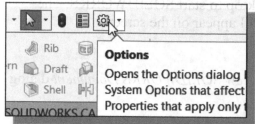

1. Select the **Options** icon from the *Menu Bar* toolbar to open the *Options* dialog box.

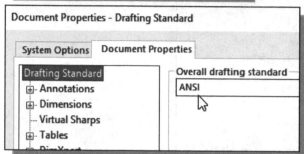

2. Switch to the **Document Properties** tab and set the *Drafting Standard* to **ANSI** as shown in the figure.

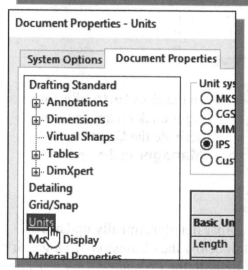

3. Click **Units** as shown in the figure.

4. Select **MKS (meter, kilogram, second)** under the *Unit system* options.

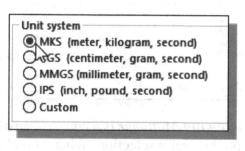

5. On your own, examine the different settings available, such as defining the degree of accuracy with which the units will be displayed.

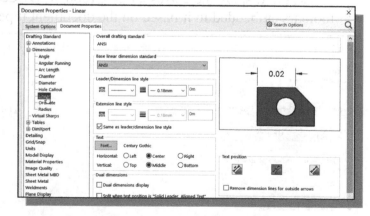

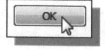

6. Click **OK** in the *Options* dialog box to accept the selected settings.

Create the CAD Model – Structural Member Approach

To create the truss system, we will first establish the locations of the center axes of the members by creating a 3D sketch. We will then use the SOLIDWORKS **Structural Member tool** to use the predefined profiles for the cross sections of the truss system.

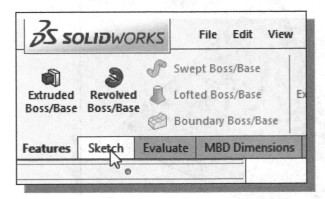

1. Click the **Sketch** tab in the *Command Manager* as shown.

❖ Note the *Sketch* toolbar is displayed in the *Command Manager* area with different sketching tools.

2. Activate the **3D Sketch** command, which is located in the *Sketch* toolbar, to create a new sketch.

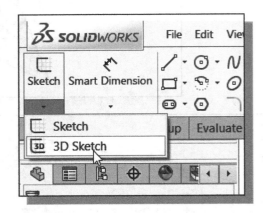

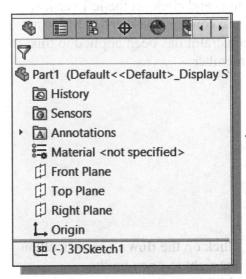

❖ Note that in the *Feature Manager*, the *3D Sketch1* item confirms the sketch type.

3. Click the **Line** icon in the *Sketch* toolbar as shown.

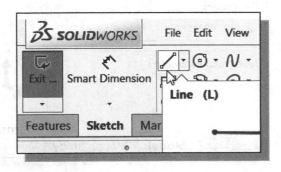

4. Start the line segment at the left side; create the vertical segment from the bottom up and create the five connected line segments as shown. Note the extra member is created to aid the 3D construction of the necessary members.

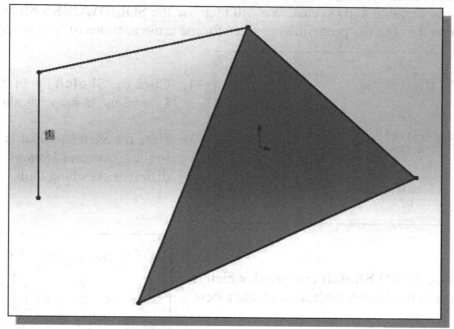

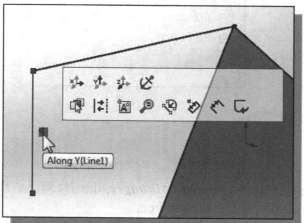

5. On your own, turn on the display of relations and click on the left vertical line and confirm the **Vertical (Along Y)** constraint has been applied to this left member.

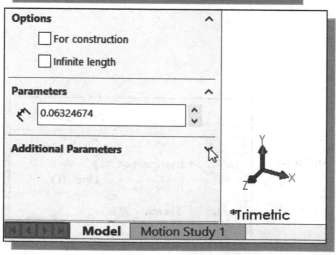

6. Click on the **down arrow** icon on the right to open up the *Additional Parameters* panel, which contains the coordinates of the endpoints of the selected line.

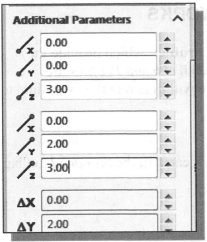

7. Adjust the coordinates of the two endpoints to **0,0,3** and **0,2,3** as shown. This is member **BE** of the truss system as shown in the below figure.

8. Click **Close Dialog** to exit the *Line Properties* dialog box.

9. On your own, continue to adjust the other endpoints to match with the following coordinates: **A(5,4,3)**, **B(0,2,3)**, **C(5,0,0)**, **D(5,0,6)**, **E(0,0,3)**.

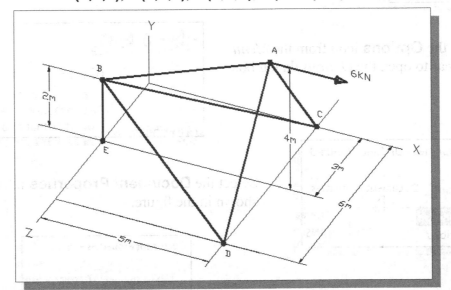

10. On your own, create the additional members **BC**, **BD** and delete line **CD**, the extra piece in the system. (Double check the coordinates before proceeding to the next step.)

11. Click the **Exit sketch** icon near the upper right corner of the graphics window as shown.

12. On your own, save the current model using file name **Truss3D**.

Create a New Weldment Profile in SOLIDWORKS

In SOLIDWORKS, all structural members use **weldment profiles**, which provide the definitions of the properties of the associated cross sections. *Weldment Profiles* are identified by **Standard**, **Type**, and **Size**. In this section, we will illustrate the procedure to create new profiles and add them to the existing library.

1. Click on the **New** icon, located in the *Standard* toolbar as shown.

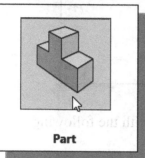

2. Select **Part** by clicking on the first icon in the *New SOLIDWORKS Document* dialog box as shown.

3. Click on the **OK** button to accept the settings.

4. Click on the **Options** icon from the *Menu Bar* toolbar to open the *Options* dialog box.

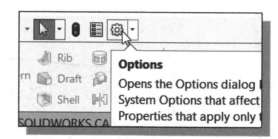

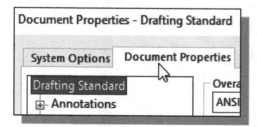

5. Select the **Document Properties** tab as shown in the figure.

6. Click **Units** as shown in the figure.

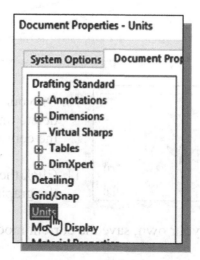

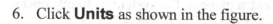

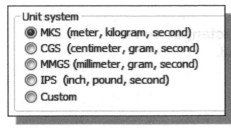

7. Select **MKS (meter, kilogram, second)** under the *Unit system* options.

Type	Unit	Decimals	Fraction
Basic Units			
Length	meters	.12	
Dual Dimension Length	millimeters	.12 / .123	
Angle	degrees	.1234 / .1234as	
Mass/Section Properties		.123456	

8. On your own, set the *Length decimals* to **four digits** as shown.

9. Click **OK** in the *Options* dialog box to accept the selected settings.

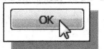

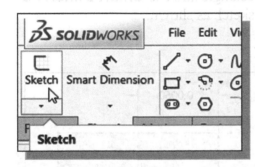

10. Create a new 2D sketch by clicking the **Sketch** icon, in the *Sketch* toolbar, to create a new sketch.

11. Click the **Front Plane**, in the graphics area, to set the sketching plane of our sketch as shown.

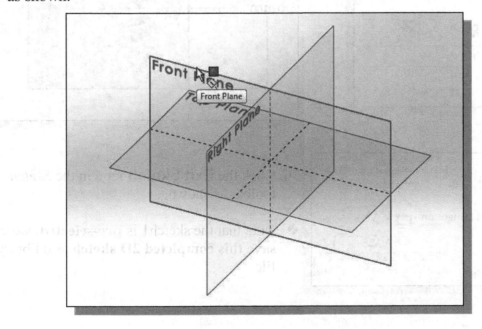

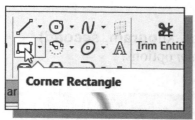

12. Click the **Corner Rectangle** icon to activate the **Rectangle** command.

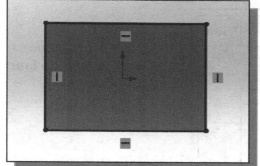

13. On your own, create a rectangle roughly centered at the origin of the world coordinate system.

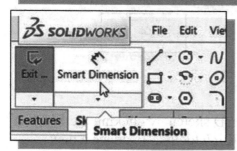

14. On your own, apply the **Equal** constraint to the sides of the rectangle and use **Smart Dimension** to adjust the sketch as shown.

15. On your own, create location dimensions using the equations as shown.

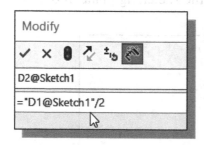

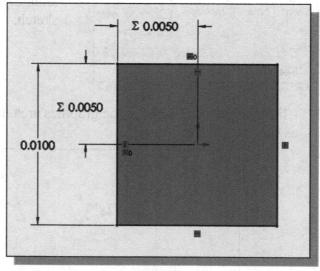

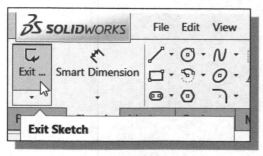

16. Click the **Exit Sketch** icon in the *Sketch* toolbar as shown.

❖ Note that the sketch1 is **pre-selected**, we will **save this completed 2D sketch** as a library file.

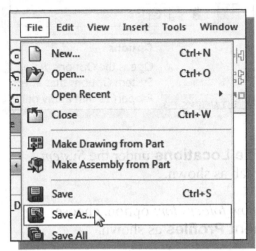

17. In the pull-down menu, select **Save As** to save our 2D sketch.

18. Choose the **Lib Feat Part** type, which is for a *Library* file, and *enter* the profile name as ***square.SLDLFP***.

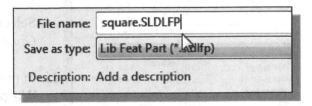

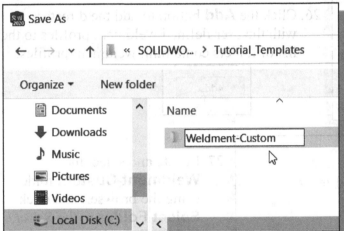

19. On your own, create and switch to a new weldment profiles folder. Such as:

C:\SOLIDWORKS Data \Tutorial_Templates \Weldment-Custom

- Hint: Use the right-mouse-click in the folder list area.

20. Switch to the new **Weldment-Custom** folder and create a new folder using the name **ISO-Custom**.

21. Switch to the new **ISO-Custom** folder and create a new folder using the name **Solid**.

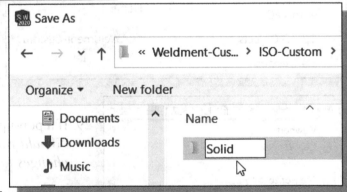

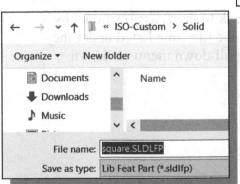

22. Switch to the new **Solid** folder and *save* the profile ***square.SLDLFP.***

23. Click on the **Options** icon from the *Menu Bar* toolbar to open the *Options* dialog box.

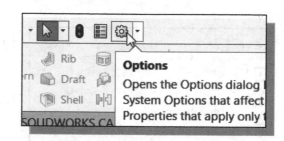

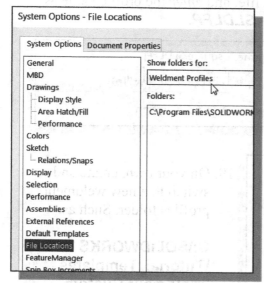

24. Select **File Locations** under the *System Options* tab as shown.

25. In the *Show folders for:* option, choose **Weldment Profiles** as shown.

26. Click the **Add** button to add the directory with the user-defined weldment profiles to the list of folders containing weldment profiles.

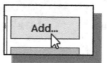

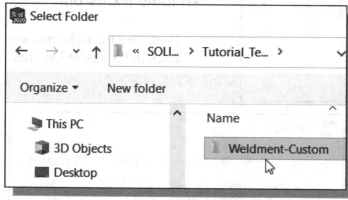

27. Locate and select the **Weldment-Custom** folder using the browser, and click **Select Folder** in the *Select Folder* dialog box.

28. Select **OK** in the *System Options* dialog box to exit.

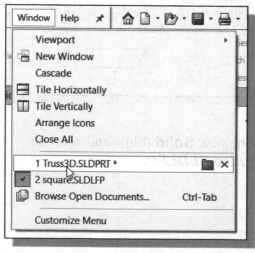

29. If a pop-up window appears with the question *"Would you like to make the following changes to your search paths?"* click **Yes**

30. Switch back to the *Truss3D* model by selecting the filename listed under the *Window* pull-down menu as shown.

Create Structural Members using the New Profile

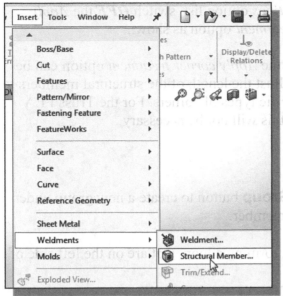

1. Select **Insert → Weldments → Structural Member** in the pull-down menu as shown.

❖ Note a *Weldment* feature is automatically added in the *Model History Tree* when the **Structural Member** command is activated.

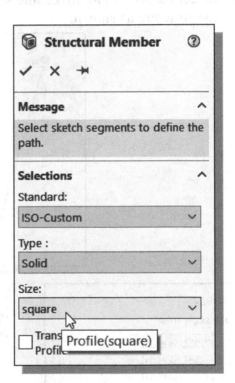

2. In the *Structural Member Property Manager* dialog box, set the *Standard* option to **ISO-Custom** as shown.

3. In the *Structural Member Property Manager* dialog box, set the *Type* option to **Solid** as shown.

4. In the *Structural Member Property Manager* dialog box, set the *Size* option to **square** as shown.

❖ Note that the selected **square profile** was defined in the previous section. The profile is stored under the folders that define its *standard*, *type* and *size*.

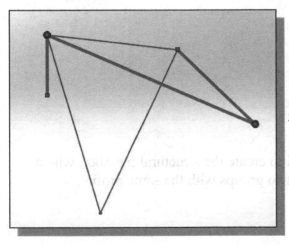

5. Select the **three line segments** that form the right side of the structure as shown.

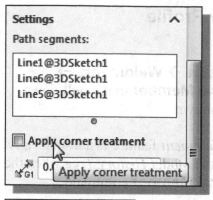

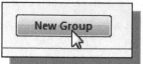

6. In the *Settings* option list, switch **OFF** the *Apply corner treatment* option as shown.

❖ Note that the *Apply corner treatment* option can be used to adjust the length of the structural members to form specific types of corners. For the Truss FEA analysis, this will not be necessary.

7. Click on the **New Group** button to create a new group under the same structural member.

8. Select the remaining **three line segments**, forming a loop, that are on the left side of the structure as shown.

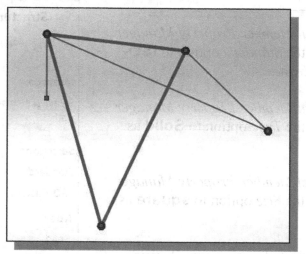

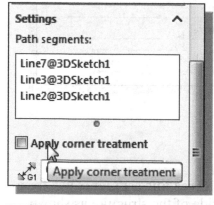

9. In the *Settings* option list, switch **OFF** the *Apply corner treatment* option as shown.

10. Click **OK** to create the structural member, which contains two groups with the same profile.

Edit the Dimensions of the New Profile

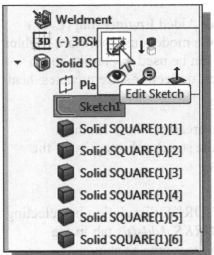

1. In the *Feature Manager*, click on the [**Triangle**] symbol in front of the *Structural Member* to expand the list.

2. Click once with the left-mouse-button on the associated **sketch**.

3. Select **Edit Sketch** in the option list as shown.

4. Click on the **Normal To** icon to rotate the viewing direction perpendicular to the selected 2D sketch.

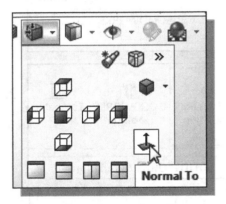

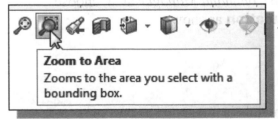

5. On your own, use the **Zoom to Area** command to adjust the view of the 2D sketch.

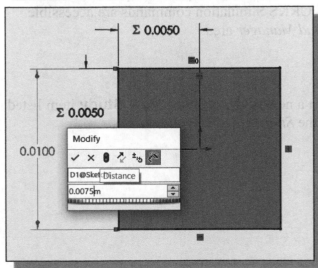

6. Double-click, with the left-mouse-button, on the height dimension and modify the value to **0.0075m** as shown.

7. On your own, confirm the adjustment of the width dimension and **exit** the *Sketch* mode.

Activate the SOLIDWORKS Simulation Module

SOLIDWORKS Simulation is a multi-discipline Computer Aided Engineering (CAE) tool that enables users to simulate the physical behavior of a model, and therefore enables users to improve the design. SOLIDWORKS Simulation can be used to predict how a design will behave in the real world by calculating stresses, deflections, frequencies, heat transfer paths, etc.

The SOLIDWORKS Simulation product line features two areas of Finite Element Analysis: **Structure** and **Thermal**. *Structure* focuses on the structural integrity of the design, and *thermal* evaluates heat-transfer characteristics.

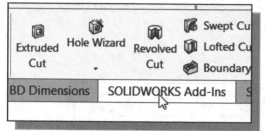

1. Start SOLIDWORKS Simulation by selecting the ***SOLIDWORKS Add-Ins*** tab in the *Command Manager* area as shown.

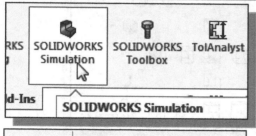

2. In the *Command Manager* toolbar, choose **SOLIDWORKS Simulation** as shown.

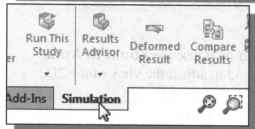

3. In the *Command Manager* area, choose **Simulation** as shown.

❖   Note that the SOLIDWORKS Simulation module is integrated as part of SOLIDWORKS. All of the SOLIDWORKS Simulation commands are accessible through the icon panel in the *Command Manager* area.

4. To start a new study, click the **New Study** item listed under the *Study Advisor* as shown.

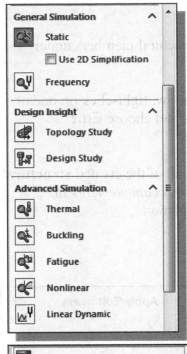

5. Select **Static** as the type of analysis to be performed with SOLIDWORKS Simulation.

❖ Note that different types of analyses are available, which include both structural static and dynamic analyses, as well as the thermal analysis.

6. Click **OK** to start the definition of a structural static analysis.

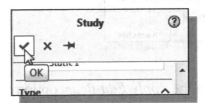

❖ In the *Feature Manager* area, note that a new panel, the *FEA Study* window, is displayed with all the key items listed.

❖ Also note that the **Static 1** tab is activated, which indicates the use of the FEA model.

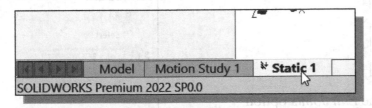

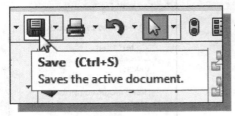

7. On your own, save a copy of the current model as **Truss3D**.

Setting Up the Truss Elements

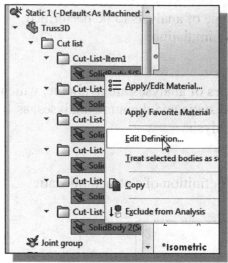

1. Preselect all of the structural members under *Truss3d.*

2. In the *FEA Study* window, right-click on one of the pre-selected items and choose **Edit Definition** as shown.

❖ Note that, by default, all of the created **structural members** are treated as *beam elements* in SOLIDWORKS Simulation.

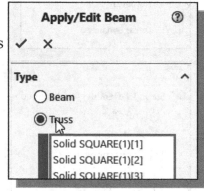

3. Select **Truss** in the *Apply/Edit Beam* option window as shown.

4. Click **OK** to accept the setting.

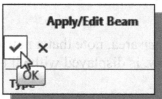

5. In the *FEA Study* window, right-click once on *Joint Group* to bring up the option list and select **Edit** as shown.

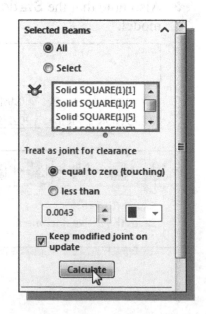

6. Select **Calculate** in the *Edit Joints* option window as shown.

❖ Note that five nodes are present in the truss system.

7. Click **OK** to accept the setting.

Assign the Element Material Property

Next, we will set up the *Material Property* for the elements. The *Material Property* contains the general material information, such as *Modulus of Elasticity, Poisson's Ratio,* etc. that is necessary for the FEA analysis.

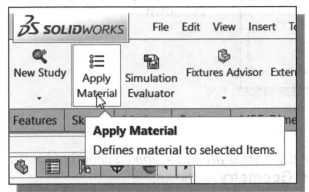

1. Choose the **Apply Material** option from the pull-down menu as shown.

❖ Note the default list of materials, which are available in the pre-defined SOLIDWORKS Simulation material library, is displayed.

2. Select **Alloy Steel** in the *Material* list as shown.

3. Confirm the **Units** option to display **SI – N/m^2 (Pa)**.

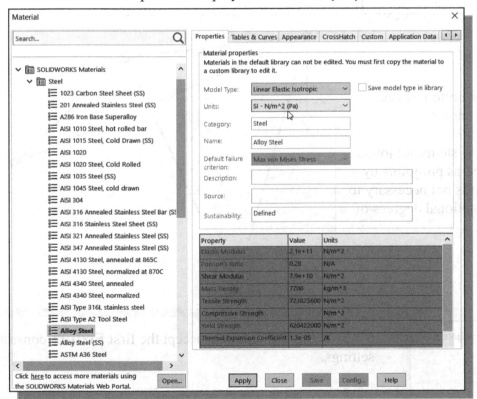

4. Click **Apply** to assign the material property, then click **Close** to exit the Material Assignment command.

Applying Boundary Conditions – Constraints

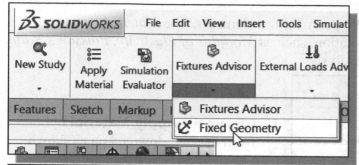

1. Choose **Fixed Geometry** by clicking the icon in the toolbar as shown.

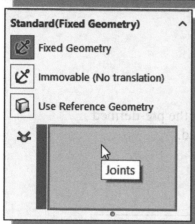

2. Confirm the *Standard(Fixed Geometry)* option is set to **Fixed Geometry** as shown.

3. Activate the *Joints* list option box by clicking on the inside of the **Joints** list box as shown.

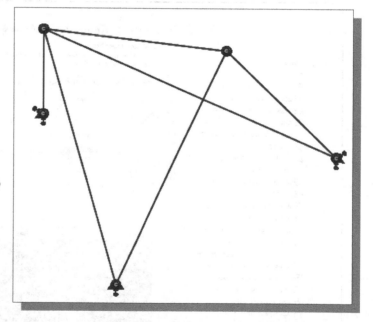

4. Select the bottom **three nodes** as shown.

❖ For truss systems, all joints are treated as pin-joints by default; it is not necessary to fix the rotational degrees of freedom.

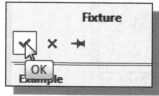

5. Click on the **OK** button to accept the first **Fixture** constraint settings.

❖ The displayed three small arrows indicate the associated node is fixed in all directions.

Applying the External Load

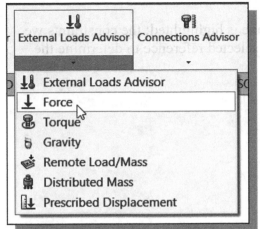

1. Choose **External Loads → Force** by clicking the icon in the toolbar as shown.

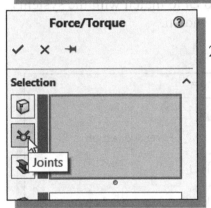

2. Change the *Force/Torque* option to **Joints** as shown.

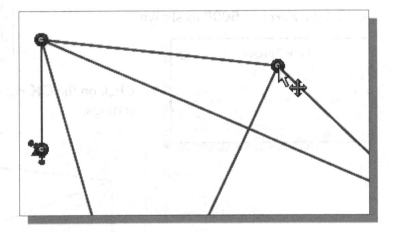

3. Select the top right node point as shown.

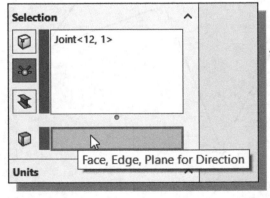

4. Activate the *Direction Reference* list option box by clicking on the inside of the **Reference** list box as shown.

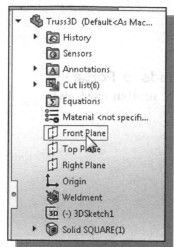

5. Select *Front Plane* as the *direction reference* as shown.

❖ Note the selected plane is highlighted; the constraints we set will be using the selected reference to determine the constraint direction.

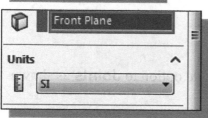

6. Set the *Units* option to **SI** to match with the systems units we are using, as shown.

7. Click on the **Along Plane Dir 1** icon to activate the force direction.

8. Set the *Force* to **6000** as shown.

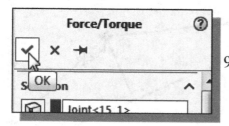

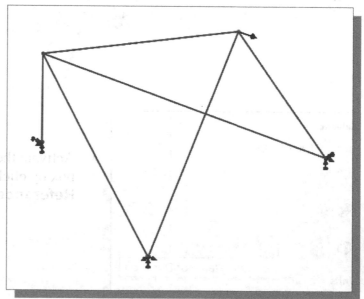

9. Click on the **OK** button to accept the *Force/Torque* settings.

Create the FEA Mesh and Run the Solver

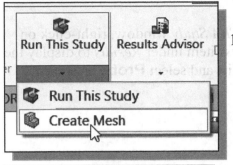

1. Choose **Create Mesh** by clicking the icon under *Run This Study* in the toolbar as shown.

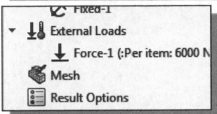

❖ Note the **Mesh** icon has changed, which indicates the FEA Mesh has been created.

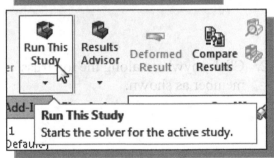

2. Click on the **Run This Study** button to start the *FEA Solver* to calculate the results.

❖ Note the stress results are displayed when the *Solver* has completed the FEA calculations. On your own, switch the stress display to **Axial** stress.

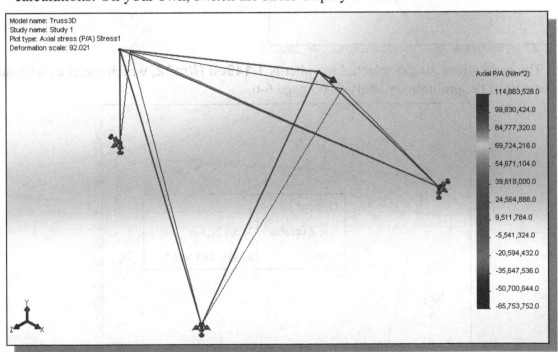

Using the Probe Option to View Individual Stress

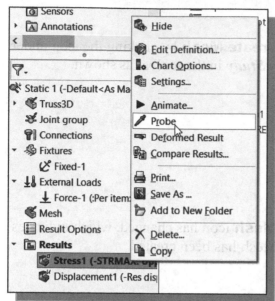

1. In the *FEA Study* window, right-click on the ***Stress*** item under *Results* to display the option list and select **Probe** as shown.

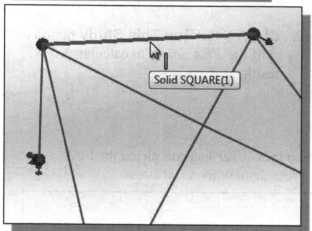

2. Click anywhere along the top truss member as shown.

❖ The axial stress for the selected member is **1.149e8 N/m^2**, which matches with the result of the preliminary analysis on page 6-6.

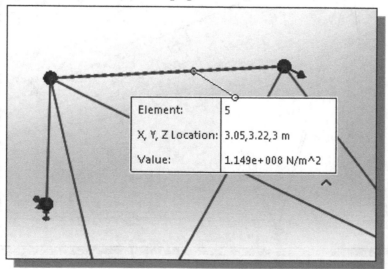

Element:	5
X, Y, Z Location:	3.05,3.22,3 m
Value:	1.149e+008 N/m^2

Viewing the Internal Loads of All members

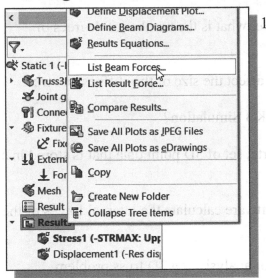

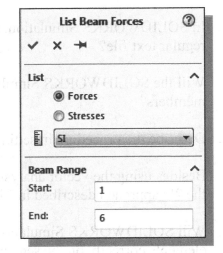

1. In the *FEA Study* window, right-click on the **Results** item to display the option list and select **List Beam Forces** as shown.

2. Confirm the list option is set to **Forces**.

3. Set the *Units* to **SI** as shown.

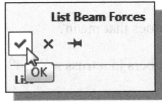

4. Click **OK** to display the results.

❖ On your own, compare the FEA results to the results from the preliminary analysis.

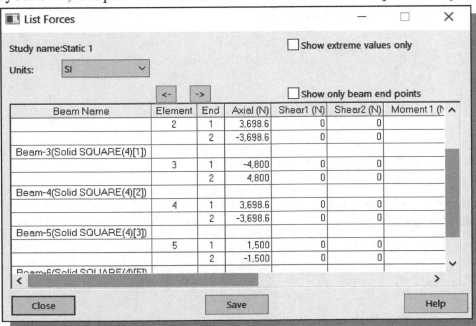

Beam Name	Element	End	Axial (N)	Shear1 (N)	Shear2 (N)	Moment 1 (N
	2	1	3,698.6	0	0	
		2	-3,698.6	0	0	
Beam-3(Solid SQUARE(4)[1])						
	3	1	-4,800	0	0	
		2	4,800	0	0	
Beam-4(Solid SQUARE(4)[2])						
	4	1	3,698.6	0	0	
		2	-3,698.6	0	0	
Beam-5(Solid SQUARE(4)[3])						
	5	1	1,500	0	0	
		2	-1,500	0	0	
Beam-6(Solid SQUARE(4)[5])						

Questions:

1. For a truss element in three-dimensional space, what is the number of degrees of freedom for the element?

2. How does the number of degrees of freedom affect the size of the stiffness matrix?

3. How do we create 3D points in SOLIDWORKS Simulation?

4. In SOLIDWORKS Simulation, can we import a set of 3D point data that is stored in a regular text file?

5. Will the SOLIDWORKS Simulation FEA software calculate the internal forces of the members?

6. Describe the procedure in setting up the vector analysis of a 3D truss problem.

7. Besides using the vector analysis procedure, can we solve the example problem by the 2D approach described in Chapter 4?

8. Will SOLIDWORKS Simulation display the orientation of cross sections of truss elements correctly on the screen?

9. Truss members are also known as two-force members. What does that mean?

10. Can we use several different material properties for truss members in a truss system?

Exercises: Determine the normal stress in each member of the truss structures.

1. All joints are ball-joints; bottom 3 joints are fixed to the floor.

    Material: Steel

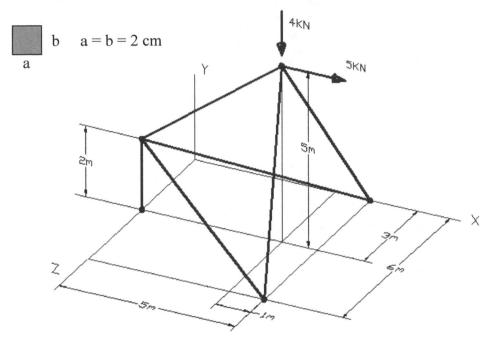

2. All joints are ball-joints, and joints D, F, F are fixed to the floor. (Hint: On your your own, create a circular weldment profile; see page 9-9 for additional help if needed.)

    Material: Steel
    Diameter: 1 in.

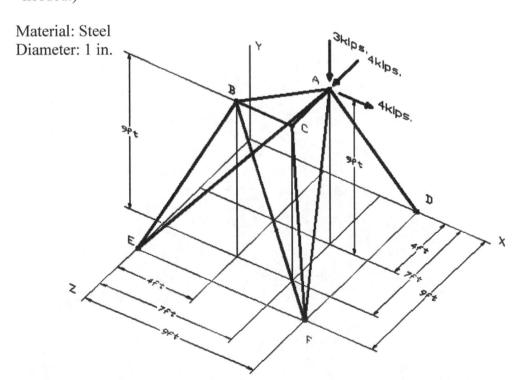

Notes:

Chapter 7
Basic Beam Analysis

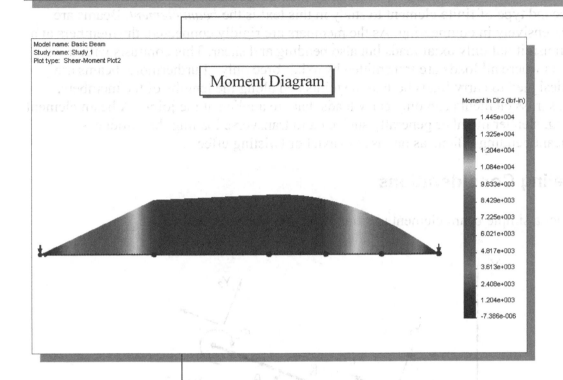

Model name: Basic Beam
Study name: Study 1
Plot type: Shear-Moment Plot2

Moment Diagram

Moment in Dir2 (lbf-in)

1.445e+004
1.325e+004
1.204e+004
1.084e+004
9.633e+003
8.429e+003
7.225e+003
6.021e+003
4.817e+003
3.613e+003
2.408e+003
1.204e+003
-7.386e-006

Learning Objectives

♦ **Understand the basic assumptions for Beam elements.**

♦ **Apply Constraints and Forces on FE Beam Models.**

♦ **Apply Distributed Loads on Beams.**

♦ **Create and Display a Shear Diagram.**

♦ **Create and Display a Moment Diagram.**

♦ **Perform Basic Beam Analysis using SOLIDWORKS Simulation.**

Introduction

The truss element discussed in the previous chapters does have a practical application in structural analysis, but it is a very limiting element since it can only transmit axial loads. The second type of finite element to study in this text is the *beam element*. Beams are used extensively in engineering. As the members are rigidly connected, the members at a joint transmit not only axial loads but also bending and shear. This contrasts with truss elements where all loads are transmitted by axial force only. Furthermore, beams are often designed to carry loads both at the joints and along the lengths of the members, whereas truss elements can only carry loads that are applied at the joints. A beam element is a long, slender member generally subjected to transverse loading that produces significant bending effects as opposed to axial or twisting effects.

Modeling Considerations

Consider a simple beam element positioned in a two-dimensional space:

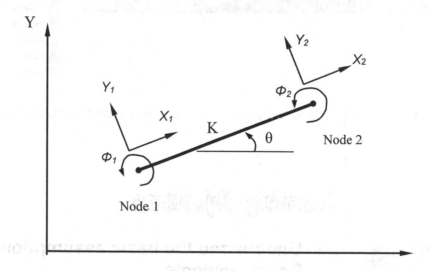

A beam element positioned in two-dimensional space is typically modeled to possess three nodal displacements (two translational and one rotational) at each node: three degrees of freedom at each node. Each element therefore possesses six degrees of freedom. A finite element analysis using beam elements typically provides a solution to the displacements, reaction forces, and moments at each node. The formulation of the beam element is based on the elastic beam theory, which implies the beam element is initially straight, linearly elastic, and loads (forces and moments) are applied at the ends. Therefore, in modeling considerations, place nodes at all locations where concentrated forces and moments are applied. For a distributed load, most finite element procedures replace the distributed load with an equivalent load set, which is applied to the nodes available along the beam span. Accordingly, in modeling considerations, place more nodes along the beam spans with distributed loads to lessen the errors.

Problem Statement

Determine the maximum normal stress developed in the steel member.

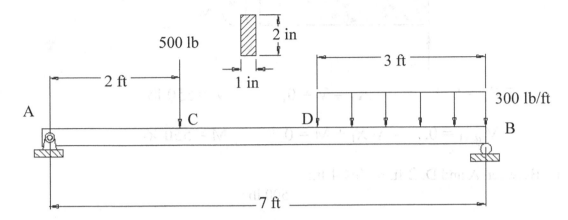

Preliminary Analysis

Free Body Diagram of the member:

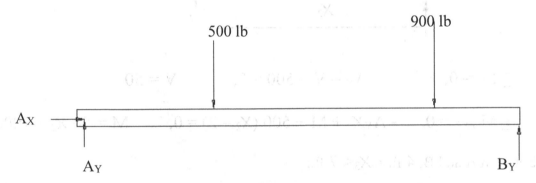

Applying the equations of equilibrium:

$$\sum M_{@A} = 0 = B_Y \times 7 - 500 \times 2 - 900 \times 5.5$$
$$\sum F_X = 0 = A_X$$
$$\sum F_Y = 0 = A_Y + B_Y - 500 - 900$$

Therefore,

$$B_Y = 850 \text{ lb}, A_X = 0 \text{ and } A_Y = 550 \text{ lb}$$

Next, construct the shear and moment diagrams for the beam. Although this is not necessary for most FEA analyses, we will use this example to build up our confidence with SOLIDWORKS Simulation's analysis results.

♦ Between A and C, 0 ft. $< X_1 < 2$ ft.:

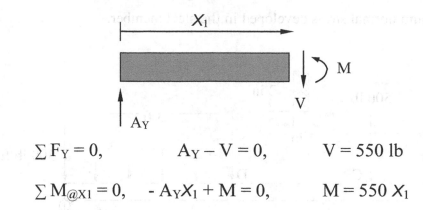

$\sum F_Y = 0$, $A_Y - V = 0$, $V = 550$ lb

$\sum M_{@X1} = 0$, $- A_Y X_1 + M = 0$, $M = 550\ X_1$

♦ Between A and D, 2 ft. $< X_2 < 4$ ft.:

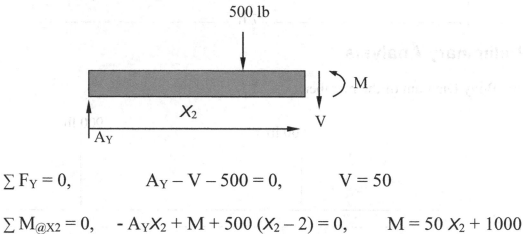

$\sum F_Y = 0$, $A_Y - V - 500 = 0$, $V = 50$

$\sum M_{@X2} = 0$, $- A_Y X_2 + M + 500\ (X_2 - 2) = 0$, $M = 50\ X_2 + 1000$

♦ Between A and B, 4 ft. $< X_3 < 7$ ft.:

500 lb

$\sum F_Y = 0$, $A_Y - V - 500 - 300\ (X_3 - 4) = 0$, $V = (-300\ X_3 + 1250)$
$\sum M_{@X3} = 0$, $-A_Y X_3 + M + 500\ (X_3 - 2) + 300\ (X_3 - 4)\ (X_3 - 4)/2 = 0$

$$M = -150\ X_3^2 + 1250\ X_3 - 1400$$

Shear and Moment diagrams:

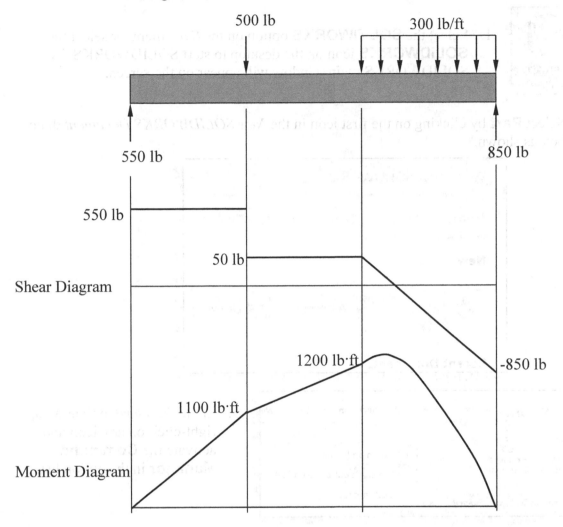

500 lb

300 lb/ft

550 lb   850 lb

550 lb

50 lb

Shear Diagram

1200 lb·ft   -850 lb

1100 lb·ft

Moment Diagram

Determine the maximum normal stress developed in the beam:

$$V = (-300 X_3 + 1250) = 0, \quad X_3 = 4.16 \text{ ft.}$$

$$M = -150 X_3^2 + 1250 X_3 - 1400 = 1204 \text{ ft-lb} = 14450 \text{ in-lb}$$

Therefore,

$$\sigma_{max} = \frac{MC}{I} = 1204 \, (h/2)/(bh^3/12) = 150.5 \text{ lb/ft}^2 = 21675 \text{ psi}$$

Start SOLIDWORKS

1. Select the **SOLIDWORKS** option on the *Start* menu or select the **SOLIDWORKS** icon on the desktop to start SOLIDWORKS. The SOLIDWORKS main window will appear on the screen.

2. Select **Part** by clicking on the first icon in the *New SOLIDWORKS Document* dialog box as shown.

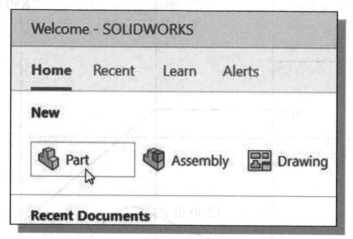

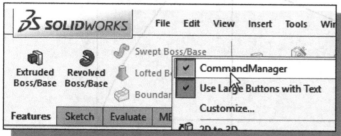

3. In the *Standard* toolbar area, right-click on any icon and activate the **Command Manager** in the option list.

❖ The **Command Manager** is a context-sensitive toolbar that dynamically updates based on the user's selection. When you click a tab below the *Command Manager*, it updates to display the corresponding toolbar. By default, the *Command Manager* has toolbars embedded in it based on the document type.

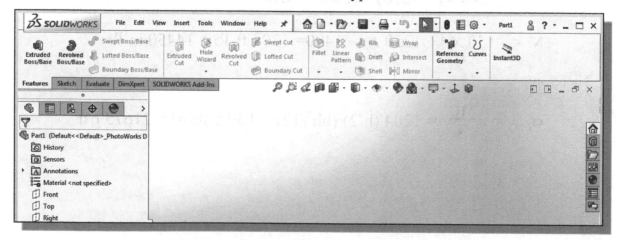

Units Setup

When starting a new CAD file, the first thing we should do is choose the units we would like to use. We will use the English setting (inches) for this example.

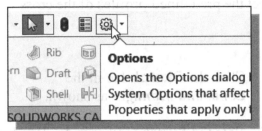

1. Select the **Options** icon from the *Menu Bar* toolbar to open the *Options* dialog box.

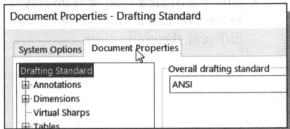

2. Switch to the **Document Properties** tab and set the *Drafting Standard* to **ANSI** as shown in the figure.

3. Click **Units** as shown in the figure.

4. Select **IPS (inch, pound, second)** under the *Unit system* options.

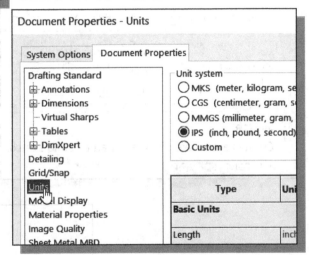

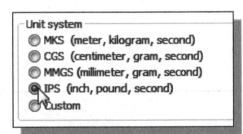

5. On your own, examine the different settings available, such as defining the degree of accuracy with which the units will be displayed.

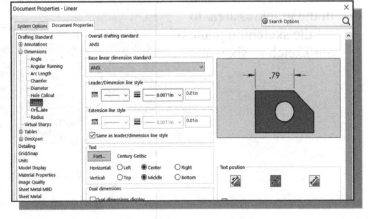

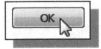

6. Click **OK** in the *Options* dialog box to accept the selected settings.

Create the CAD Model – Structural Member approach

To create the nine member truss system, we will first establish the locations of the nine center axes of the members by creating a 2D sketch. We will then use the SOLIDWORKS **Structural Member** tool to select the predefined profiles of the cross sections of the system.

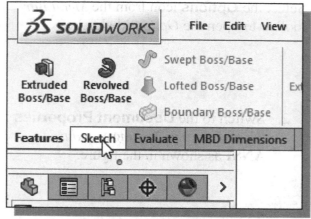

1. Click the **Sketch** tab in the *Command Manager* as shown.

❖ Note the *Sketch* toolbar is displayed in the *Command Manager* area with different sketching tools.

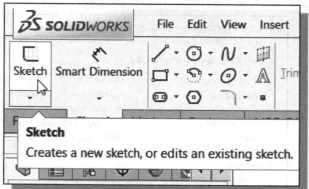

2. Click the **Sketch** icon, in the *Sketch* toolbar, to create a new sketch.

3. Click the **Front Plane**, in the graphics area, to set the sketching plane of our sketch as shown.

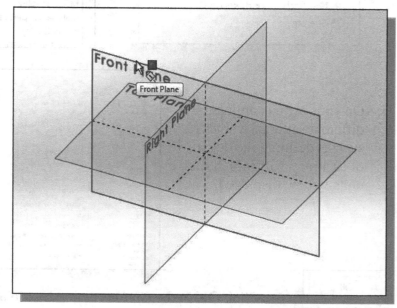

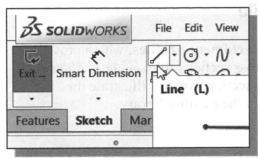

4. Click the **Line** icon in the *Sketch* toolbar as shown.

5. Start the line segment at the origin and create the **two connected line segments** as shown.

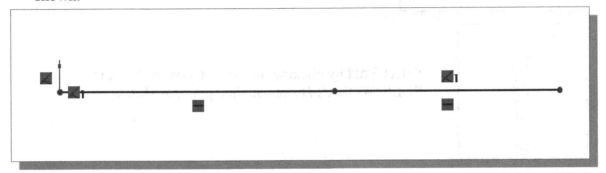

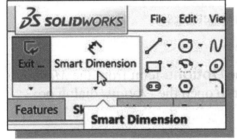

6. On your own, use the **Smart Dimension** command to adjust the sketch as shown.

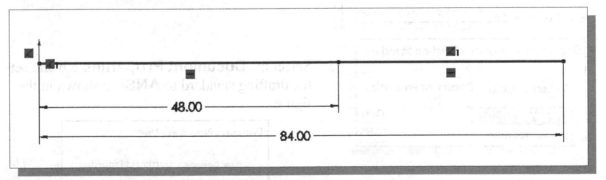

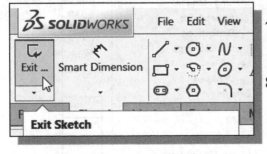

7. Click the **Exit Sketch** icon in the *Sketch* toolbar as shown.

8. On your own, save the current model using the file name **Basic-Beam**.

Create a Rectangular Weldment Profile

❖ In SOLIDWORKS, all structural members use **weldment profiles**, which provide the definitions of the properties of the associated cross sections. *Weldment Profiles* are identified by **Standard**, **Type**, and **Size**. In this section, we will illustrate the procedure to create new profiles and add them to the existing library.

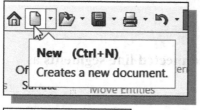

1. Click on the **New** icon, located in the *Standard* toolbar as shown.

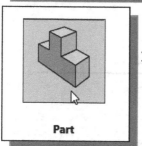

2. Select **Part** by clicking on the first icon in the *New* SOLIDWORKS *Document* dialog box as shown.

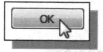

3. Click on the **OK** button to accept the settings.

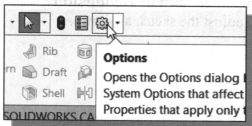

4. Click on the **Options** icon from the *Menu Bar* toolbar to open the *Options* dialog box.

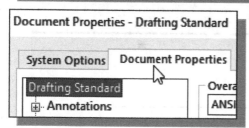

5. Select the **Document Properties** tab and set the drafting standard to **ANSI** as shown in the figure.

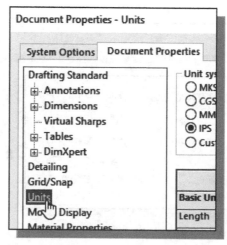

6. Click **Units** as shown in the figure.

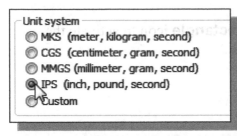

7. Select **IPS (inch, pound, second)** under the *Unit system* options.

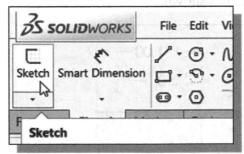

8. On your own, confirm the length decimals to two digits as shown.

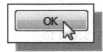

9. Click **OK** in the *Options* dialog box to accept the selected settings.

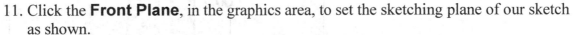

10. Create a new 2D sketch by clicking the **2D Sketch** icon, in the *Sketch* toolbar.

11. Click the **Front Plane**, in the graphics area, to set the sketching plane of our sketch as shown.

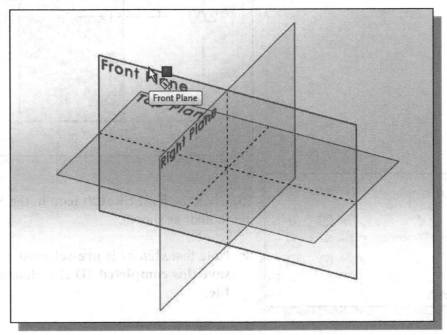

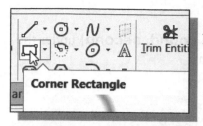

12. Click the **Corner Rectangle** icon to activate the Rectangle command.

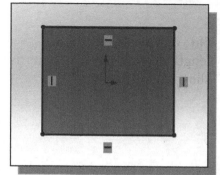

13. On your own, create a rectangle roughly centered at the origin of the world coordinate system.

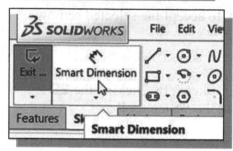

14. On your own, use the **Smart Dimension** to adjust the sketch as shown.

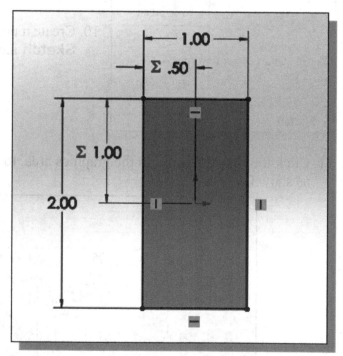

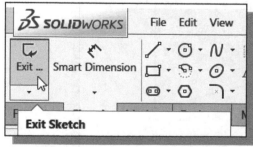

15. Click the **Exit Sketch** icon in the *Sketch* toolbar as shown.

❖ Note that *sketch1* is **pre-selected**, we will **save this completed 2D sketch** as a **library file**.

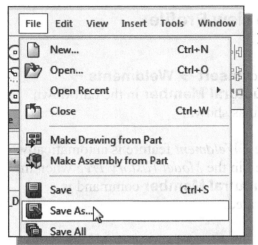

16. In the *pull-down menu*, select **Save As** to save our 2D sketch.

17. Choose the **SLDLFP** type, which is for a *Library* file, and *Save* the profile as **Rectangular.SLDLFP**.

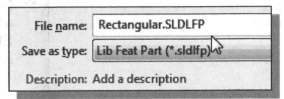

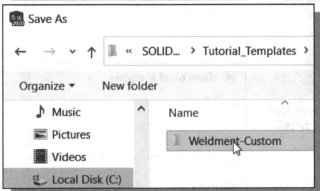

18. On your own, switch to the new weldment profiles folder that you created in chapter 6, such as:

C:\SOLIDWORKS Data \Tutorial_Templates \Weldment-Custom

19. Switch to the **Weldment-Custom** folder and create a new folder using the name **ANSI-Custom**.

20. Switch to the new **ANSI-Custom** folder and create a new folder using the name **Solid**.

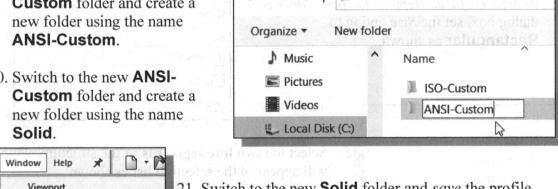

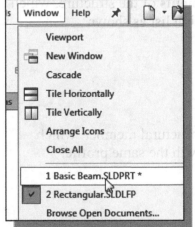

21. Switch to the new **Solid** folder and *save* the profile **Rectangular.SLDLFP.**

22. Switch back to the *Basic Beam* model by selecting the filename listed under the *Window* pull-down menu as shown.

Create Structural Members Using the New Profile

1. Select **Insert → Weldments → Structural Member** in the pull-down menu as shown.

❖ Note a *Weldment* feature is automatically added in the *Model History Tree* when the **Structural Member** command is activated.

2. In the *Structural Member Property Manager* dialog box, set the *Standard* option to **ansi-custom** as shown.

3. In the *Structural Member Property Manager* dialog box, set the *Type* option to **solid** as shown.

4. In the *Structural Member Property Manager* dialog box, set the *Size* option to **Rectangular** as shown.

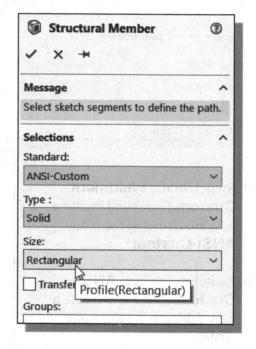

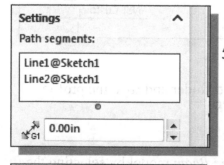

5. Select the **two line segments** of the structure, which will appear in the selection list as shown.

6. Click **OK** to create the structural member, which contains two segments with the same profile.

Adjust the Orientation of the Profile

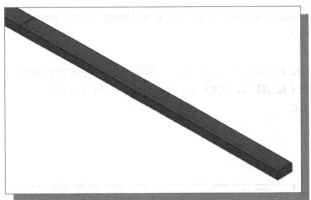

The current *weldment profile* of the beam is aligned perpendicular to the world X axis, but its orientation needs to be rotated 90 degrees for our analysis.

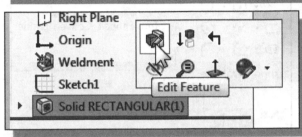

1. In the *Feature Manager*, click on **Structural Member1** and select **Edit Feature** to enter the *Edit Feature* mode.

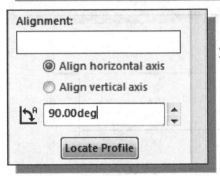

2. Scroll down to the bottom and enter **90.00** as the rotation value.

3. Click **OK** to create the structural member, which contains two segments with the same profile.

4. On your own, use the **Zoom/View** commands and confirm the orientation of the profile is as shown.

Add a Datum Point for the Concentrated Load

In SOLIDWORKS Simulation, to apply a concentrated point load, a datum point must exist on the outside surface of the solid model.

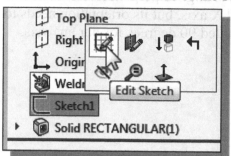

1. In the *Feature Manager*, click on **Sketch1** and select **Edit Sketch** to enter the *Edit Sketch* mode.

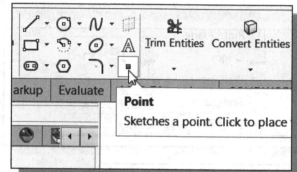

2. In the *Sketch* toolbar, click **Point** to activate the command.

3. Click a location to the left side on the **left line segment** as shown. We will position the point at the *midpoint* of the first segment by adding a location dimension.

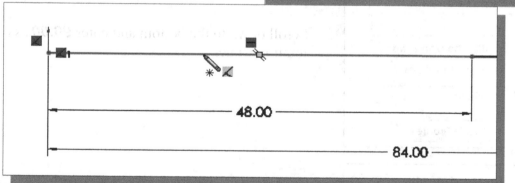

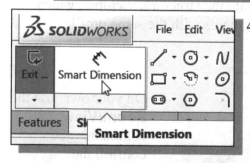

4. On your own, use the **Smart Dimension** to adjust the sketch as shown.

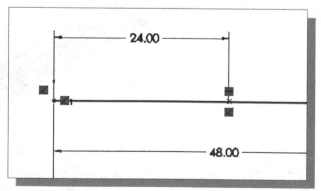

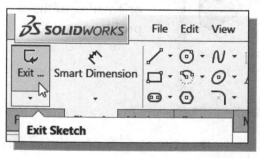

5. Click the **Exit sketch** icon in the *Sketch* toolbar as shown.

6. Set the display to **Isometric view** by using the control in the display panel.

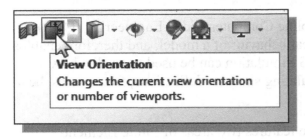

7. In the *Feature* toolbar, select **Point** from the *Reference Geometry* list as shown.

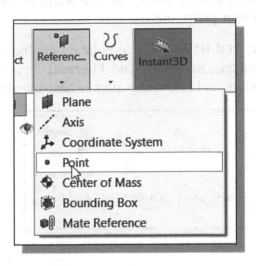

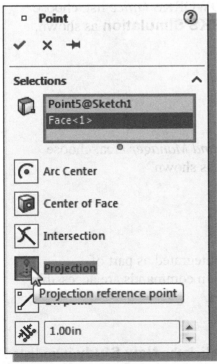

8. In the *Point* dialog box, switch on the **Projection** option as shown.

9. Select the **top surface** of the beam and the **point** we added in Sketch1. (Use the dynamic zoom function if needed.)

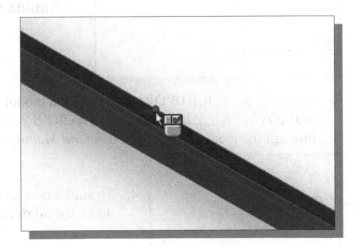

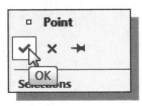

10. Click **OK** to accept the selection and create the point.

Activate the SOLIDWORKS Simulation Module

SOLIDWORKS Simulation is a multi-discipline Computer Aided Engineering (CAE) tool that enables users to simulate the physical behavior of a model, and therefore enables users to improve the design. SOLIDWORKS Simulation can be used to predict how a design will behave in the real world by calculating stresses, deflections, frequencies, heat transfer paths, etc.

The SOLIDWORKS Simulation product line features two areas of Finite Element Analysis: **Structure** and **Thermal**. *Structure* focuses on the structural integrity of the design, and *thermal* evaluates heat-transfer characteristics.

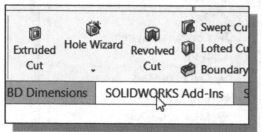

1. Start SOLIDWORKS Simulation by selecting the **SOLIDWORKS Simulation** tab in the *Command Manager* area as shown.

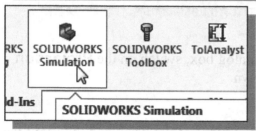

2. In the SOLIDWORKS *Office* list, choose **SOLIDWORKS Simulation** as shown.

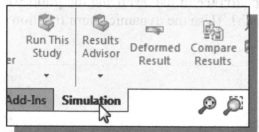

3. In the *Command Manager* area, choose **Simulation** as shown.

❖ Note that the SOLIDWORKS Simulation module is integrated as part of SOLIDWORKS. All of the SOLIDWORKS Simulation commands are accessible through the icon panel in the *Command Manager* area.

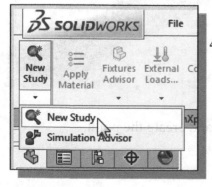

4. To start a new study, click the **New Study** item listed under the *Study Advisor* as shown.

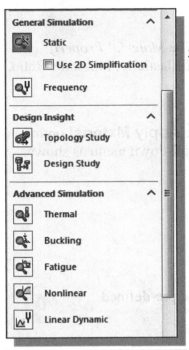

5. Select **Static** as the type of analysis to be performed with SOLIDWORKS Simulation.

❖ Note that different types of analyses are available, which include both structural static and dynamic analyses, as well as the thermal analysis.

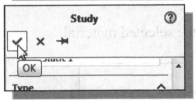

6. Click **OK** to start the definition of a structural static analysis.

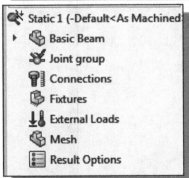

❖ In the *Feature Manager* area, note that a new panel, the *FEA Static* window, is displayed with all the key items listed.

❖ Also note that the **Static 1** tab is activated, which indicates the use of the FEA model.

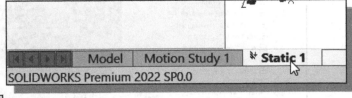

7. On your own, save a copy of the current model.

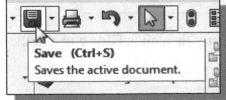

Assign the Element Material Property

Next, we will set up the *Material Property* for the elements. The *Material Property* contains the general material information, such as Modulus of Elasticity, Poisson's Ratio, etc. that is necessary for the FEA analysis.

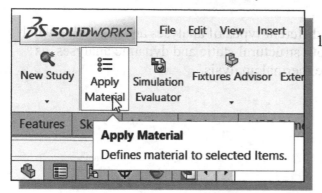

1. Choose the **Apply Material** option from the pull-down menu as shown.

❖ Note the default list of materials, which are available in the pre-defined SOLIDWORKS Simulation material library, is displayed.

2. Select **Alloy Steel** in the *Material* list as shown.

3. Set the **Units** option to display **English (IPS)** to make the selected material available for use in the current FEA model.

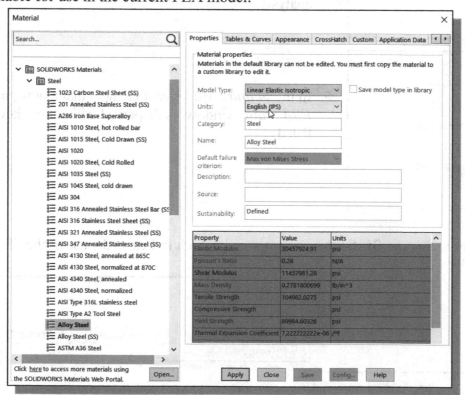

4. Click **Apply** to assign the material property then click **Close** to exit the Material Assignment command.

Apply Boundary Conditions – Constraints

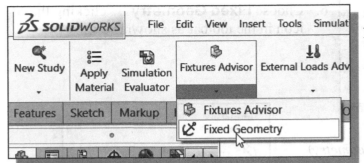

1. Choose **Fixed Geometry** by clicking the icon in the toolbar as shown.

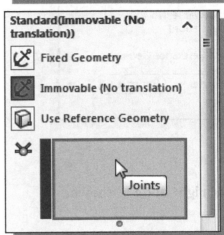

2. In the *Standard* option list, choose **Immovable (No translation)** to set up a **Pin** support at this point.

3. Activate the *Joints* list option box by clicking on the inside of the **Joints** list box as shown.

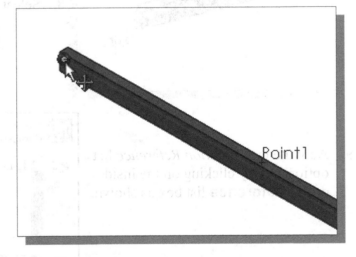

4. Select the **left node** as shown in the figure.

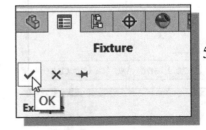

5. Click on the **OK** button to accept the first **Fixture** constraint settings.

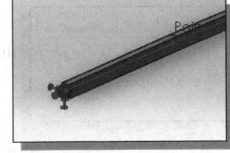

❖ The small arrows indicate constraints have been applied to the associated node in all three directions.

❖ For the node on the right, we will apply a separate roller constraint.

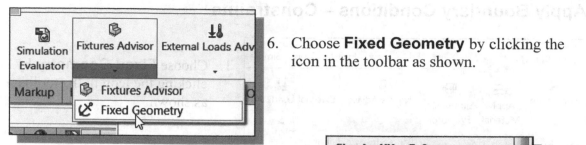

6. Choose **Fixed Geometry** by clicking the icon in the toolbar as shown.

7. Change the *Standard (Fixed Geometry)* option to **Use Reference Geometry** as shown.

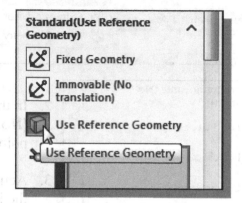

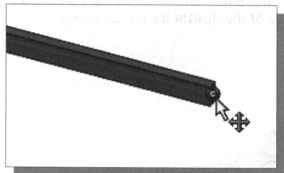

8. Select the **right node** as shown.

9. Activate the *Direction Reference* list option box by clicking on the inside of the **Reference** list box as shown.

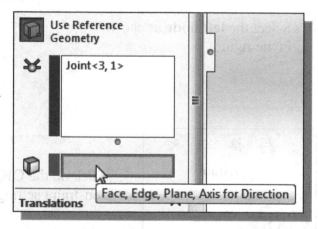

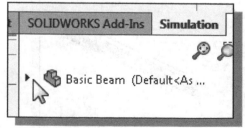

10. Expand the *Model History Tree*, located at the top left corner of the graphics window, by clicking on the **[triangle]** symbol as shown.

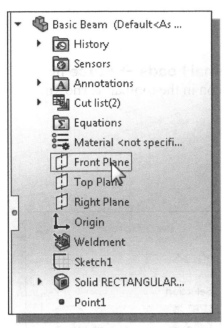

11. Select **Front Plane** as the *direction reference* as shown.

❖ Note the selected plane is highlighted; the constraints we set will be using the selected reference to determine the constraint direction.

12. Set the distance measurement to **inches** to match with the systems units we are using.

13. In the *Translations* constraints list, click on the **Normal to Plane** and **Along Plane Dir 2** icons to activate the constraint.

14. Set the *Normal to Plane* and *Along Plane Dir 2 distances* to **0** as shown.

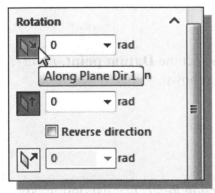

15. Click on the **Along Plane Dir 1** and **Along Plane Dir 2** icons to activate the rotation constraints.

16. Set the *Along Plane Dir 1* and *Along Plane Dir 2 angles* to **0** as shown.

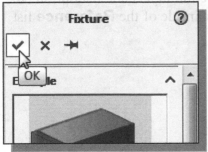

17. Click on the **OK** button to accept the second **Displacement** constraint settings.

❖ Note that no additional constraint is needed for the point near the center of the beam. For beam systems, all joints are treated as **fixed** by default, locking all six degrees of freedom.

Apply the Concentrated Point Load

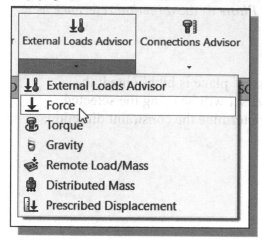

1. Choose **External Loads → Force** by clicking the icon in the toolbar as shown.

2. Confirm the *Force/Torque* option is set to **Points** as shown.

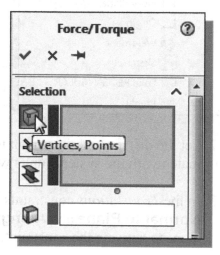

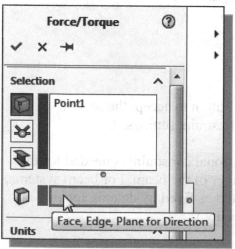

3. Select the **Datum point**, *Point1*, as shown.

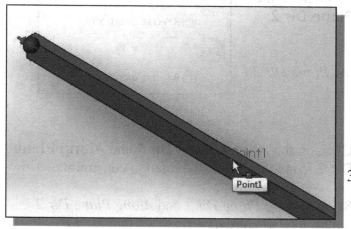

4. Activate the *Direction Reference* list option box by clicking on the inside of the **Reference** list box as shown.

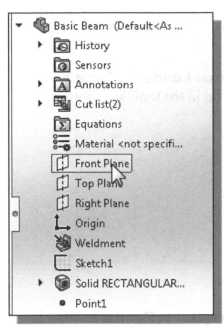

5. Select *Front Plane* as the *direction reference* as shown.

❖ Note the selected plane is highlighted; the constraints we set will be using the selected reference to determine the constraint direction.

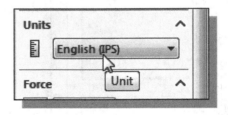

6. Set the *Units* option to **IPS** to match with the systems units we are using, as shown.

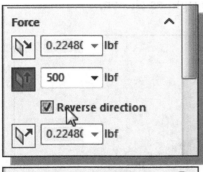

7. Click on the **Along Plane Dir 2** icon to activate the force direction.

8. Set the *Force* to **500** as shown.

9. Activate the **Reverse direction** option.

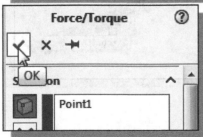

10. Click on the **OK** button to accept the *Force/Torque* settings.

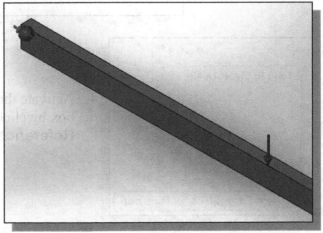

Apply the Distributed Load

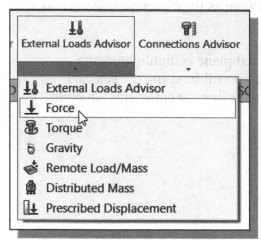

1. Choose **External Loads → Force** by clicking the icon in the toolbar as shown.

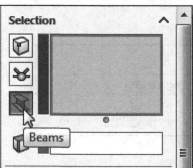

2. Set the *Force/Torque* option to **Beams** as shown.

3. Select the **Beam** on the right as shown.

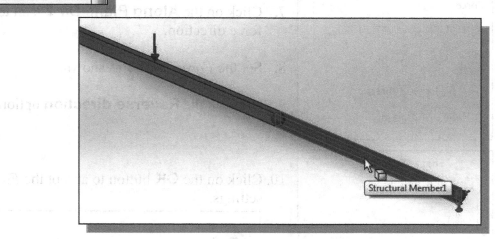

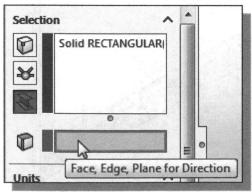

4. Activate the *Direction Reference* list option box by clicking on the inside of the **Reference** list box as shown.

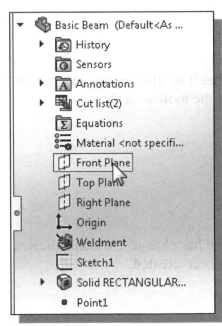

5. Select *Front Plane* as the *direction reference* as shown.

❖ Note the selected plane is highlighted; the constraints we set will be using the selected reference to determine the constraint direction.

6. Set the *Units* option to **IPS** to match with the systems units we are using, as shown.

7. Activate the **Per unit length** option as shown.

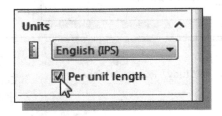

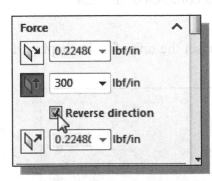

8. Click on the **Along Plane Dir 2** icon to activate the force direction.

9. Set the *Force* to **300** as shown.

10. Activate the **Reverse direction** option.

11. Click on the **OK** button to accept the *Force/Torque* settings.

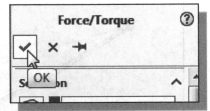

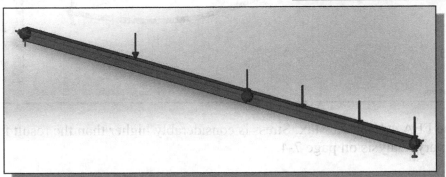

Create the FEA Mesh and Run the Solver

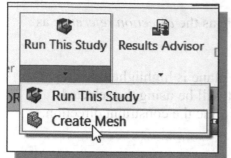

1. Choose **Create Mesh** by clicking the icon under *Run This Study* in the toolbar as shown.

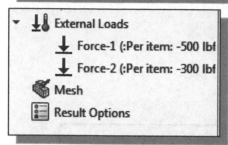

❖ Note the **Mesh** icon has changed, which indicates the FEA Mesh has been created.

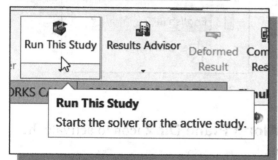

2. Click on the **Run This Study** button to start the *FEA Solver* to calculate the results.

❖ Note the stress results are displayed when the *Solver* has completed the FEA calculations.

3. On your own, set the units for the displayed stress to **psi**.

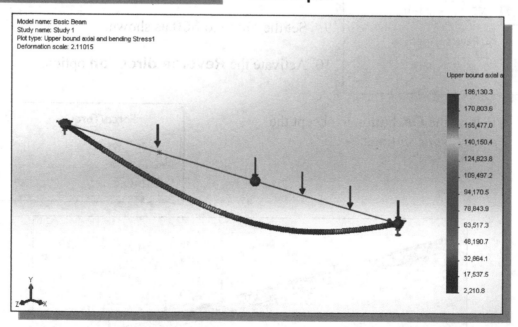

❖ Note the FEA calculated Max. Stress is considerably higher than the result from the preliminary analysis on page 7-4.

What Went Wrong?

With the discrepancy between the FEA result and the preliminary analysis, let's examine the FEA Shear diagram, which is also available in SOLIDWORKS Simulation.

1. In the *Static1* list area, right-mouse-click once on *Result* and choose **Define Beam Diagrams** as shown.

2. Choose **Shear Force in Dir 1** and set the units to **lbf** under the *Display* option list as shown.

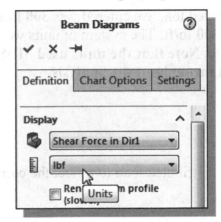

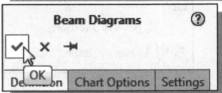

3. Click **OK** to accept the settings and view the shear diagram.

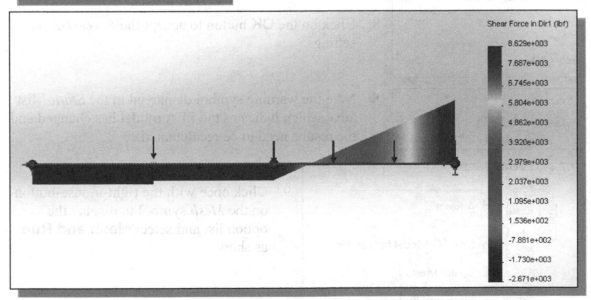

❖ The shear diagram shows the system has a fairly large reaction force, **8629 lb**, on the right side of the beam. Note the total load of the system should be 1400 lb.

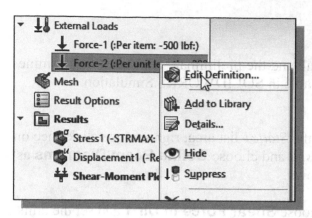

4. Select the **Distributed load**, **Force-2** in the *Static1* list area.

5. Click once with the right-mouse-button to display the option list and select **Edit Definition** as shown.

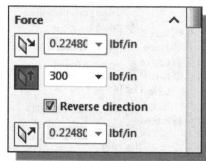

6. The distributed load we entered was **300 lb/in**, and it should be **300 lb/ft**. The system of units we are using is *inches-lbf*. **Note that the units used MUST be consistent throughout the analysis.**

7. Adjust the distributed load to reflect the correct unit as shown.

    300 lb/ft = (300/12) lb/in = 25 lb/in

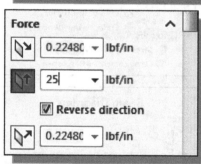

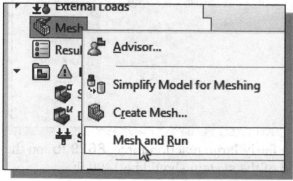

8. Click on the **OK** button to accept the *Force/Torque* settings.

❖ Note the warning symbol displayed in the *Static1* list area, which indicates the FEA model has changed and the results need to be recalculated.

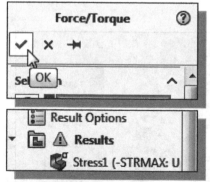

9. Click once with the right-mouse-button on the *Mesh* symbol to display the option list and select **Mesh and Run** as shown.

❖ Note the FEA calculated Max. Stress, **21674**, now matches perfectly to the result from the preliminary analysis on page 7-4.

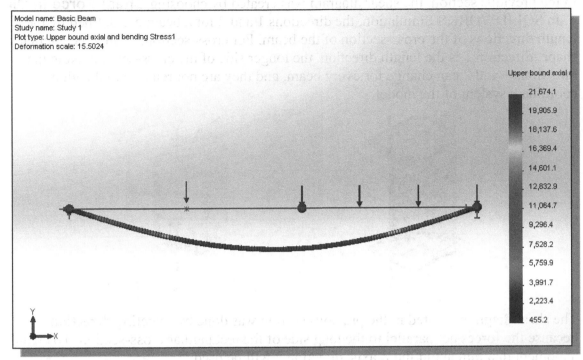

10. Click once with the right-mouse-button on the *Shear-Moment-Plot1* item to display the option list and select **Show** as shown.

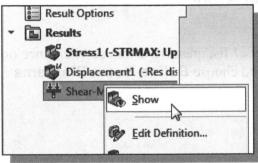

❖ Note the *Shear* diagram matches with the results from the preliminary analysis done at the beginning of the chapter.

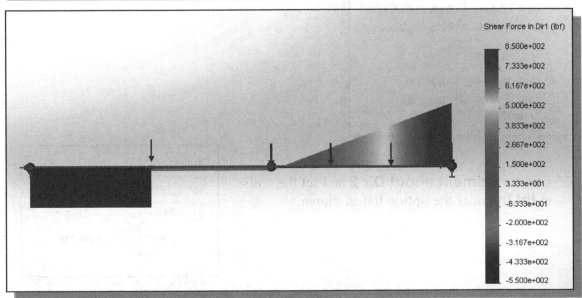

Directions 1 and 2 in Shear and Moment Diagrams

In the previous section, the shear diagram was created by choosing **Shear Force in Dir 1**. In SOLIDWORKS Simulation, the directions 1 and 2 for a beam are defined by the length directions of the cross-section of the beam. For cross-sections with a rectangular shape, direction 1 is the length direction, the longer side of the cross-section. Note that directions 1 and 2 can change for every beam, and they are not relative to the global coordinate system of the model.

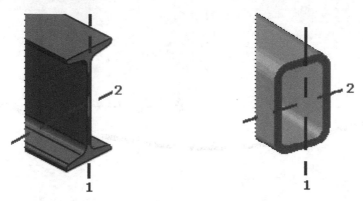

The shear diagram created in the previous section was done by selecting **direction 1** because the forces act parallel to the long side of the rectangular cross-section. To create a moment diagram about the Z axis, direction 2 will be used.

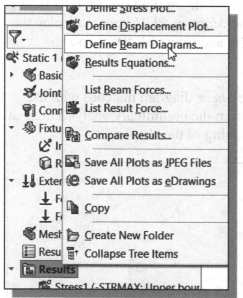

1. In the *Static1* list area, right-mouse-click once on **Result** and choose **Define Beam Diagrams** as shown.

2. Choose **Moment about Dir 2** and set the units to **lbf.in** under the option list as shown.

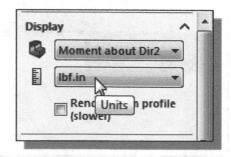

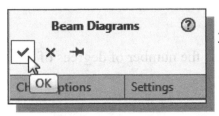

3. Click **OK** to accept the settings and view the shear diagram.

4. Switch to the bottom view by selecting the **Bottom** view icon in the standard view display option list as shown.

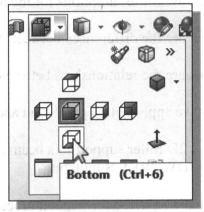

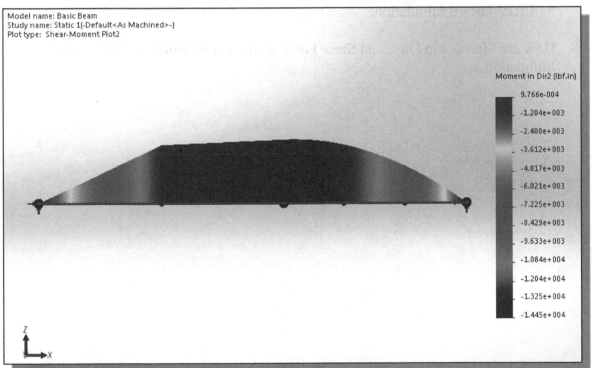

Model name: Basic Beam
Study name: Static 1(-Default<As Machined>-)
Plot type: Shear-Moment Plot2

Moment in Dir2 (lbf.in)

9.766e-004
-1.204e+003
-2.408e+003
-3.612e+003
-4.817e+003
-6.021e+003
-7.225e+003
-8.429e+003
-9.633e+003
-1.084e+004
-1.204e+004
-1.325e+004
-1.445e+004

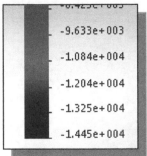

-9.633e+003
-1.084e+004
-1.204e+004
-1.325e+004
-1.445e+004

❖ The FEA result of the maximum moment is **-1.445e+4**, which matches the result of the preliminary analysis on page 7-4 except for the negative sign. Note that most FEA packages do not use the Shear-Moment diagram sign convention that is commonly used (as outlined in the preliminary analysis section at the beginning of this chapter).

Questions:

1. For a beam element in three-dimensional space, what is the number of degrees of freedom it possesses?

2. What are the assumptions for the beam element?

3. What are the differences between Truss Element and Beam Element?

4. What are the relationships between the shear diagram and the moment diagram?

5. Can we apply both a point load and a distributed load to the same beam element?

6. For a 2D roller support in a beam system, how should we set the constraints in SOLIDWORKS Simulation?

7. List and describe the general procedure to display the moment diagram in SOLIDWORKS Simulation.

8. How are Moment in Dir 2 and Shear Force in Dir 1 determined in SOLIDWORKS Simulation?

Exercises:

Determine the maximum stress produced by the loads and create the shear and moment diagram.

1. Material: Steel, Diameter 2.5 in.

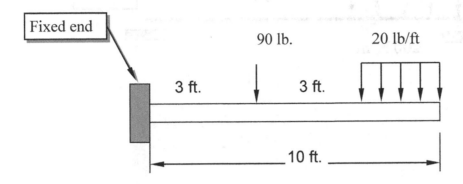

2. Material: Aluminum

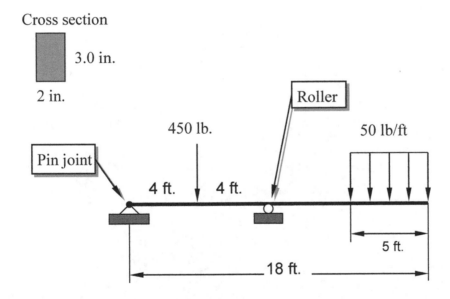

3.   Material: Steel, ∅ 3 cm.

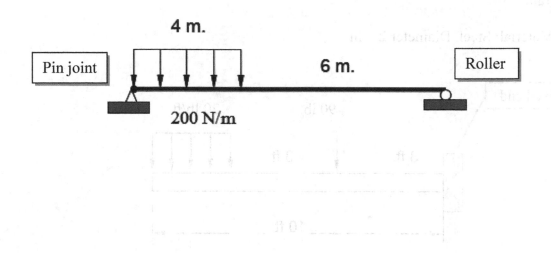

Chapter 8
Beam Analysis Tools

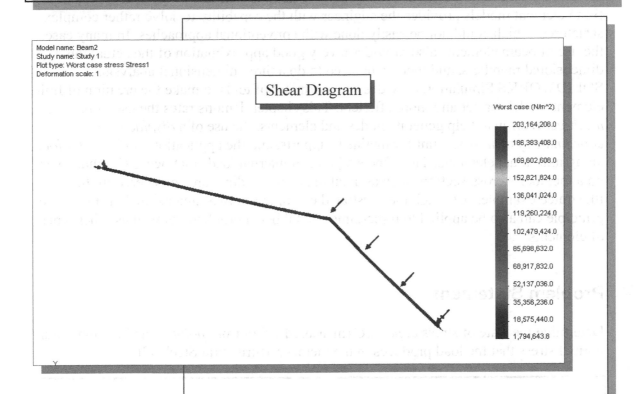

Model name: Beam2
Study name: Study 1
Plot type: Worst case stress Stress1
Deformation scale: 1

Shear Diagram

Worst case (N/m^2)

203,164,208.0
186,383,408.0
169,602,608.0
152,821,824.0
136,041,024.0
119,260,224.0
102,479,424.0
85,698,632.0
68,917,832.0
52,137,036.0
35,356,236.0
18,575,440.0
1,794,643.8

Learning Objectives

◆ **Analyze Structures with Inclined supports.**
◆ **Create FE constraints using the Reference direction options.**
◆ **Use Datum points to view FEA results at specific locations.**
◆ **Analyze Structures with Combined Stresses.**
◆ **View the Results with the Animation options.**

Introduction

The beam element is the most common of all structural elements as it is commonly used in buildings, towers, bridges and many other structures. The use of beam elements in finite element models provides the engineer with the capability to solve rather complex structures, which could not be easily done with conventional approaches. In many cases, the use of beam elements also provides a very good approximation of the actual three-dimensional members, and there is no need to do a three-dimensional analysis. SOLIDWORKS Simulation provides an assortment of tools to make the creation of finite element models easier and more efficient. This chapter demonstrates the use of *automatic mesh-generation* to help generate nodes and elements, the use of a *displacement coordinate system* to account for inclined supports, and the application of *elemental loads* along individual elements. The effects of several internal loads that occur simultaneously on a member's cross-section, such as axial load and bending, are considered in the illustrated example. Although the illustrated example is a two-dimensional problem, the principle can also be applied to three-dimensional beam problems as well as other types of elements.

Problem Statement

Determine the state of stress at *point C* (measured 1.5 m from *point A*) and the maximum normal stress that the load produces in the member. (Structural Steel A36)

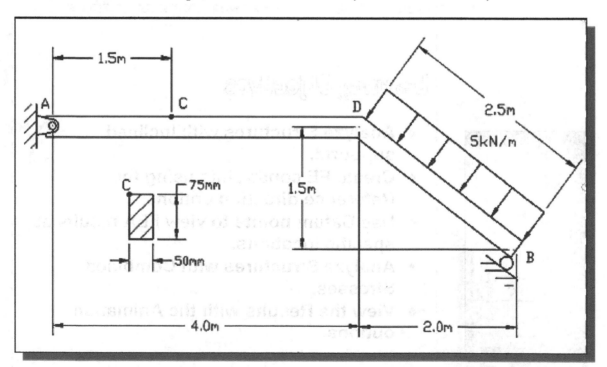

Preliminary Analysis

Free Body Diagram of the member:

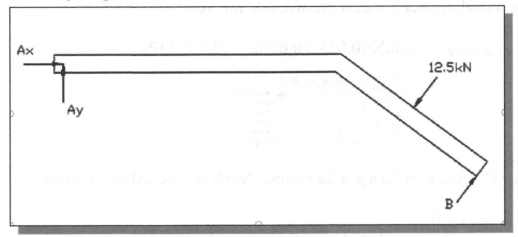

Applying the equations of equilibrium:

$$\Sigma M_{@A} = 0, \ \Sigma F_X = 0, \ \Sigma F_Y = 0,$$

Therefore B = 9.76 kN, A_X = 1.64 kN and A_Y = 2.19 kN

Consider a segment of the horizontal portion of the member:

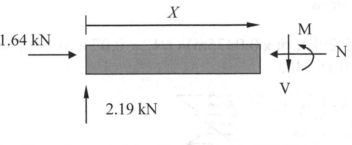

$$\Sigma F_X = 0, \quad 1.64 - N = 0, N = 1.64 \text{ kN}$$
$$\Sigma F_Y = 0, \quad 2.19 - V = 0, V = 2.19 \text{ kN}$$
$$\Sigma M_{@X} = 0, \ -A_Y X + M = 0, \quad M = 2.19 \ X \text{ kN-m}$$

At *point C*, X = 1.5 m and

N = 1.64 kN, V = 2.19 kN, M = 3.29 kN-m

➤ The state of stress at *point C* can be determined by using the principle of superposition. The stress distribution due to each loading is first determined, and then the distributions are superimposed to determine the resultant stress distribution. The principle of superposition can be used for this purpose provided a linear relationship exists between the stress and the loads.

Stress Components

Normal Force:
      The normal stress at C is a compressive uniform stress.

$$\sigma_{normal\_force} = 1.64kN/(0.075 \times 0.05)m^2 = 0.437 \text{ MPa}$$

Shear Force:
      Point C is located at the top of the member. No shear stress existed at *point C*.

$$\tau_{shear\_force} = 0$$

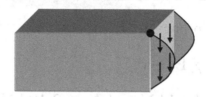

Bending Moment:
      Point C is located at 37.5 mm from the neutral axis. The normal stress at C is a compressive uniform stress.

$$\sigma_{bending\_moment} = (3.29kN\text{-}m \times 0.0375m)/(1/12 \times 0.05 \times (0.075)^3)m^4$$
$$= 70.16 \text{ MPa}$$

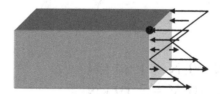

Superposition:
      The shear stress is zero and combining the normal stresses gives a compressive stress at *point C*:

$$\sigma_C = 0.437 \text{ MPa} + 70.16 \text{ MPa} = 70.6 \text{ MPa}$$

Examine the horizontal segment of the member:

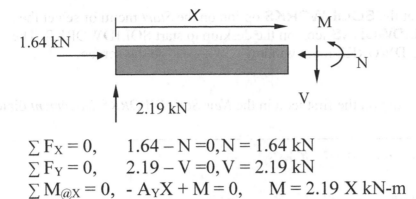

$$\Sigma \, F_X = 0, \quad 1.64 - N = 0, N = 1.64 \text{ kN}$$
$$\Sigma \, F_Y = 0, \quad 2.19 - V = 0, V = 2.19 \text{ kN}$$
$$\Sigma \, M_{@X} = 0, \quad - A_Y X + M = 0, \quad M = 2.19 \text{ X kN-m}$$

The maximum normal stress for the horizontal segment will occur at *point D*, where X = 4m.

$$\sigma_{normal\_force} = 1.64 \text{kN}/(0.05 \times 0.075) \text{m}^2 = 0.437 \text{ MPa}$$

$$\sigma_{bending\_moment} = (8.76 \text{kN-m} \times 0.0375 \text{m})/(1/12 \times 0.05 \times (0.075)^3) \text{m}^4$$
$$= 186.9 \text{MPa}$$

$$\sigma_{max@D} = 0.437 \text{ MPa} + 186.9 \text{ MPa} = 187.34 \text{ MPa}$$

➢ Does the above calculation provide us the maximum normal stress developed in the structure? To be sure, it would be necessary to check the stress distribution along the inclined segment of the structure. We will rely on the SOLIDWORKS *Simulation* solutions to find the maximum stress developed. The above calculation (the state of stress at *point C* and *point D*) will serve as a check of the SOLIDWORKS *Simulation* FEA solutions.

Start SOLIDWORKS

1. Select the **SOLIDWORKS** option on the *Start* menu or select the **SOLIDWORKS** icon on the desktop to start SOLIDWORKS. The SOLIDWORKS main window will appear on the screen.

2. Select **Part** by clicking on the first icon in the *New SOLIDWORKS Document* dialog box as shown.

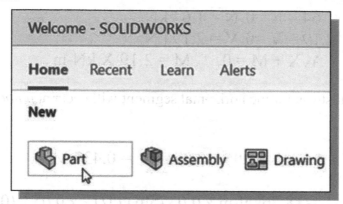

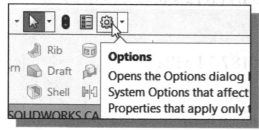

3. Select the **Options** icon from the *Menu* toolbar to open the *Options* dialog box.

4. Switch to the **Document Properties** tab and reset the *Drafting Standard* to **ANSI** as shown in the figure.

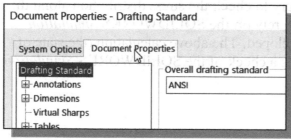

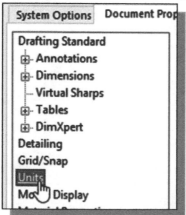

5. Set the *Unit system* to **MKS (meter, kilogram, second)**.

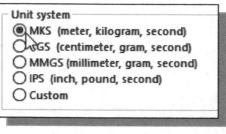

Create the CAD Model – Structural Member Approach

To create the Beam system, we will first establish the locations of the two center axes of the two sections by creating a 2D sketch. We will then use the SOLIDWORKS **Structural Member** tool to select the predefined profiles of the cross sections of the system.

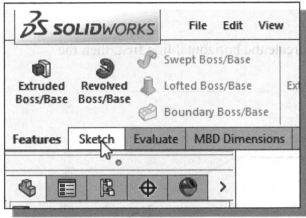

1. Click the **Sketch** tab in the *Command Manager* as shown.

   ❖ Note the *Sketch* toolbar is displayed in the *Command Manager* area with different sketching tools.

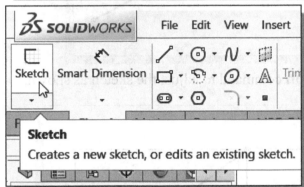

2. Click the **Sketch** icon, in the *Sketch* toolbar, to create a new sketch.

3. Click the **Front Plane**, in the graphics area, to set the sketching plane of our sketch as shown.

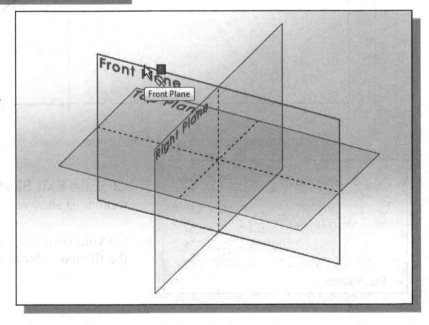

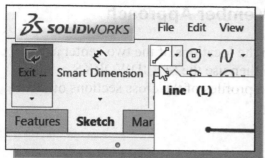

4. Click the **Line** icon in the *Sketch* toolbar as shown.

5. Start the line segment at the origin and create the horizontal line first, then the inclined line as shown.

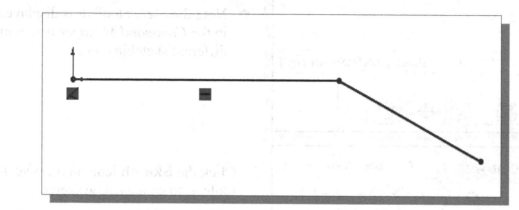

6. On your own, use the **Smart Dimension** command to adjust the sketch as shown.

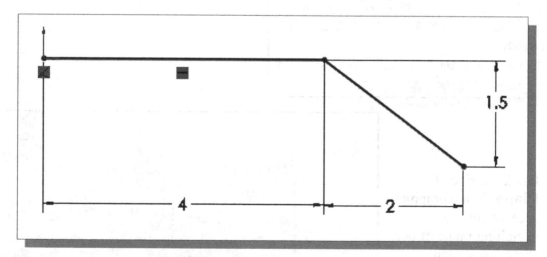

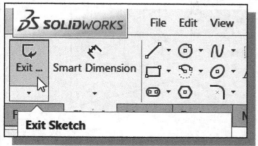

7. Click the **Exit Sketch** icon in the *Sketch* toolbar as shown.

8. On your own, save the current model using the file name **Beam2**.

Create a Rectangular Weldment Profile

In SOLIDWORKS, all structural members use **weldment profiles**, which provide the definitions of the properties of the associated cross sections. *Weldment Profiles* are identified by **Standard**, **Type** and **Size**. In this section, we will illustrate the procedure to create new profiles and add them to the existing library.

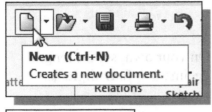

1. Click on the **New** icon, located in the *Standard* toolbar as shown.

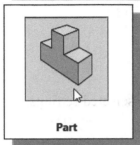

2. Select **Part** by clicking on the first icon in the *New* SOLIDWORKS *Document* dialog box as shown.

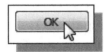

3. Click on the **OK** button to accept the settings.

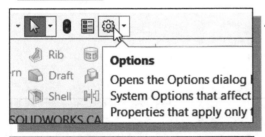

4. Click on the **Options** icon from the *Menu* toolbar to open the *Options* dialog box.

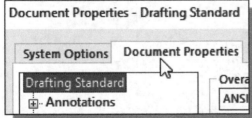

5. Select the **Document Properties** tab as shown in the figure.

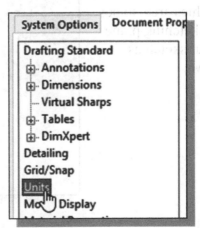

6. Click **Units** as shown in the figure.

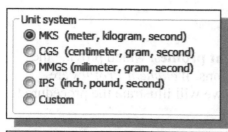

7. Select **MKS (meter, kilogram, second)** under the *Unit system* options.

Type	Unit	Decimals	Fractions
Basic Units			
Length	meters	.1234	
Dual Dimension Length	inches	.12	
Angle	degrees	.123	
		.1234	
Mass/Section Properties		.123456	
Length	meters	.1234567	
Mass	kilograms	.12345678	

8. On your own, set the length decimals to **four digits** as shown.

9. Click **OK** in the *Options* dialog box to accept the selected settings.

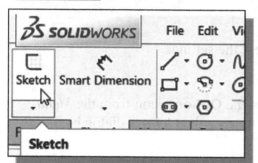

10. Create a new 2D sketch by clicking the **2D Sketch** icon in the *Sketch* toolbar.

11. Click the **Front Plane**, in the graphics area, to set the sketching plane of our sketch as shown.

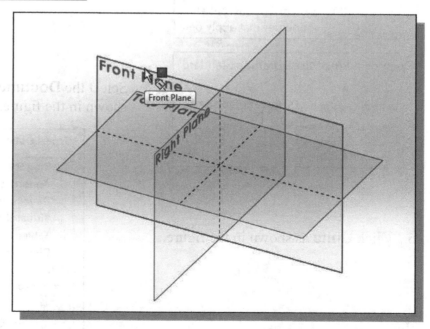

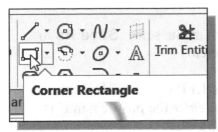

12. Click the **Corner Rectangle** icon to activate the **Rectangle** command.

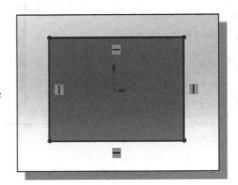

13. On your own, create a rectangle roughly centered at the origin of the world coordinate system.

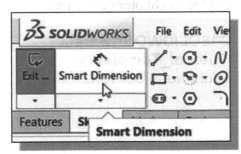

14. On your own, use the **Smart Dimension** to adjust the sketch as shown.

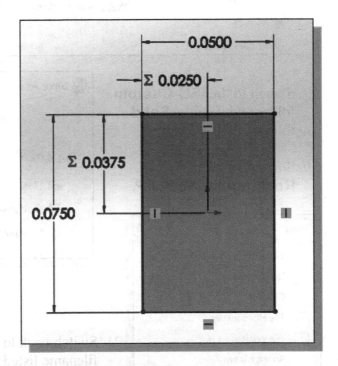

15. Click the **Exit Sketch** icon in the *Sketch* toolbar as shown.

❖ Note that *sketch1* is **pre-selected**; we will **save this completed 2D sketch** as a **library file**.

16. In the pull-down menu, select **Save As** to save our 2D sketch.

17. Choose the **SLDLFP** type, which is for a *Library* file, and *enter* the profile name as **Rectangular.SLDLFP**.

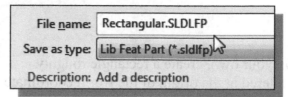

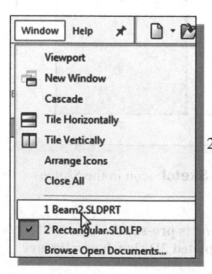

18. On your own, switch to the new weldment profiles folder that you created in chapter 6, such as:

    C:\SOLIDWORKS Data \Tutorial_Templates \Weldment-Custom.

19. Switch to the **ISO-Custom** folder and then the **Solid** folder.

20. S*ave* the profile **Rectangular.*SLDLFP.***

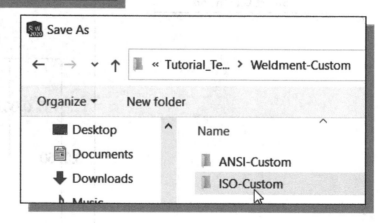

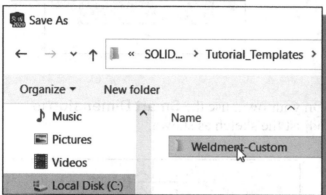

21. Switch back to the **Beam2** model by selecting the filename listed under the **Window** pull-down menu as shown.

Create Structural Members Using the New Profile

1. Select **Insert → Weldments → Structural Member** in the pull-down menu as shown.

❖ Note a *Weldment* feature is automatically added in the *Model History Tree* when the **Structural Member** command is activated.

2. In the *Structural Member PropertyManager* dialog box, set the *Standard* option to **iso** as shown.

3. In the *Structural Member PropertyManager* dialog box, set the *Type* option to **solid** as shown.

4. In the *Structural Member PropertyManager* dialog box, set the *Size* option to **Rectangular** as shown.

5. Select the **two line segments** of the structure as shown.

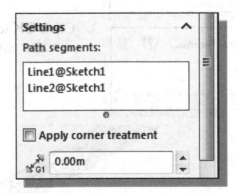

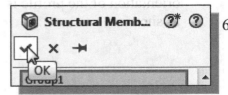

6. Click **OK** to create the structural member, which contains two segments with the same profile.

Adjust the Orientation of the Profile

The current *weldment profile* of the beam is aligned perpendicular to the beam direction, but its orientation needs to be rotated 90 degrees for our analysis.

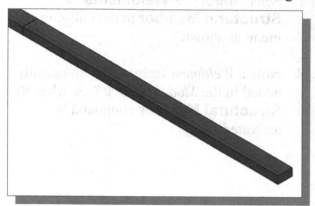

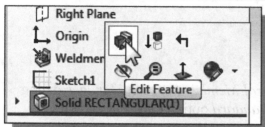

1. In the *Feature Manager*, click on **Solid RECTANGULAR(1)** and select **Edit Feature** to enter the *Edit Feature* mode.

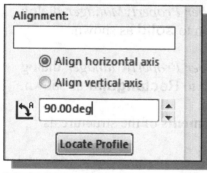

2. Scroll down to the bottom and enter **90.00** as the rotation value.

3. Click **OK** to create the structural member, which contains two segments with the same profile.

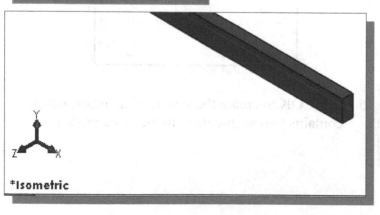

4. On your own, use the **Zoom/View** commands and confirm the orientation of the profile is as shown.

Add a Datum Point for the 1.5m location

We will place a datum point at 1.5m away from the left end of the system; the point is used as a reference point for viewing the FEA results.

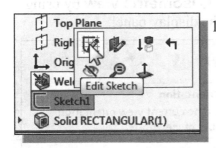

1. In the *Feature Manager*, click on **Sketch1** and select **Edit Sketch** to enter the *Edit Sketch* mode.

2. In the *Sketch* toolbar, click **Point** to activate the command.

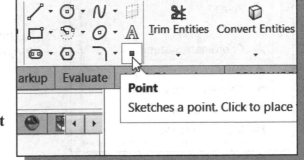

3. Click a location on the **left line segment** as shown.

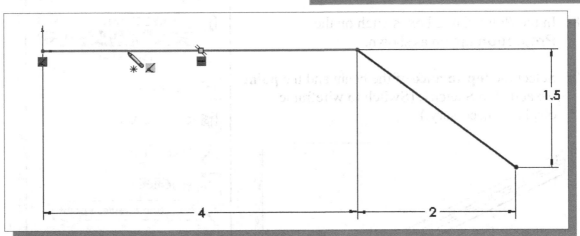

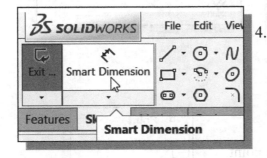

4. On your own, use the **Smart Dimension** to adjust the sketch as shown.

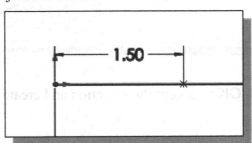

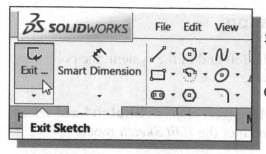

5. Click the **Exit sketch** icon in the *Sketch* toolbar as shown.

6. Set the display to **Isometric view** by using the control in the display panel.

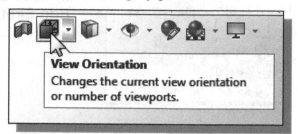

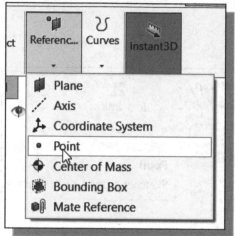

7. In the *Feature* toolbar, select **Point** from the *Reference Geometry* list as shown.

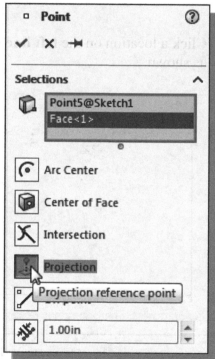

8. In the *Point* dialog box, switch on the **Projection** option as shown.

9. Select the **top surface** of the beam and the **point** we added in Sketch1. (Switch to wireframe display if necessary.)

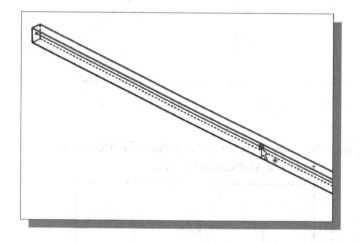

10. Click **OK** to accept the selection and create the point.

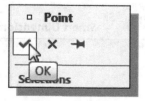

Activate the SOLIDWORKS Simulation Module

SOLIDWORKS Simulation is a multi-discipline Computer Aided Engineering (CAE) tool that enables users to simulate the physical behavior of a model, and therefore enables users to improve the design. SOLIDWORKS Simulation can be used to predict how a design will behave in the real world by calculating stresses, deflections, frequencies, heat transfer paths, etc.

The SOLIDWORKS Simulation product line features two areas of Finite Element Analysis: **Structure** and **Thermal**. *Structure* focuses on the structural integrity of the design, and *thermal* evaluates heat-transfer characteristics.

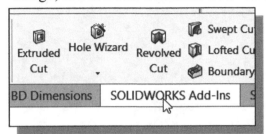

1. Start SOLIDWORKS Simulation by selecting the **SOLIDWORKS Add-Ins** tab in the *Command Manager* area as shown.

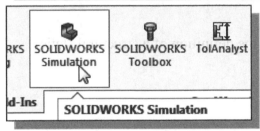

2. In the SOLIDWORKS *Office* list, choose **SOLIDWORKS Simulation** as shown.

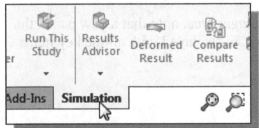

3. In the *Command Manager* area, choose **Simulation** as shown.

❖ Note that the SOLIDWORKS Simulation module is integrated as part of SOLIDWORKS. All of the SOLIDWORKS Simulation commands are accessible through the icon panel in the *Command Manager* area.

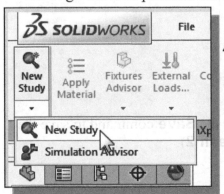

4. To start a new study, click the **New Study** item listed under the *Study Advisor* as shown.

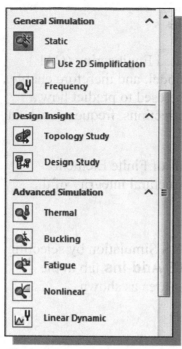

5. Select **Static** as the type of analysis to be performed with SOLIDWORKS Simulation.

❖ Note that different types of analyses are available, which include both structural static and dynamic analyses, as well as the thermal analysis.

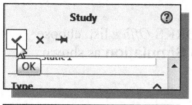

6. Click **OK** to start the definition of a structural static analysis.

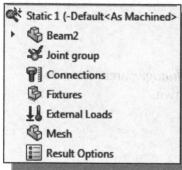

❖ In the *Feature Manager* area, note that a new panel, the *FEA Static* window, is displayed with all the key items listed.

❖ Also, note the **Static 1** tab is activated, which indicates the use of the FEA model.

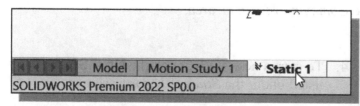

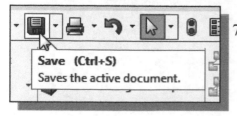

7. On your own, use the **Save** command to save the current model (**Beam2**).

Assign the Element Material Property

➢ Next, we will set up the *Material Property* for the elements. The *Material Property* contains the general material information, such as Modulus of Elasticity, Poisson's Ratio, etc. that is necessary for the FEA analysis.

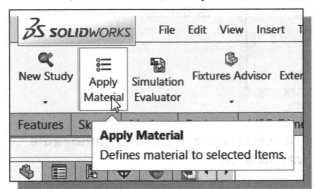

1. Choose **Apply Materials** option from the pull-down menu as shown.

❖ Note the default list of materials, which are available in the pre-defined SOLIDWORKS Simulation material library, is displayed.

2. Select **Alloy Steel** in the *Material* list as shown.

3. Confirm the **Units** option to display **SI – N/m^2 (PA)**.

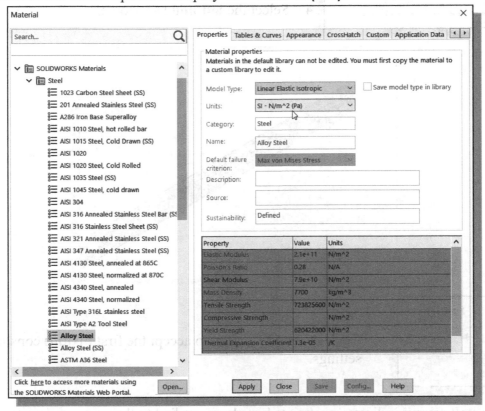

4. Click **Apply** to assign the material property then click **Close** to exit the Material Assignment command.

Apply Boundary Conditions – Constraints

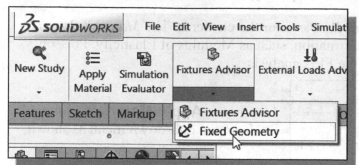

1. Choose **Fixed Geometry** by clicking the icon in the toolbar as shown.

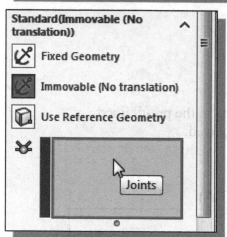

2. In the *Standard* option list, choose **Immovable (No translation)**.

❖ For Beam systems, all joints are treated as **fixed** by default, locking all six degrees of freedom.

3. Activate the *Joints* list option box by clicking on the inside of the **Joints** list box as shown.

4. Select the **left node** as shown.

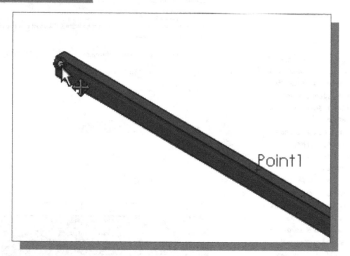

Point1

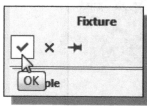

5. Click on the **OK** button to accept the first Fixture constraint settings.

❖ The small arrows indicate constraints have been applied to the associated node.

❖ For the support on the inclined surface, we will apply a separate roller constraint.

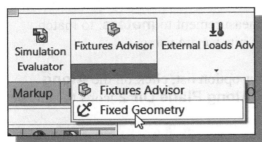

6. Choose **Fixed Geometry** by clicking the icon in the toolbar as shown.

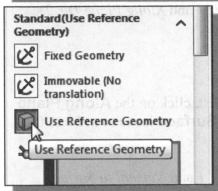

7. Change the *Standard(Fixed Geometry)* option to **Use Reference Geometry** as shown.

8. Select the **right node**, the node that is on the inclined section.

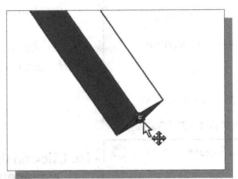

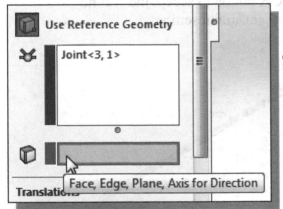

9. Activate the *Direction Reference* list option box by clicking on the inside of the **Reference** list box as shown.

10. **Zoom-in** on the right end of the beam and select the **small surface** at the end as shown.

❖ Note the selected plane is highlighted; the constraints we set will be using the selected reference to determine the constraint direction.

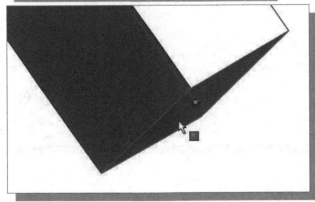

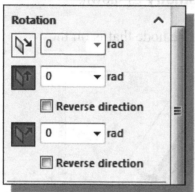

11. Set the distance measurement to **meters**, to match with the systems units we are using.

12. In the *Translations* option list, click on the **Along Plane Dir 1** and **Along Plane Dir 2** icons to activate the constraint.

13. Set the *Along Plane Dir 1* and *Along Plane Dir 2* distances to **0** as shown.

14. In the *Rotation* option list, click on the **Along Plane Dir 2** and **Normal to Surface** icons to activate the constraint.

15. Set the *Along Plane Dir 1* and *Normal to Surface* distances to **0 rad** as shown.

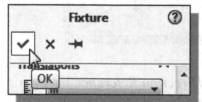

16. Click on the **OK** button to accept the second **Displacement** constraint settings.

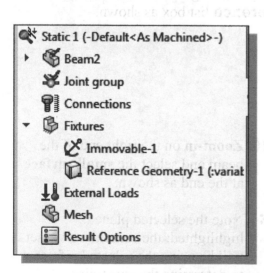

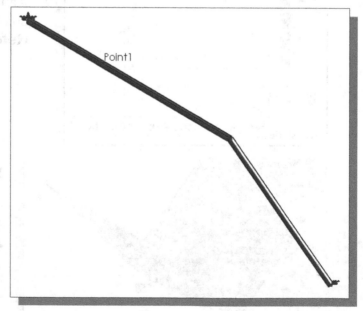

Apply the Distributed Load

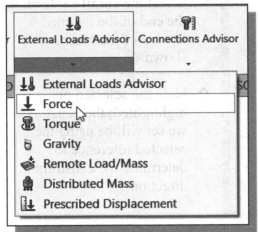

1. Choose **External Loads → Force** by clicking the icon in the toolbar as shown.

2. Set the *Force/Torque* option to **Beams** as shown.

❖ The *Beams* option allows us to apply loads along the entire beam member.

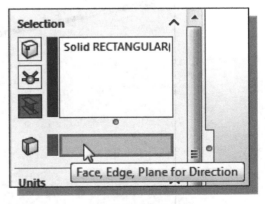

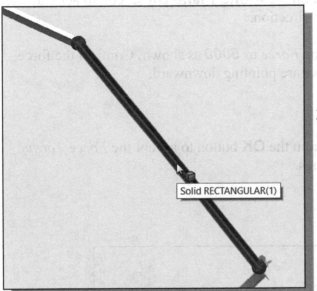

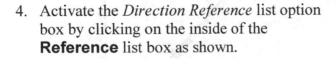

3. Select the **beam** on the right as shown.

4. Activate the *Direction Reference* list option box by clicking on the inside of the **Reference** list box as shown.

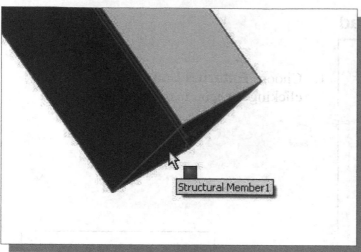

5. Select the **small surface**, at the end of the inclined section of the beam, as shown.

❖ Note the selected plane is highlighted; the constraints we set will be using the selected reference to determine the constraint direction.

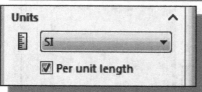

6. Set the *Units* option to **SI** to match with the systems units we are using as shown.

7. Activate the **Per unit length** option as shown.

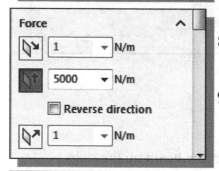

8. Click on the **Along Plane Dir 2** icon to activate the force direction.

9. Set the *Force* to **5000** as shown. Confirm the force arrows are pointing downward.

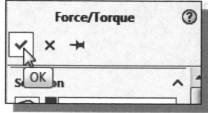

10. Click on the **OK** button to accept the *Force/Torque* settings.

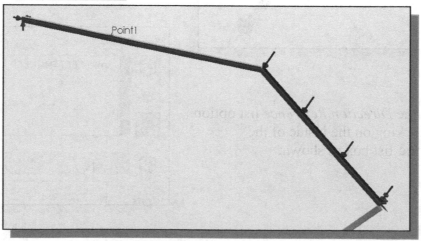

Create the FEA Mesh and Run the Solver

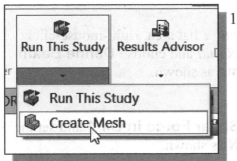

1. Choose **Create Mesh** by clicking the icon under *Run This Study* in the toolbar as shown.

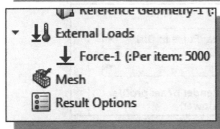

❖ Note the **Mesh** icon has changed, which indicates the FEA Mesh has been created.

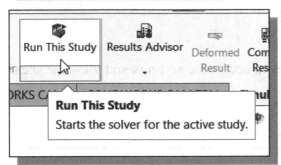

2. Click on the **Run This Study** button to start the *FEA Solver* to calculate the results.

❖ Note the stress results are displayed when the *Solver* has completed the FEA calculations.

❖ Note the FEA calculated Max. Stress matches the result from the preliminary analysis on page 8-4.

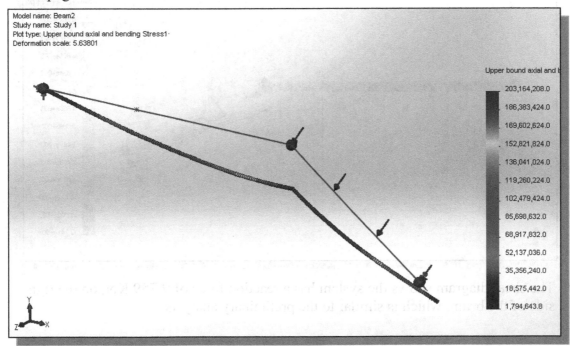

Model name: Beam2
Study name: Study 1
Plot type: Upper bound axial and bending Stress1
Deformation scale: 5.63801

Upper bound axial and t

| |
| 203,164,208.0 |
| 186,383,424.0 |
| 169,602,624.0 |
| 152,821,824.0 |
| 136,041,024.0 |
| 119,260,224.0 |
| 102,479,424.0 |
| 85,698,632.0 |
| 68,917,832.0 |
| 52,137,036.0 |
| 35,356,240.0 |
| 18,575,442.0 |
| 1,794,643.8 |

Shear and Moment Diagrams

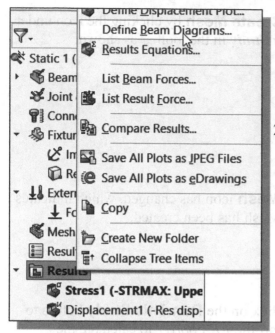

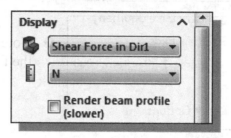

1. In the *Static 1* list area, right-mouse-click once on *Result* and choose **Define Beam Diagrams** as shown.

2. Choose **Shear Force in Dir 1** and set the units to **N** as shown.

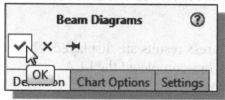

3. Click **OK** to accept the settings and view the shear diagram.

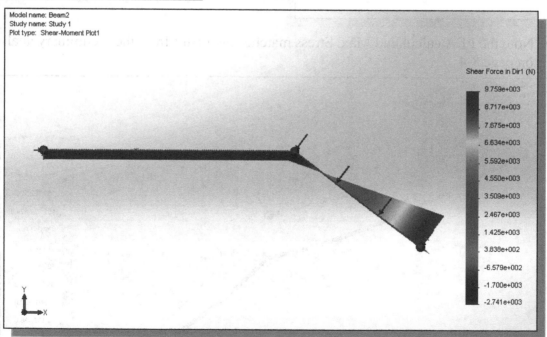

❖ The shear diagram shows the system has a reaction force of 9.759 KN, on the right side of the beam, which is similar to the preliminary analysis.

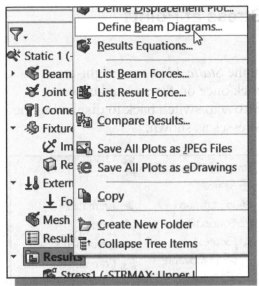

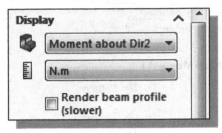

4. In the *Static 1* list area, right-mouse-click once on *Result* and choose **Define Beam Diagrams** as shown.

5. Choose **Moment about Dir2** and set the units to **N.m** as shown.

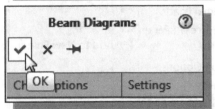

6. Click **OK** to accept the settings and view the shear diagram. Rotate the display to view the moment diagram.

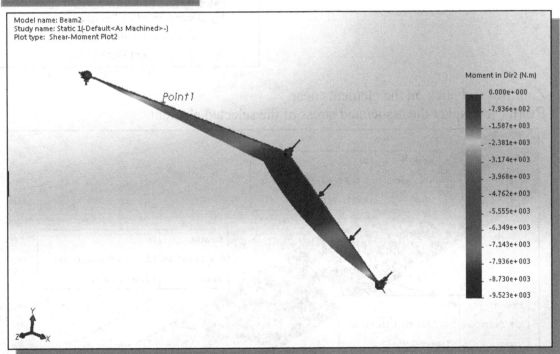

Model name: Beam2
Study name: Static 1(-Default<As Machined>-)
Plot type: Shear-Moment Plot2

Point1

Moment in Dir2 (N.m)

0.000e+000
-7.936e+002
-1.587e+003
-2.381e+003
-3.174e+003
-3.968e+003
-4.762e+003
-5.555e+003
-6.349e+003
-7.143e+003
-7.936e+003
-8.730e+003
-9.523e+003

-7.143e+03
-7.936e+03
-8.730e+03
-9.523e+03

❖ The FEA result of the maximum moment is **-9.523e+3**, which is higher than the result of the preliminary analysis (point D) on page 8-5.

Using the Probe Option to Examine Stress at Point1

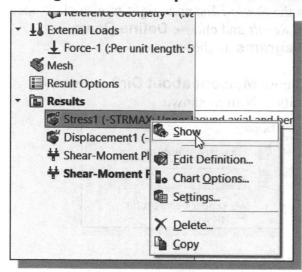

1. In the *Static 1* list area, right-mouse-click once on *Stress1* and choose **Show** to switch back to displaying the Stresses as shown.

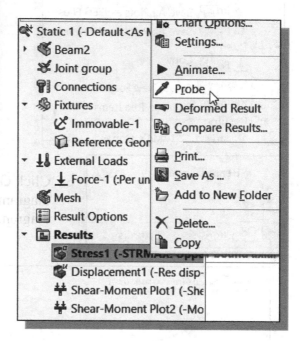

2. Right-mouse-click once on *Stress1* and choose **Probe** to switch to the *Probe* option as shown.

3. Zoom-in and click on the elements near Point1 to display the associated stress of the selected element.

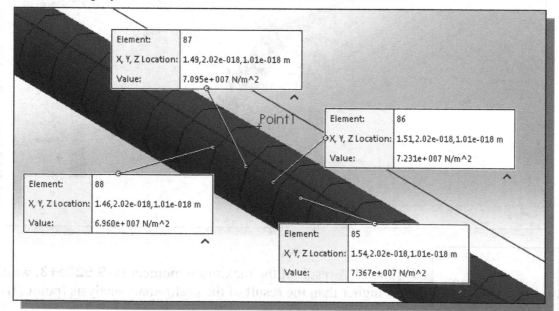

❖ Note the preliminary analysis done on page 8-4 calculated the stress as **70.6 MPa** at Point1.

Questions:

1. How do we account for inclined supports and apply angled loads on individual elements in SOLIDWORKS Simulation?

2. Will SOLIDWORKS Simulation calculate and display the reaction forces that are located at inclined supports?

3. List two of the standard beam cross section types that are available in SOLIDWORKS Simulation.

4. In SOLIDWORKS Simulation, list and describe the two options to establish beam elements you have learned so far.

5. Are the SOLIDWORKS Simulation analysis results stored inside the SOLIDWORKS part files?

6. How do we animate the SOLIDWORKS Simulation analysis results?

Exercises:

Determine the maximum stress produced by the loads and create the shear and moment diagrams for the structures.

1. Fixed-end support at A.
    Material: Steel
            Diameter 1.0 in.

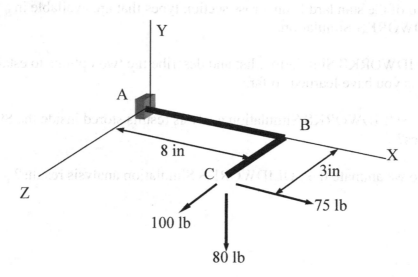

2. A Pin-Joint support and a roller support.
        Material: Aluminum Alloy 6061 T6

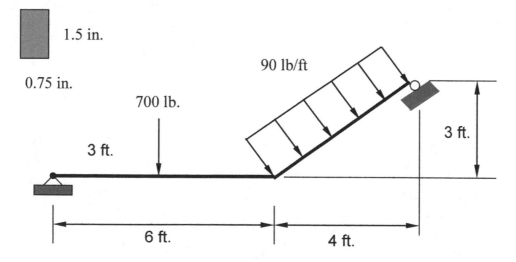

3.   A Pin-Joint support and a roller support.

   Material: Steel

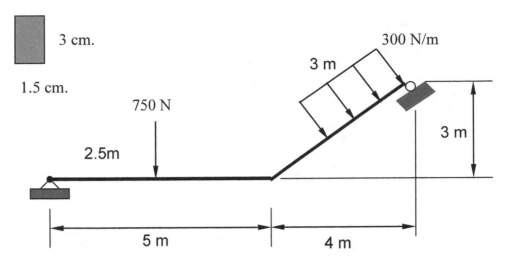

Notes:

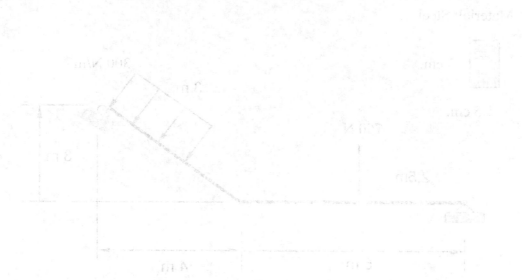

Chapter 9
Statically Indeterminate Structures

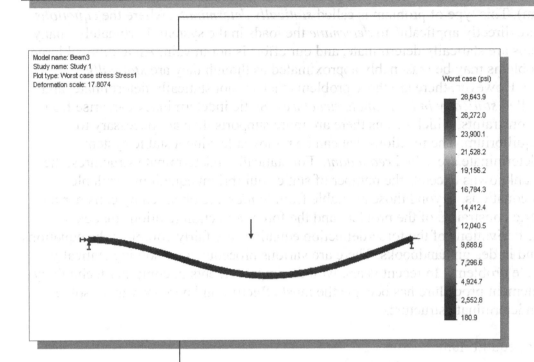

Model name: Beam3
Study name: Study 1
Plot type: Worst case stress Stress1
Deformation scale: 17.8074

Worst case (psi)

28,643.9
26,272.0
23,900.1
21,528.2
19,156.2
16,784.3
14,412.4
12,040.5
9,668.6
7,296.6
4,924.7
2,552.8
180.9

Learning Objectives

♦ **Perform Statically Indeterminate Beam Analysis.**

♦ **Understand and Apply the Principle of Superposition.**

♦ **Identify Statically Indeterminate Structures.**

♦ **Apply and Modify Boundary Conditions on Beams.**

♦ **Generate Shear and Moment diagrams.**

Introduction

Up to this point, we have dealt with a very convenient type of problem in which the external reactions can be solved by using the *equations of Statics* (Equations of Equilibrium). This type of problem is called *statically determinate*, where the *equations of Statics* are directly applicable to *determine* the loads in the system. Fortunately, many real problems are statically determinate, and our effort is not in vain. An additional large class of problems may be reasonably approximated as though they are *statically determinate*. However, there are those problems that are not statically determinate, and these are called *statically indeterminate structures*. Static indeterminacy can arise from redundant constraints, which means there are more supports than are necessary to maintain equilibrium. The reactions that can be removed leaving a stable system statically determinate are called *redundant*. For statically indeterminate structures, the number of unknowns exceeds the number of static equilibrium equations available. Additional equations, beyond those available from *Statics*, are obtained by considering the geometric constraints of the problem and the force-deflection relations that exist. Most of the derivations of the force-deflection equations are fairly complex; the equations can be found in design handbooks. There are various procedures for solving statically indeterminate problems. In recent years, with the improvements in computer technology, the finite element procedure has become the most effective and popular way to solve statically indeterminate structures.

Equations of equilibrium:

$$\sum M = 0$$
$$\sum F = 0$$

Statically determinate structure:

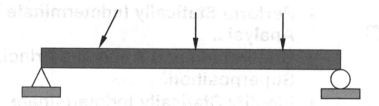

Statically indeterminate structure:

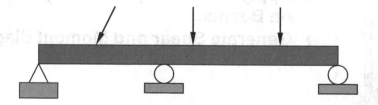

Problem Statement

Determine the reactions at point A and point B. Also determine the maximum normal stress that the loading produces in the Aluminum (6061) member (Diameter 2").

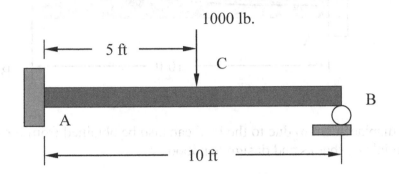

Preliminary Analysis

The reactions at point A and B can be determined by using the principle of superposition. By removing the roller support at point B, the displacement at point B due to the loading is first determined. Since the displacement at point B should be zero, the reaction at point B must cause an upward displacement of the same amount. The displacements are superimposed to determine the reaction. The *principle of superposition* can be used for this purpose, and once the reaction at B is determined, the other reactions are determined from the equations of equilibrium.

1

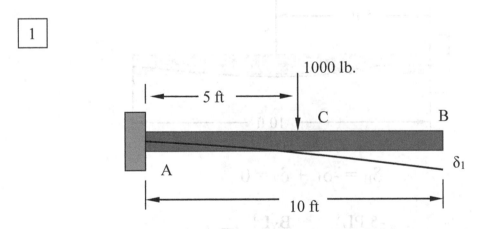

The displacement δ_1 can be obtained from most strength of materials textbooks and design handbooks:

$$\delta_1 = \frac{5\,PL^3}{48\,EI}$$

2

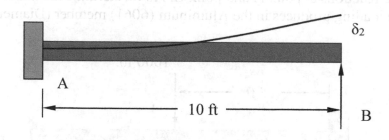

δ_2

A

10 ft

B

The displacement δ_2 due to the load can also be obtained from most strength of materials textbooks and design handbooks:

$$\delta_2 = \frac{B_Y L^3}{3 E I}$$

3   (Superposition of 1 and 2)

1000 lb.

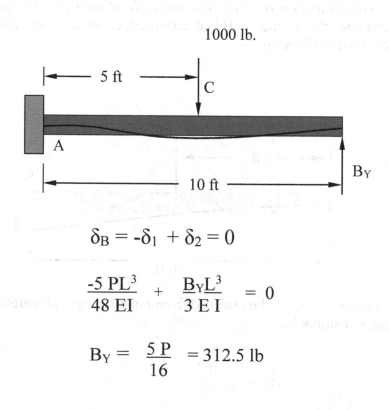

5 ft

C

A

B$_Y$

10 ft

$$\delta_B = -\delta_1 + \delta_2 = 0$$

$$\frac{-5 PL^3}{48 EI} + \frac{B_Y L^3}{3 E I} = 0$$

$$B_Y = \frac{5 P}{16} = 312.5 \text{ lb}$$

Free Body Diagram of the system:

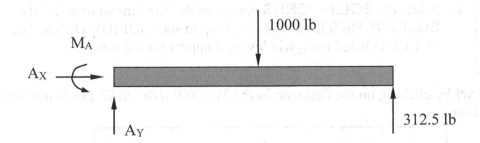

The reactions at A can now be solved:

$$\sum F_X = 0, \quad A_X = 0$$

$$\sum F_Y = 0, \quad 312.5 - 1000 + A_Y = 0, \quad A_Y = 687.5 \text{ lb.}$$

$$\sum M_{@A} = 0, \quad 312.5 \times 10 - 1000 \times 5 + M_A = 0, \quad M_A = 1875 \text{ ft-lb}$$

The maximum normal stress at point A is from the bending stress:

$$\sigma = \frac{MC}{I}$$

for the circular cross section:

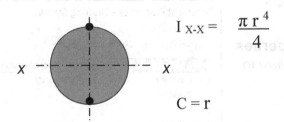

$$I_{X-X} = \frac{\pi r^4}{4}$$

$$C = r$$

Therefore,

$$\sigma_A = \frac{MC}{I} = \frac{4 M r}{\pi r^4} = 4.125 \times 10^6 \text{ lb/ft}^2 = 28648 \text{ lb/in}^2$$

Start SOLIDWORKS

1. Select the **SOLIDWORKS** option on the *Start* menu or select the **SOLIDWORKS** icon on the desktop to start SOLIDWORKS. The SOLIDWORKS main window will appear on the screen.

2. Select **Part** by clicking on the first icon in the *New SOLIDWORKS Document* dialog box as shown.

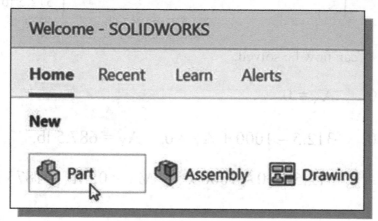

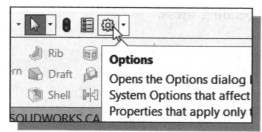

3. Select the **Options** icon from the *Menu* toolbar to open the *Options* dialog box.

4. Switch to the **Document Properties** tab and reset the *Drafting Standard* to **ANSI** as shown in the figure.

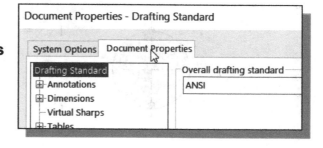

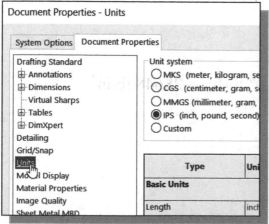

5. On your own, set the *Unit system* to **IPS (inch, pound, second)** as shown.

Create the CAD Model

To create the nine member truss system, we will first establish the locations of the nine center axes of the members by creating a 2D sketch. We will then use the SOLIDWORKS **Structural Member** tool to select the predefined profiles of the cross sections of the system.

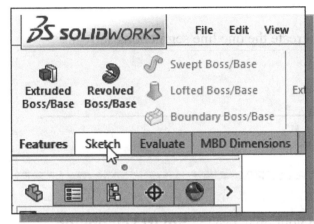

1. Click the **Sketch** tab in the *Command Manager* as shown.

❖ Note the *Sketch* toolbar is displayed in the *Command Manager* area with different sketching tools.

2. Click the **Sketch** icon, in the *Sketch* toolbar, to create a new sketch.

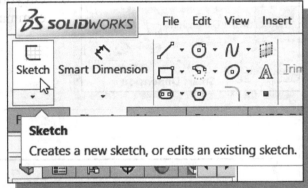

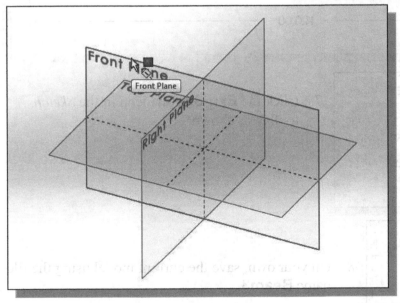

3. Click the **Front Plane**, in the graphics area, as shown.

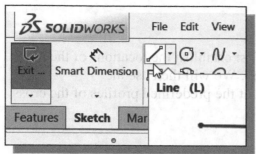

4. Click the **Line** icon in the *Sketch* toolbar as shown.

5. Start the line segment at the origin and create the one line segment as shown.

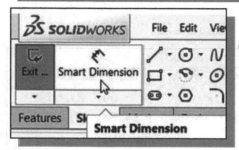

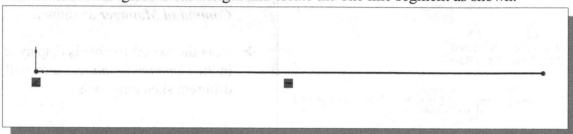

6. On your own, use the **Smart Dimension** command to adjust the sketch as shown.

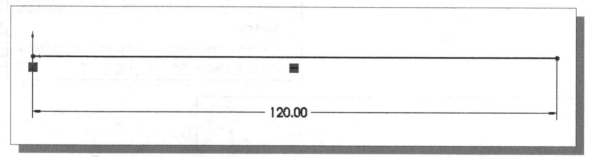

120.00

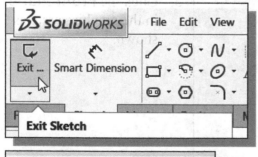

7. Click the **Exit Sketch** icon in the *Sketch* toolbar as shown.

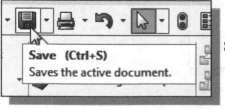

8. On your own, save the current model using the file name **Beam3**.

Create a Circular Weldment Profile

In SOLIDWORKS, all structural members use **weldment profiles**, which provide the definitions of the properties of the associated cross sections. *Weldment Profiles* are identified by **Standard**, **Type** and **Size**. In this section, we will illustrate the procedure to create new profiles and add them to the existing library.

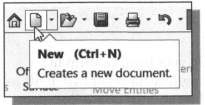

1. Click on the **New** icon, located in the *Standard* toolbar as shown.

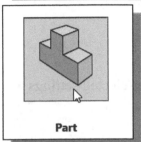

Part

2. Select **Part** by clicking on the first icon in the *New* SOLIDWORKS *Document* dialog box as shown.

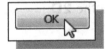

3. Click on the **OK** button to accept the settings.

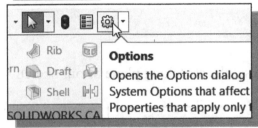

4. Click on the **Options** icon from the *Menu Bar* toolbar to open the *Options* dialog box.

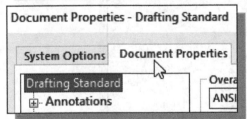

5. Select the **Document Properties** tab and set the drafting standard to **ANSI** as shown in the figure.

6. Click **Units** as shown in the figure.

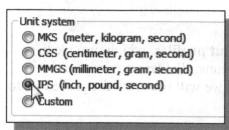

7. Select **IPS (inch, pound, second)** under the *Unit system* options.

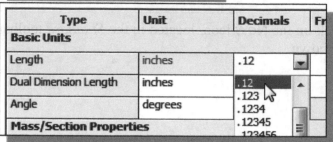

8. On your own, confirm the length decimals to two digits as shown.

9. Click **OK** in the *Options* dialog box to accept the selected settings.

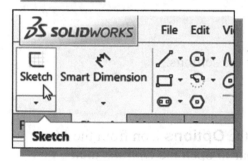

10. Create a new 2D sketch by clicking the **2D Sketch** icon, in the *Sketch* toolbar.

11. Click the **Front Plane**, in the graphics area, to set the sketching plane of our sketch as shown.

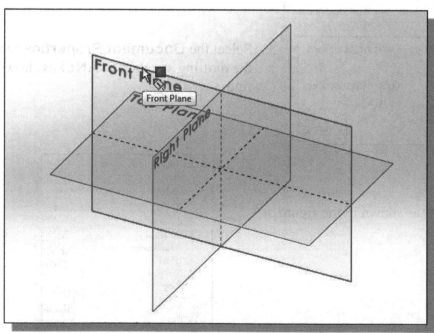

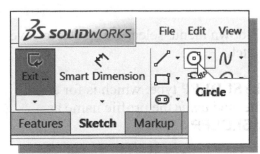

12. Click the **Circle** icon to activate the **Circle** command.

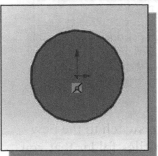

13. On your own, create a circle roughly centered at the origin of the world coordinate system.

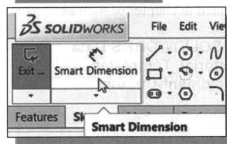

14. On your own, use the **Smart Dimension** to adjust the sketch as shown.

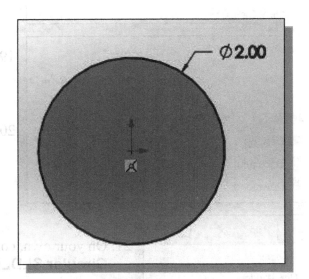

15. Click the **Exit Sketch** icon in the *Sketch* toolbar as shown.

❖ Note the dimensions disappeared from the screen. For the *weldment profile*, the file should only contain one 2D sketch feature.

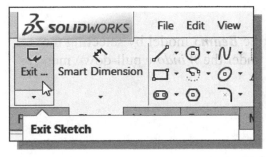

16. In the pull-down menu, select **Save As** to save our 2D sketch.

17. Choose the **SLDLFP** type, which is for a *Library* file, and *enter* the profile name as **Circular.SLDLFP**.

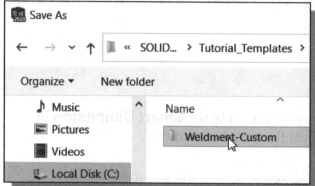

18. On your own, switch to the new weldment profiles folder that you created in chapter 6, such as:

**C:\SOLIDWORKS Data
\Tutorial_Templates
\Weldment-Custom**

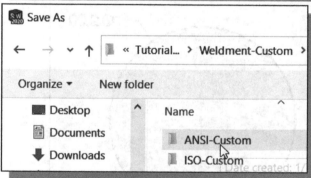

19. Switch to the **Weldment-Custom** folder and create a new folder using the name **ANSI-Custom**.

20. Switch to the **ANSI-Custom** folder and then switch to the **Solid** folder.

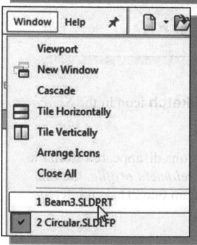

21. On your own, continue and *Save* the profile **Circular.SLDLFP**.

22. Switch back to the *Beam3* model by selecting the filename listed under the *Window* pull-down menu as shown.

Create Structural Members Using the New Profile

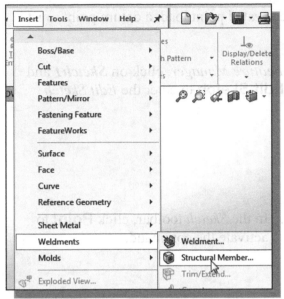

1. Select **Insert → Weldments → Structural Member** in the pull-down menu as shown.

❖ Note a *Weldment* feature is automatically added in the *Model History Tree* when the **Structural Member** command is activated.

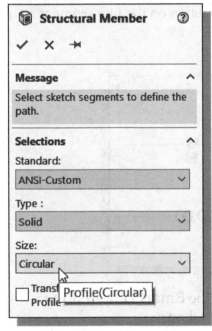

2. In the *Structural Member Property Manager* dialog box, set the *Standard* option to **ANSI-Custom** as shown.

3. In the *Structural Member Property Manager* dialog box, set the *Type* option to **Solid** as shown.

4. In the *Structural Member Property Manager* dialog box, set the *Size* option to **Circular** as shown.

5. Select the **line segment** of the structure.

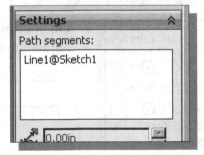

6. Click **OK** to create the structural member, which contains two segments with the same profile.

Add a Datum Point for the Concentrated Load

In SOLIDWORKS Simulation, to apply a concentrated point load, a datum point must exist on the outside surface of the solid model.

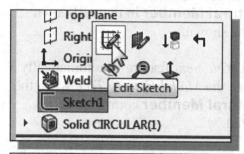

1. In the *Feature Manager*, click on **Sketch1** and select **Edit Sketch** to enter the *Edit Sketch* mode.

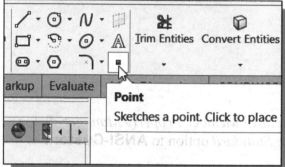

2. In the *Sketch* toolbar, click **Point** to activate the command.

3. Click a location on the line segment as shown.

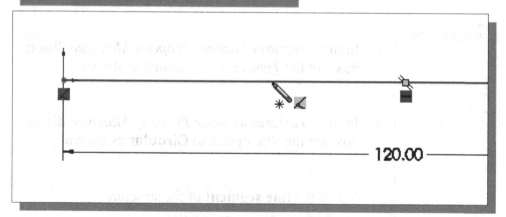

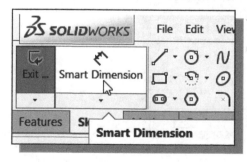

4. On your own, use the **Smart Dimension** to adjust the sketch as shown.

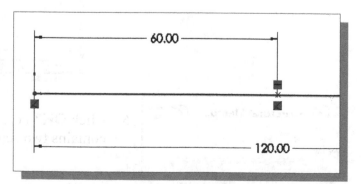

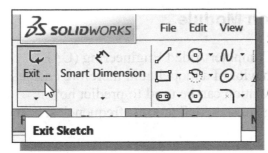

5. Click the **Exit sketch** icon in the *Sketch* toolbar as shown.

6. Set the display to **Isometric view** by using the control in the display panel.

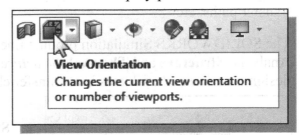

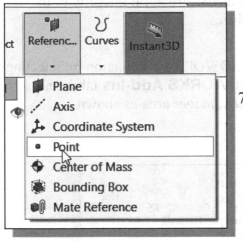

7. In the *Feature* toolbar, select **Point** from the *Reference Geometry* list as shown.

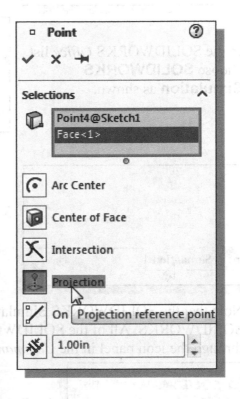

8. In the *Point* dialog box, switch on the **Projection** option as shown.

9. Select the **circular surface** of the beam and the **point** we added in *Sketch1*.

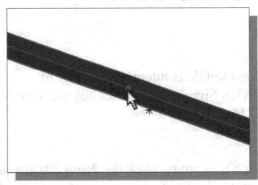

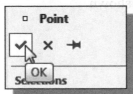

10. Click **OK** to accept the selection and create the point.

Activate the SOLIDWORKS Simulation Module

SOLIDWORKS Simulation is a multi-discipline Computer Aided Engineering (CAE) tool that enables users to simulate the physical behavior of a model, and therefore enables users to improve the design. SOLIDWORKS Simulation can be used to predict how a design will behave in the real world by calculating stresses, deflections, frequencies, heat transfer paths, etc.

The SOLIDWORKS Simulation product line features two areas of Finite Element Analysis: **Structure** and **Thermal**. *Structure* focuses on the structural integrity of the design, and *thermal* evaluates heat-transfer characteristics.

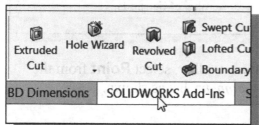

1. Start SOLIDWORKS Simulation by selecting the **SOLIDWORKS Add-Ins** tab in the *Command Manager* area as shown.

2. In the SOLIDWORKS *Office* list, choose **SOLIDWORKS Simulation** as shown.

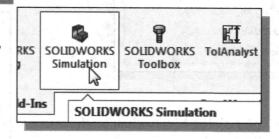

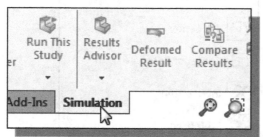

3. In the *Command Manager* area, choose **Simulation** as shown.

❖ Note that the SOLIDWORKS Simulation module is integrated as part of SOLIDWORKS. All of the SOLIDWORKS Simulation commands are accessible through the icon panel in the *Command Manager* area.

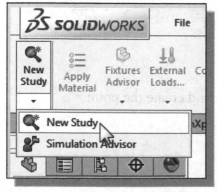

4. To start a new study, click the **New Study** item listed under the *Study Advisor* as shown.

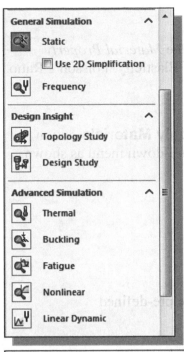

5. Select **Static** as the type of analysis to be performed with SOLIDWORKS Simulation.

❖ Note that different types of analyses are available, which include both structural static and dynamic analyses, as well as the thermal analysis.

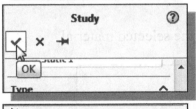

6. Click **OK** to start the definition of a structural static analysis.

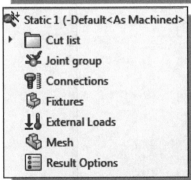

❖ In the *Feature Manager* area, note that a new panel, the *FEA Static* window, is displayed with all the key items listed.

❖ Also note the **Static 1** tab is activated, which indicates the use of the FEA model.

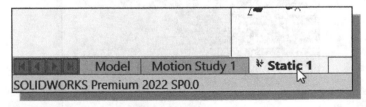

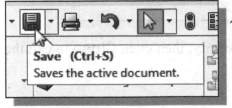

7. On your own, use the **Save** command to save the current model (**Beam3**).

Assign the Element Material Property

Next, we will set up the *Material Property* for the elements. The *Material Property* contains the general material information, such as Modulus of Elasticity, Poisson's Ratio, etc. that is necessary for the FEA analysis.

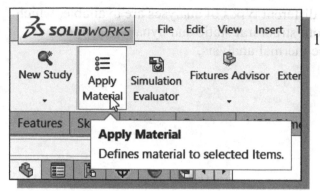

1. Choose **Apply Materials** option from the pull-down menu as shown.

❖ Note the default list of materials, which are available in the pre-defined SOLIDWORKS Simulation material library, is displayed.

2. Select **Alloy Steel** in the *Material* list as shown.

3. Set the **Units** option to display **English (IPS)** to make the selected material available for use in the current FEA model.

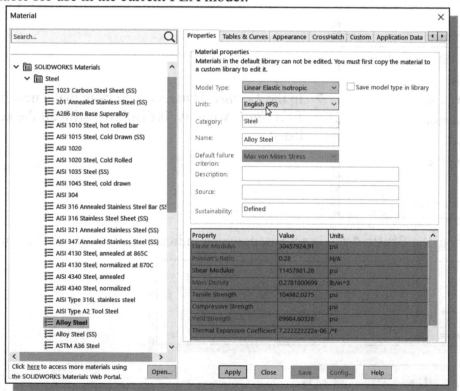

4. Click **Apply** to assign the material property, then click **Close** to exit the Material Assignment command.

Apply Boundary Conditions – Constraints

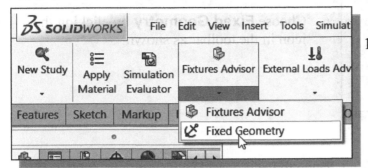

1. Choose **Fixed Geometry** by clicking the icon in the toolbar as shown.

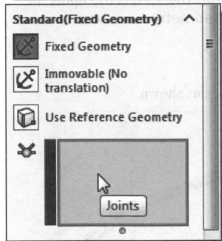

2. In the *Standard* option list, choose **Fixed Geometry**.

❖ For Beam systems, all joints are treated as **fixed** by default, locking all six degrees of freedom.

3. Activate the *Joints* list option box by clicking on the inside of the **Joints** list box as shown.

4. Select the **left node** as shown.

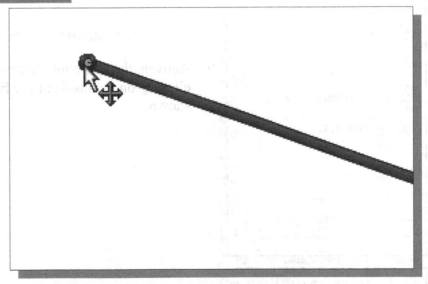

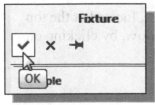

5. Click on the **OK** button to accept the first **Fixture** constraint settings.

❖ The small arrows indicate constraints have been applied to the associated node.
❖ For the node on the right, we will apply a separate roller constraint.

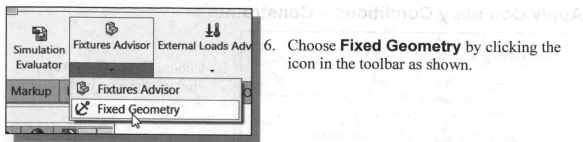

6. Choose **Fixed Geometry** by clicking the icon in the toolbar as shown.

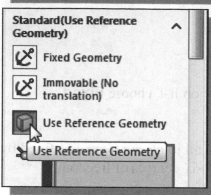

7. Change the *Standard(Fixed Geometry)* option to **Use Reference Geometry** as shown.

8. Select the **right node** as shown.

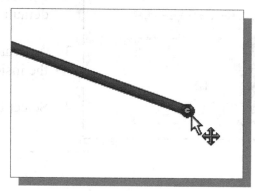

9. Activate the *Direction Reference* list option box by clicking on the inside of the **Reference** list box as shown.

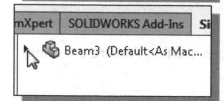

10. Expand the *Model History Tree*, located at the top left corner of the graphics window, by clicking on the **triangle** symbol as shown.

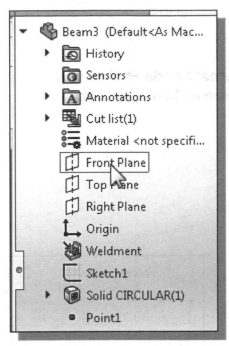

11. Select **Front Plane** as the *direction reference* as shown.

❖ Note the selected plane is highlighted; the constraints we set will be using the selected reference to determine the constraint direction.

12. Set the distance measurement to **inches** to match with the systems units we are using.

13. In the *Translation* constraints list, click on the **Normal to Plane** and **Along Plane Dir 2** icons to activate the constraint.

14. Set the *Normal to Plane* and *Along Plane Dir 2 distances* to **0** as shown.

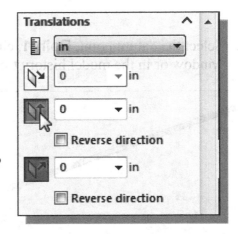

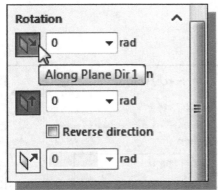

15. Click on the **Along Plane Dir 1** and **Along Plane Dir 2** icons to activate the constraint.

16. Set the *Along Plane Dir 1* and *Along Plane Dir 2 angles* to **0** as shown.

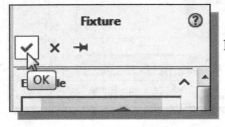

17. Click on the **OK** button to accept the second Displacement constraint settings.

Apply the Concentrated Point Load

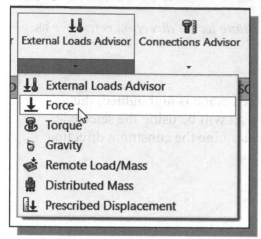

1. Choose **External Loads → Force** by clicking the icon in the toolbar as shown.

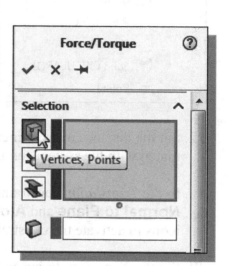

2. Confirm the *Force/Torque* option is set to **Points** as shown.

3. Select the datum point, **Point1**, either in the graphics window or in the model history tree.

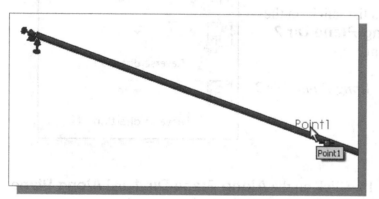

4. Activate the *Direction Reference* list option box by clicking on the inside of the **Reference** list box as shown.

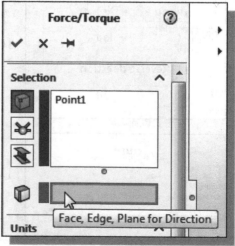

5. Select **Front Plane** as the *direction reference* as shown.

❖ Note the selected plane is highlighted; the constraints we set will be using the selected reference to determine the constraint direction.

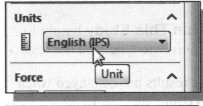

6. Set the *Units* option to **IPS** to match with the systems units we are using, as shown.

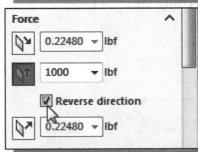

7. Click on the **Along Plane Dir 2** icon to activate the force direction.

8. Set the *Force* to **1000** as shown.

9. Activate the **Reverse direction** option.

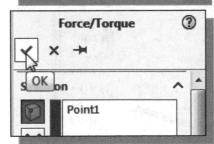

10. Click on the **OK** button to accept the *Force/Torque* settings.

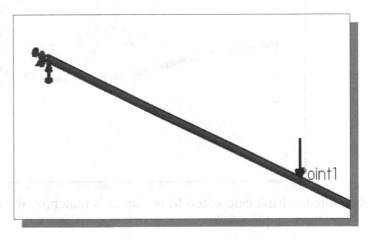

Create the FEA Mesh and Run the Solver

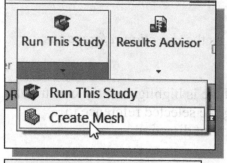

1. Choose **Create Mesh** by clicking the icon under *Run This Study* in the toolbar as shown.

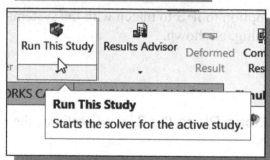

❖ Note the **Mesh** icon has changed, which indicates the FEA Mesh has been created.

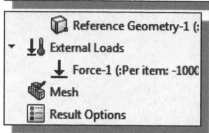

2. Click on the **Run This Study** button to start the *FEA Solver* to calculate the results.

❖ Note the stress results are displayed when the *Solver* has completed the FEA calculations.

3. On your own, set the units for the displayed stress to **psi**.

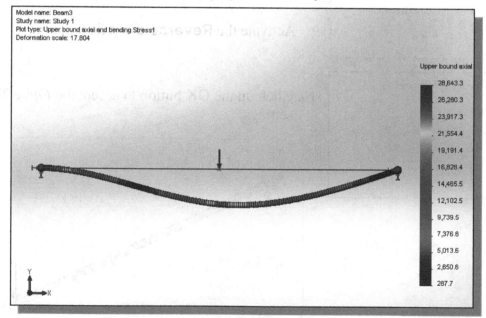

❖ Note the FEA calculated Max. Stress is matching the result from the preliminary analysis on page 9-5.

Viewing the Internal Loads of All members

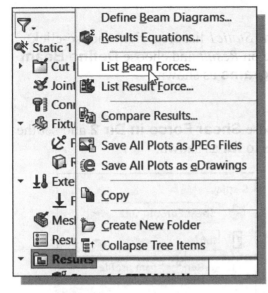

1. In the *FEA Static* window, click once with the right-mouse-button on the *Results* item to display the option list and select **List Beam Forces** as shown.

2. Confirm the list option is set to **Forces**.

3. Set the *Units* to **English (IPS)** as shown.

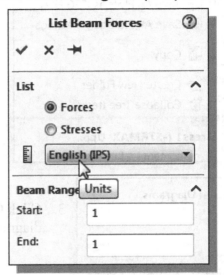

4. Click **OK** to display the results.

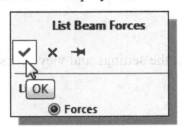

❖ Note that although there is only one structural member, SOLIDWORKS Simulation actually created and calculated 166 elements.

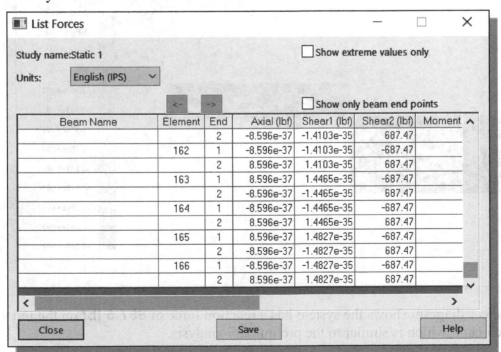

Beam Name	Element	End	Axial (lbf)	Shear1 (lbf)	Shear2 (lbf)	Moment
		2	-8.596e-37	-1.4103e-35	687.47	
	162	1	-8.596e-37	-1.4103e-35	-687.47	
		2	8.596e-37	1.4103e-35	687.47	
	163	1	8.596e-37	1.4465e-35	-687.47	
		2	-8.596e-37	-1.4465e-35	687.47	
	164	1	-8.596e-37	-1.4465e-35	-687.47	
		2	8.596e-37	1.4465e-35	687.47	
	165	1	8.596e-37	1.4827e-35	-687.47	
		2	-8.596e-37	-1.4827e-35	687.47	
	166	1	-8.596e-37	-1.4827e-35	-687.47	
		2	8.596e-37	1.4827e-35	687.47	

Study name:Static 1

Units: English (IPS)

Show extreme values only

Show only beam end points

Close   Save   Help

Shear and Moment Diagrams

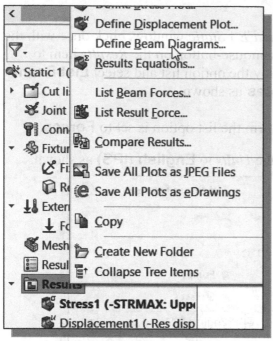

1. In the *Static1* list area, right-mouse-click once on *Result* and choose **Define Beam Diagrams** as shown.

2. Choose **Shear Force in Dir 2** and set the units to **lbf** as shown.

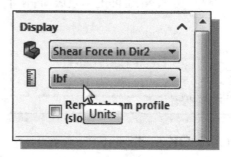

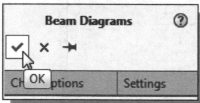

3. Click **OK** to accept the settings and view the shear diagram.

❖ The shear diagram shows the system has a reaction force of **687.5 lbf** on the left side of the beam, which is similar to the preliminary analysis.

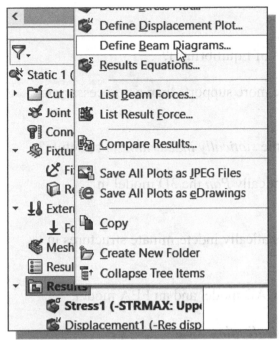

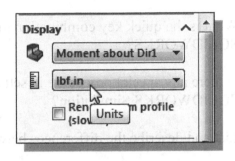

4. In the *Static1* list area, right-mouse-click once on *Result* and choose **Define Beam Diagrams** as shown.

5. Choose **Moment in Dir 1** and set the units to **lbf-in** under the *Display* option list as shown.

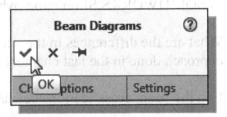

6. Click **OK** to accept the settings and view the shear diagram. Switch the display to the **Top view** to view the moment diagram.

❖ The FEA result of the maximum moment is **2.250e+4**, which matches with the result of the preliminary analysis on page 9-5.

Questions:

1. What are the equations of *Statics* (Equations of Equilibrium)?

2. What is the type of structure where there are more supports than are necessary to maintain equilibrium?

3. In the example problem, how did we solve the *statically indeterminate* structure?

4. What is the quick key combination to dynamically *Pan* the 3D model in SOLIDWORKS?

5. Is there any special procedure in setting up statically indeterminate structures in SOLIDWORKS Simulation?

6. List and describe the differences between a CAD model and an FEA model.

7. In SOLIDWORKS Simulation, what does *idealization* mean?

8. What are the differences in the way the FEA model was created in the tutorial vs. the approach done in the last chapter?

9. Can the beam cross sections be changed after beam elements have been created?

10. What is the percent error of the maximum bending stress obtained by the FEA analysis and the preliminary calculation in the tutorial?

Exercises:

Determine the maximum stress produced by the loads and create the shear and moment diagrams.

1. A cantilever beam with a roller support.

Material: Steel

3 cm

1.5 cm

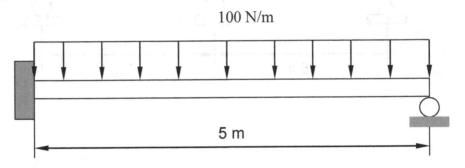

100 N/m

5 m

2. A cantilever beam with a roller support.

Material: Steel
        Diameter 2.0 in.

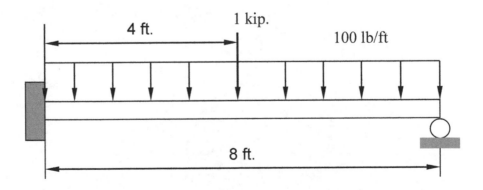

4 ft.   1 kip.   100 lb/ft

8 ft.

3. Simply Supported beam with an extra roller support.

Material: Aluminum Alloy 6061 T6

1.5 in.

1.0 in.

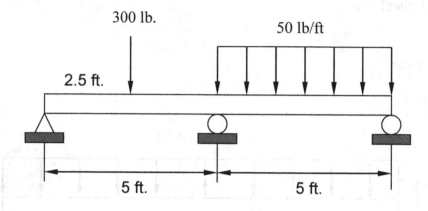

Chapter 10
Two-Dimensional Surface Analysis

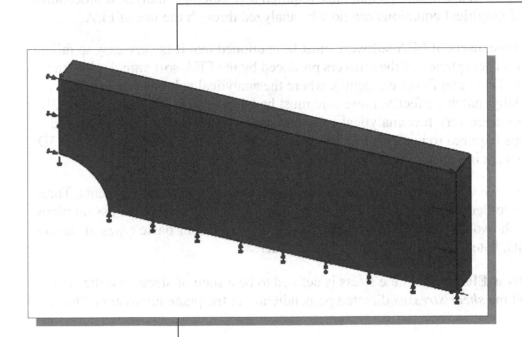

♦ **Understand the basic assumptions for 2D elements.**

♦ **Understand the Basic differences between H-Element and P-Element.**

♦ **Create and Refine 2D Surface elements.**

♦ **Understand the Stress concentration effects on Shapes.**

♦ **Perform Basic Plane Stress Analysis using SOLIDWORKS Simulation.**

Introduction

The recent developments in computer technology have triggered tremendous advancements in the development and use of 2D surface, 3D surface and 3D solid elements in FEA. Many problems that once required sophisticated analytical procedures and the use of empirical equations can now be analyzed through the use of FEA.

At the same time, users of FEA software must be cautioned that it is very easy to fall into the trap of blind acceptance of the answers produced by the FEA software. Unlike the line elements (*Truss* and *Beam* elements), where the analytical solutions and the FEA solutions usually match perfectly, more care must be taken with 2D/3D surface and 3D solid elements since very few analytical solutions can be easily obtained. On the other hand, the steps required to perform finite element analysis using 2D/3D surface and 3D solid elements are in general less complicated than that of line elements.

In this chapter, we will examine the use of two-dimensional solid finite elements. Three-dimensional problems can be reduced to two dimensions if they satisfy the assumptions associated with two-dimensional surface elements. There are four basic types of surface solid elements, listed in historical development order:

1. **Plane Stress Elements**: Plane stress is defined to be a state of stress that the *normal stress* and the *shear stresses* directed perpendicular to the plane are assumed to be zero.

2. **Plane Strain Elements**: Plane strain is defined to be a state of strain that the *normal strain* and the *shear strains* directed perpendicular to the plane are assumed to be zero.

3. **Axisymmetric Elements**: Axisymmetric structures, such as storage tanks, nozzles and nuclear containment vessels, subjected to uniform internal pressures, can be analyzed as two-dimensional systems using *Axisymmetric Elements*.

4. **Shell Elements**: Shell elements are surface elements with a thickness that is generally small compared to their other dimensions. The shell element is the most versatile type of surface elements. Unlike the plane stress elements, shell elements can have loads that are not parallel to the object surfaces.

In SOLIDWORKS Simulation, the plane stress and plane strain analyses can be performed by using the 2D surface elements. To perform a *Plane Stress/Plane Strain* analysis, a surface model or a solid model is required. This chapter demonstrates the use of the 2D surface element to solve the classical stress concentration problem – a plate with a hole in it. This project is simple enough to demonstrate the necessary steps for the finite element process, and the FEA results can be compared to the analytical solutions to assure the accuracy of the FEA procedure. Generally, members that are thin and whose loads act only in the plane can be considered to be under *plane stress*; on the other hand, members that are thick and whose loads act only in the plane can be considered to be under *plane strain*.

Problem Statement

Determine the maximum normal stress that loading produces in the 6061-T6 aluminum plate.

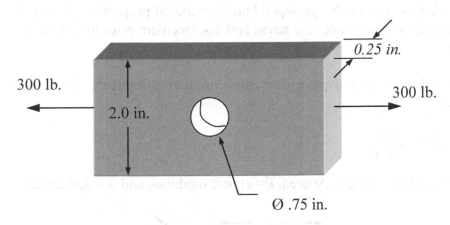

Preliminary Analysis

- **Maximum Normal Stress**

The nominal normal stress developed at the smallest cross section (through the center of the hole) in the plate is

$$\sigma_{nominal} = \frac{P}{A} = \frac{300}{(2 - 0.75) \times .25} = 960 \text{ psi.}$$

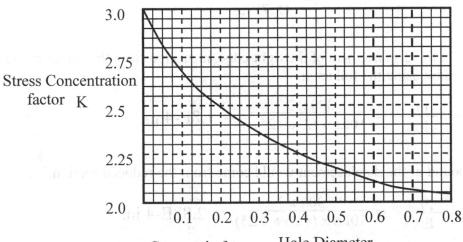

Geometric factor = .75/2 = 0.375

Stress concentration factor K is obtained from the graph, **K = 2.27**

$$\sigma_{MAX} = K \, \sigma_{nominal} = 2.27 \times 960 = 2180 \text{ psi.}$$

- **Maximum Displacement**

We will also estimate the displacement under the loading condition. For a statically determinant system the stress results depend mainly on the geometry. The material properties can be in error and still the FEA analysis comes up with the same stresses. However, the displacements always depend on the material properties. Thus, it is necessary to always estimate both the stress and displacement prior to a computer FEA analysis.

The classic one-dimensional displacement can be used to estimate the displacement of the problem:

$$\delta = \frac{PL}{EA}$$

Where P=force, L=length, A=area, E= elastic modulus, and δ = deflection.

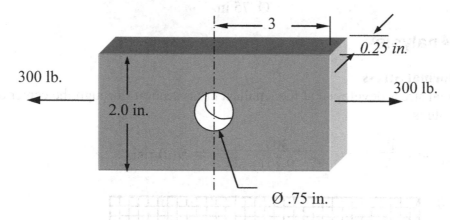

A lower bound of the displacement of the right-edge, measured from the center of the plate, is obtained by using the full area:

$$\delta_{lower} = \frac{PL}{EA} = \frac{300 \times 3}{10E6 \times (2 \times 0.25)} = 1.8E\text{-}4 \text{ in.}$$

and an upper bound of the displacement would come from the reduced section:

$$\delta_{upper} = \frac{PL}{EA} = \frac{300 \times 3}{10E6 \times (1.25 \times 0.25)} = 2.88E\text{-}4 \text{ in.}$$

but the best estimate is a sum from the two regions:

$$\delta_{average} = \frac{PL}{EA} = \frac{300 \times 0.375}{10E6 \times (1.25 \times 0.25)} + \frac{300 \times 2.625}{10E6 \times (2.0 \times 0.25)}$$

$$= 3.6E\text{-}5 + 1.58E\text{-}4 = 1.94E\text{-}4 \text{ in.}$$

Geometric Considerations of Finite Elements

For *Linear Statics analysis*, designs with symmetrical features can often be reduced to expedite the analysis.

For our plate problem, there are two planes of symmetry. Thus, we only need to create an FE model that is one-fourth of the actual system. By taking advantage of symmetry, we can use a finer subdivision of elements that can provide more accurate and faster results.

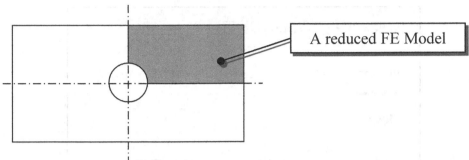

A reduced FE Model

In performing a plane stress analysis, although only the selected surface is analyzed, it is necessary to consider the constraints in all directions. For our plate model, deformations will occur along the axes of symmetry; we will therefore place roller constraints along the two center lines as shown in the figure below.

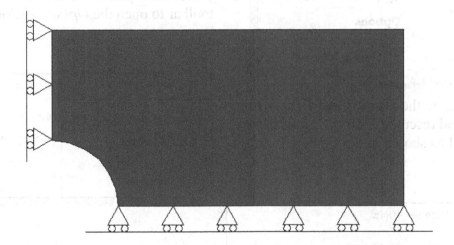

❖ One should also be cautious of using symmetrical characteristics in FEA. The symmetry characteristics of boundary conditions and loads should be considered. Also note the symmetry characteristic that is used in the *Linear Statics Analysis* does not imply similar symmetrical results in vibration or buckling modes.

Start SOLIDWORKS

1. Select the **SOLIDWORKS** option on the *Start* menu or select the **SOLIDWORKS** icon on the desktop to start SOLIDWORKS. The SOLIDWORKS main window will appear on the screen.

2. Select **Part** by clicking on the first icon in the *New SOLIDWORKS Document* dialog box as shown.

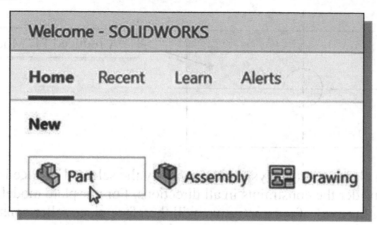

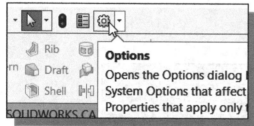

3. Select the **Options** icon from the *Menu* toolbar to open the *Options* dialog box.

4. Switch to the **Document Properties** tab and reset the *Drafting Standard* to **ANSI** as shown in the figure.

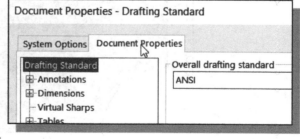

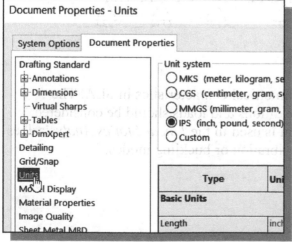

5. On your own, set the *Unit system* to **IPS (inch, pound, second)** as shown.

Create the CAD Model

To perform the surface FEA analysis, we will first create a solid model and use the front face of the solid model in SOLIDWORKS Simulation. In the next chapter, the procedure to create a surface model from a sketch is illustrated.

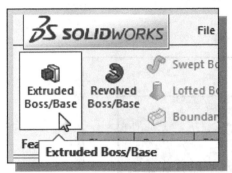

1. Click the **Extruded Boss/Base** icon, in the *Features* toolbar, to create a new extruded feature.

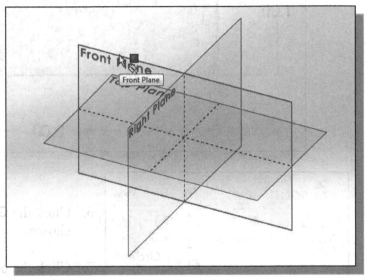

2. Click the **Front Plane**, in the graphics area, to set the sketching plane of our sketch as shown.

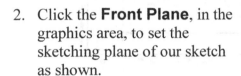

3. Click the **Rectangle** icon in the *Sketch* toolbar as shown.

4. Start the first corner of the rectangle at the origin and create the rectangle as shown.

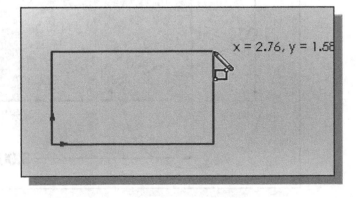

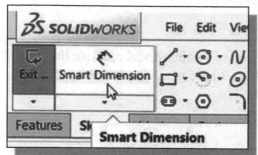

5. On your own, use the **Smart Dimension** command to adjust the sketch as shown.

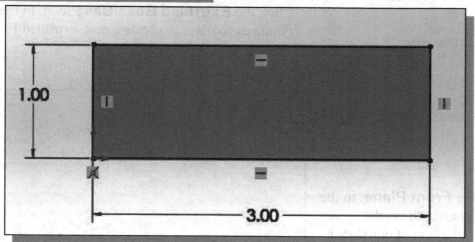

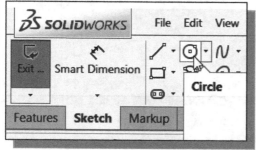

6. Click the **Circle** icon in the *Sketch* toolbar as shown.

7. Click the **origin** to align the center of the circle.

8. Create a circle of arbitrary size as shown.

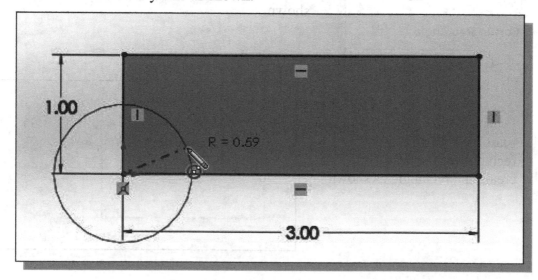

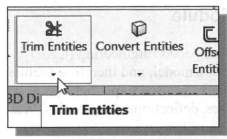

9. On your own, create and adjust the **diameter** of the circle to *0.75*.

10. Using the **Trim Entities** command, modify the sketch as shown in the figure below.

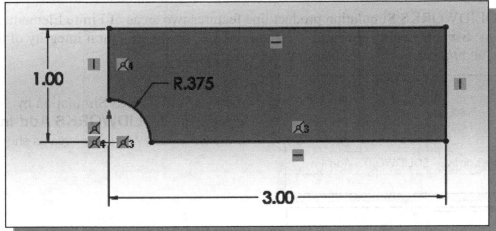

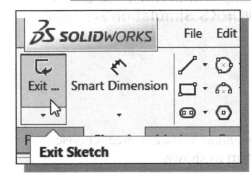

11. Click the **Exit Sketch** icon in the *Sketch* toolbar as shown.

12. On your own, create the feature by flipping the extrusion direction with a distance of **0.25** in.

13. On your own, save a copy of the current model using the file name **PlaneStress**.

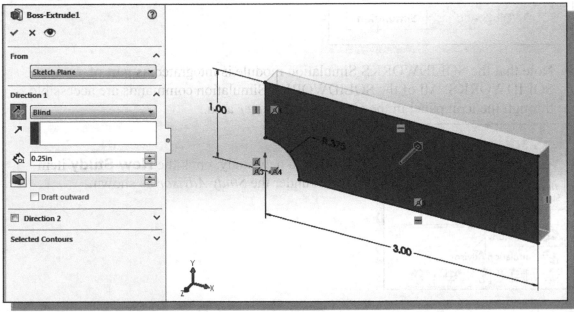

Activate the SOLIDWORKS Simulation Module

SOLIDWORKS Simulation is a multi-discipline Computer Aided Engineering (CAE) tool that enables users to simulate the physical behavior of a model, and therefore enables users to improve the design. SOLIDWORKS Simulation can be used to predict how a design will behave in the real world by calculating stresses, deflections, frequencies, heat transfer paths, etc.

The SOLIDWORKS Simulation product line features two areas of Finite Element Analysis: **Structure** and **Thermal**. *Structure* focuses on the structural integrity of the design, and *thermal* evaluates heat-transfer characteristics.

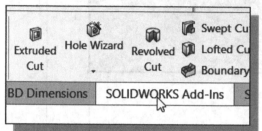

1. Start SOLIDWORKS Simulation by selecting the **SOLIDWORKS Add-Ins** tab in the *Command Manager* area as shown.

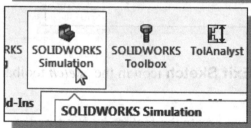

2. In the *SOLIDWORKS Office* list, choose **SOLIDWORKS Simulation** as shown.

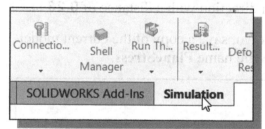

3. In the *Command Manager* area, choose **Simulation** as shown.

❖ Note that the SOLIDWORKS Simulation module is integrated as part of SOLIDWORKS. All of the SOLIDWORKS Simulation commands are accessible through the icon panel in the *Command Manager* area.

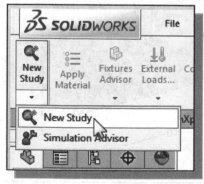

4. To start a new study, click the **New Study** item listed under the *Study Advisor* as shown.

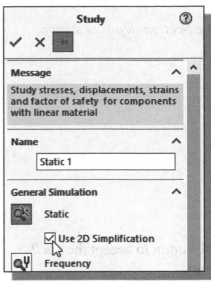

5. Select **Static** as the type of analysis to be performed with SOLIDWORKS Simulation.

6. Activate the **Use 2D Simplification** option by clicking on the check box as shown.

❖ Note that additional settings for different types of 2D analyses will be available once the *Use 2D Simplification* option is turned on.

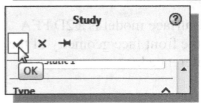

7. Click **OK** to the show additional settings for different types of 2D analyses available for a structural static analysis.

8. Confirm the 2D study type is set to Plane Stress as shown.

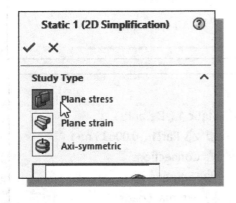

9. In the Section Definition box, note the Section Plane option is activated.

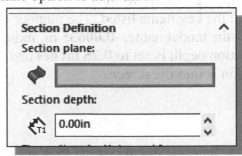

10. In the graphics area, select the **front face** of the solid model.

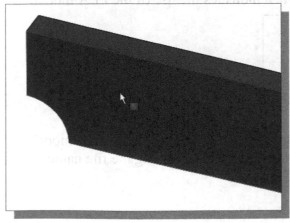

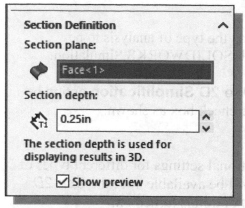

11. Enter **0.25** as the *Section depth* as shown.

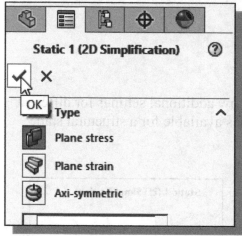

12. Click on the **OK** button to accept the *Shell Definition* settings.

❖ Note the defined surface model is a 2D FEA model, but only the front face geometry of the CAD model is referenced.

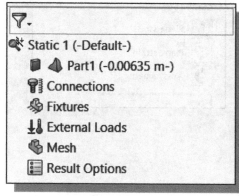

❖ In the *Feature Manager* area, note that a new panel, the *FEA Study* window, is displayed with all the key items listed. The number behind the model name, **-0.00635 m**, indicates the section depth is set to **0.25 inches** and direction is into the screen.

❖ Also note the **Static 1** tab is activated, which indicates the use of the FEA model.

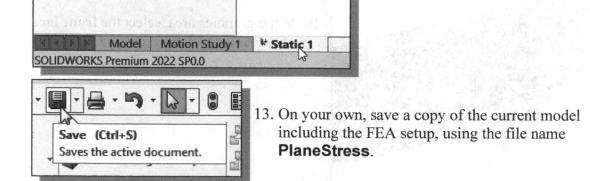

13. On your own, save a copy of the current model including the FEA setup, using the file name **PlaneStress**.

Assign the Element Material Property

Next, we will set up the *Material Property* for the elements. The *Material Property* contains the general material information, such as *Modulus of Elasticity*, *Poisson's Ratio*, etc. that is necessary for the FEA analysis.

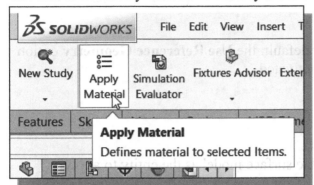

1. Choose the **Apply Materials** option as shown.

❖ Note the default list of materials, which are available in the pre-defined SOLIDWORKS Simulation material library, is displayed.

2. Select **6061-T6 Aluminum** in the *Material* list as shown.

3. Set the **Units** option to display **English (IPS)** to make the selected material available for use in the current FEA model.

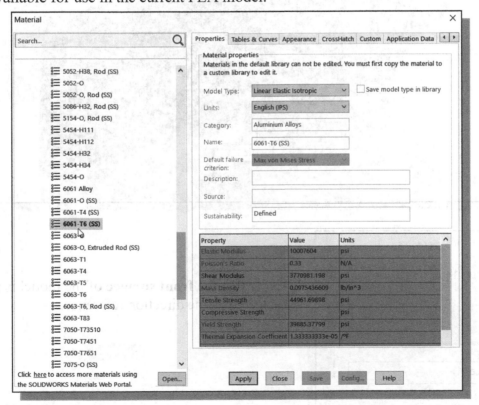

4. Click **Apply** to assign the material property, then click **Close** to exit the material assignment command.

Apply Boundary Conditions – Constraints

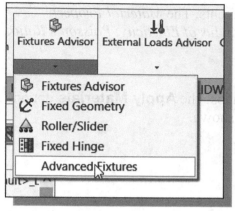

1. Choose **Advanced Fixtures** by clicking the icon in the toolbar as shown.

- By default, the **Use Reference Geometry** option is activated.

2. Select the **bottom horizontal edge** of the surface model as the entity to apply constraints.

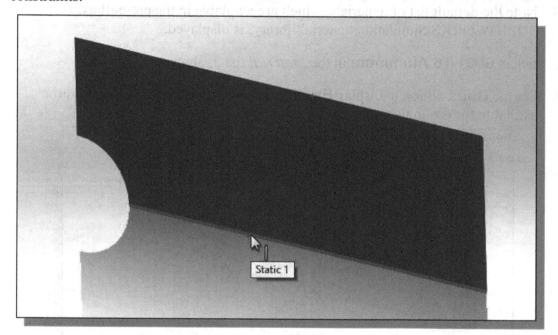

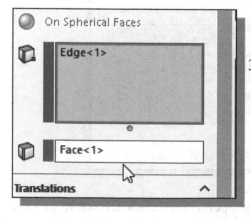

3. Note that the **front surface** of the model is pre-selected as the direction reference.

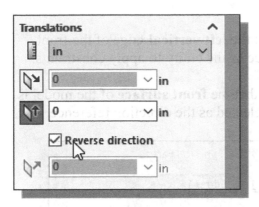

4. Set the distance measurement to **inches**, to match with the systems units we are using.

5. In the *Translation* constraints list, click on the **Along Plane Dir 2** icon to activate the constraint and set the *Along Plane Dir 2 distances* to **0** as shown.

6. Also activate the **Reverse direction** option.

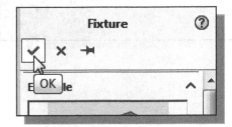

7. Click on the **OK** button to accept the first **Fixture** constraint settings.

❖ The small arrows indicate constraints have been applied to the associated edge.

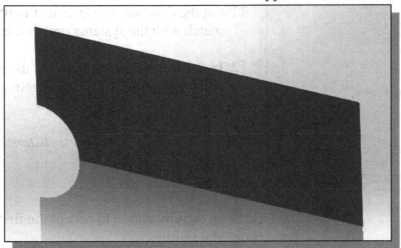

❖ For the vertical edge on the left, we will apply a separate constraint set.

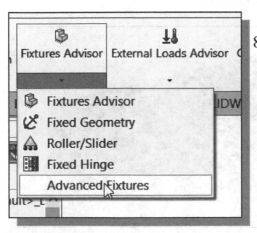

8. Choose **Advanced Fixtures** by clicking the icon in the toolbar as shown.

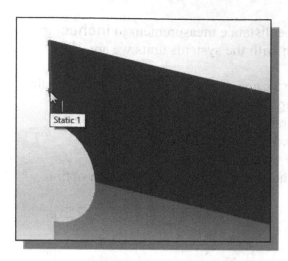

9. Select the **left vertical edge** of the front surface as the entity to apply constraints.

10. Note that the **front surface** of the model is pre-selected as the direction reference.

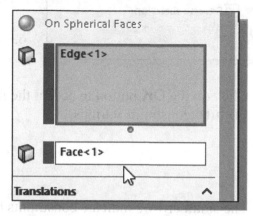

11. Set the distance measurement to **inches**, to match with the systems units we are using.

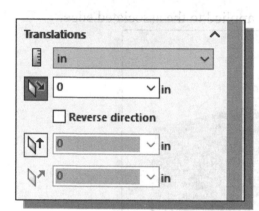

12. In the *Translation* constraints list, click on the **Along Plane Dir 1** icon to activate the constraint.

13. Set the *Along Plane Dir 1 distances* to **0** as shown.

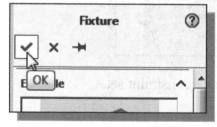

14. Click on the **OK** button to accept the first **Fixture** constraint settings.

❖ Note the applied constraints, to the two associated edges, account for the symmetry of the system.

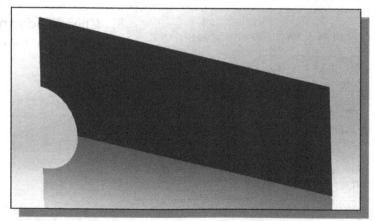

Apply the External Load

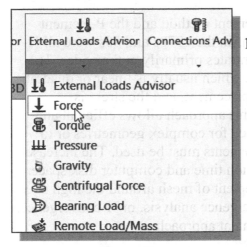

1. Choose **External Loads → Force** by clicking the icon in the toolbar as shown.

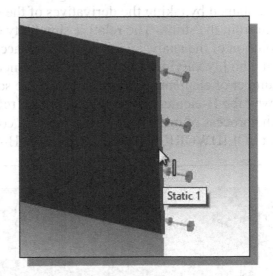

2. Select the **right edge** of the front surface as shown.

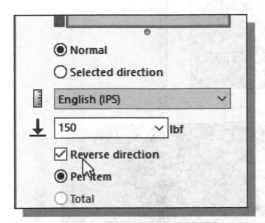

3. Set the Units to **English (IPS)**, to match with the systems units we are using.

4. Enter **150** lbf as the applied force.

5. Activate the **Reverse Direction** option by clicking on the box in front of the option.

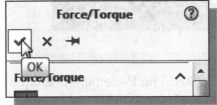

6. Click on the **OK** button to accept the *Force/Torque* settings.

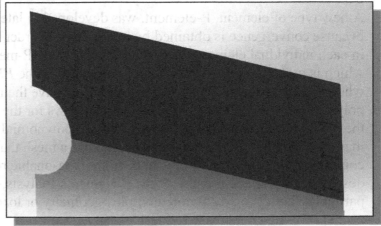

H-Element versus P- Element

There are two different approaches to FEA: the H-element method and the P-element method. Finite elements used by commercial programs in the 1970s and 80s were all H-elements. The stress analysis of the H-element concentrates primarily at the nodes. The H-element uses a low order interpolating polynomial, which usually is linear or quadratic. Strain is obtained by taking the derivatives of the displacement, and the stress is computed from the strain. The relative simplicity of this approach allows efficient and direct solution of the analysis. In order to gain accuracy for complex geometries or to represent a highly varying stress distribution, finer elements must be used. The increase in the number of elements will also increase the solution time and computer disk space. This is why the H-element method requires the refinement of mesh around the high stress areas. The process of mesh refinement is called convergence analysis, or H-convergence. Note that SOLIDWORKS Simulation uses the H-element approach.

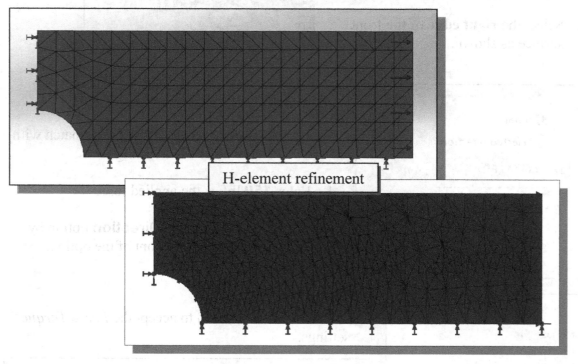

H-element refinement

A new type of element, P-element, was developed in late 1980s. The P-element is unique because convergence is obtained by increasing the order of the interpolating polynomials in each individual element. Generally speaking, the P-method uses a constant mesh, which is usually coarser than an H-element mesh. The *FEA Solver* will detect the areas where high gradients occur and those elements have their order of the interpolating polynomials automatically increased. This allows for the monitoring of expected error in the solution and then automatically increases polynomial order as needed. This is the main benefit of using P-elements. We can use a mesh that is relatively coarse, thus computational time will be low, and still get reasonable results. One should realize that despite the automatic process of the P-elements analysis, areas of the model that are of particular interest or with more complex geometry or loading could still benefit from the user specifying an increase in mesh density.

Create the first FEA Mesh – Coarse Mesh

As a rule in creating the first FEA mesh in using the H-element approach, start with a relatively small number of elements and progressively move to more refined models. The main objective of the first coarse mesh analysis is to obtain a rough idea of the overall stress distribution. In most cases, use of a complex and/or a very refined FEA model is not justifiable since it most likely provides computational accuracy at the expense of unnecessarily increased processing time.

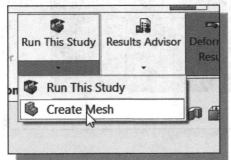

1. Choose **Create Mesh** by clicking the icon in the toolbar as shown.

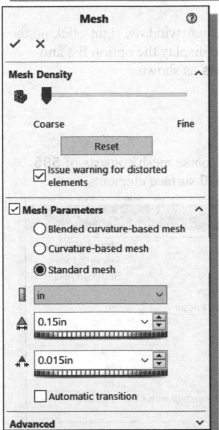

2. Switch on the **Mesh Parameters** options to show the additional control options.

3. Set the *Units* to **inches** as shown.

4. Enter **0.15** inch as the *Global Element size*.

5. Enter **0.015** inch as the *Size tolerance*.

❖ A rule of thumb to follow is the first mesh is to have at least 4 to 5 elements on the edges of the model. The shortest edge in our model is about 0.6 inches, so we will use 0.15 as the element size.

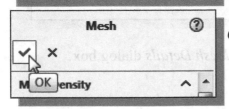

6. Click on the **OK** button to accept the *Mesh* settings.

- The *Automatic mesh* function in SOLIDWORKS Simulation has generated a fairly uniform and consistent mesh of elements.

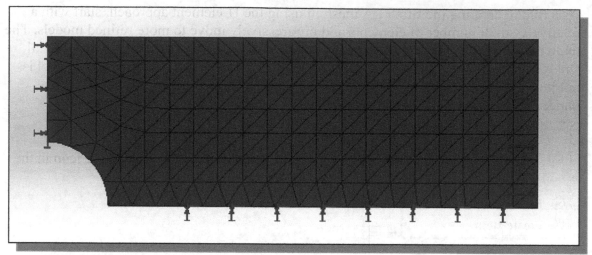

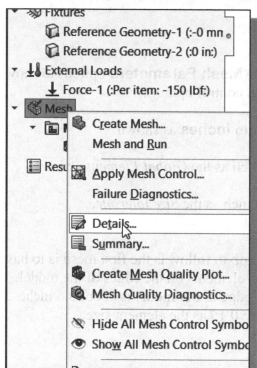

7. In the *FEA Study* window, right-click on the *Mesh* item to display the option list and select **Details** as shown.

❖ The current coarse mesh consists of **595** nodes and **270** surface elements.

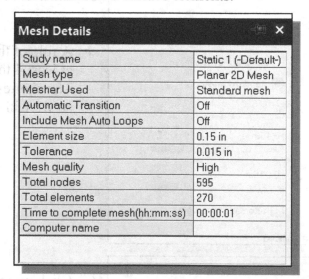

Mesh Details	
Study name	Static 1 (-Default-)
Mesh type	Planar 2D Mesh
Mesher Used	Standard mesh
Automatic Transition	Off
Include Mesh Auto Loops	Off
Element size	0.15 in
Tolerance	0.015 in
Mesh quality	High
Total nodes	595
Total elements	270
Time to complete mesh(hh:mm:ss)	00:00:01
Computer name	

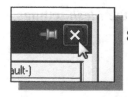

8. Click on the [**X**] icon to close the *Mesh Details* dialog box.

Run the Solver

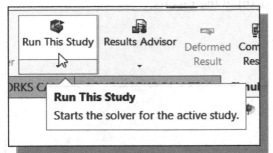

1. Click on the **Run This Study** button to start the *FEA Solver* to calculate the results.

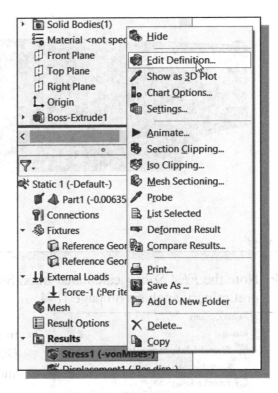

2. In the *FEA Study* window, right-click on the *Stress1* item to display the option list and select **Edit Definition** as shown.

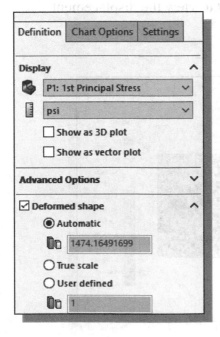

3. Set the *Stress Component* to **P1: 1ˢᵗ Principal Stress** as shown.

4. Set the *Display Units* to **psi**.

5. Set the *Deformed Shape* option to **Automatic**.

❖   The *Automatic Deformed Shape* option will enable the display of a scaled up deformation. Note that the true deformation is relatively small, and the deformation will not be visible if the display option is set to **True scale**.

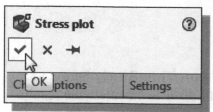

6. Click **OK** to display the results.

Model name:PlaneStress
Study name:Static 1(-Default-)
Plot type: Static nodal stress Stress1
Deformation scale: 1,474.16

P1 (psi)

2.017e+03
1.849e+03
1.681e+03
1.513e+03
1.345e+03
1.176e+03
1.008e+03
8.403e+02
6.723e+02
5.042e+02
3.361e+02
1.681e+02
0.000e+00

❖ Note the *FEA Solver* calculated Max. Stress is a bit lower than the result from the preliminary analysis on page 10-4.

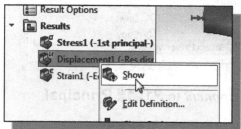

7. In the *FEA Study* window, right-click on the *Displacement1* item to display the option list and select **Show** to view the displacement.

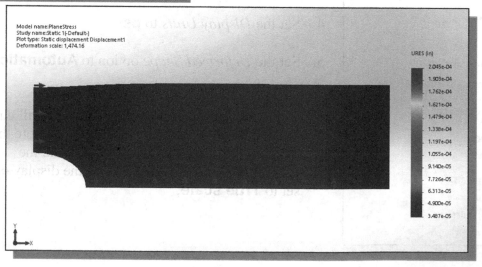

Model name:PlaneStress
Study name:Static 1(-Default-)
Plot type: Static displacement Displacement1
Deformation scale: 1,474.16

URES (in)

2.045e-04
1.903e-04
1.762e-04
1.621e-04
1.479e-04
1.338e-04
1.197e-04
1.055e-04
9.140e-05
7.726e-05
6.313e-05
4.900e-05
3.487e-05

Refinement of the FEA Mesh – Global Element Size 0.10

One method of refinement is simply to adjust the *Global Element size* to a smaller value. We will reduce the *Global Element size* to 0.10.

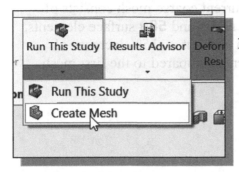

1. Choose **Create Mesh** by clicking the icon in the toolbar as shown.

2. Click on the **OK** button to proceed with deleting the old mesh and creating a new mesh of the FEA model.

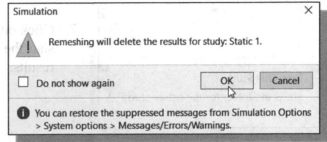

3. Enter *0.10* inch as the *Global Element size*.

4. Switch **On** the *Automatic transition* option.

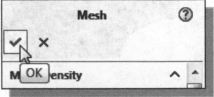

5. Click on the **OK** button to accept the *Mesh* settings and create a new mesh.

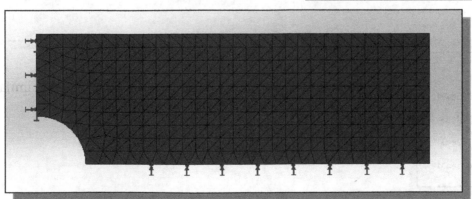

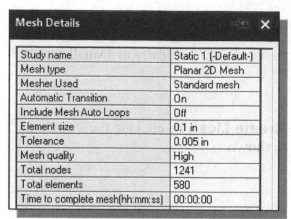

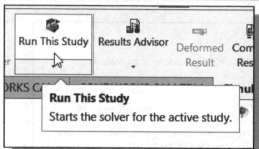

6. On your own, examine the details of the refinement.

❖ The current coarse mesh consists of **1241** nodes and **580** surface elements, which is about a 115% increase of elements compared to the first mesh.

7. Click on the **Run This Study** button to start the *FEA Solver* to calculate the results.

❖ The *FEA Solver* calculated the Max. Stress with the refinement to be **2108 psi**.

❖ The Max. Displacement is **2.045e-4**, which is very similar to the result from the first mesh.

Refinement of the FEA Mesh – Global Element Size 0.05

We will next adjust the *Global Element size* to **0.05 inch** and observe the changes in the results. In this chapter, we will demonstrate the effect of adjusting the mesh of the entire model, not just refining in the high stress area.

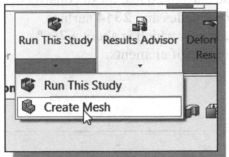

1. Choose **Create Mesh** by clicking the icon in the toolbar as shown.

2. Click on the **OK** button to proceed with deleting the old mesh and create a new mesh of the FEA model.

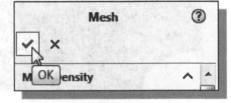

3. Enter **0.05** inch as the *Global Element size*.

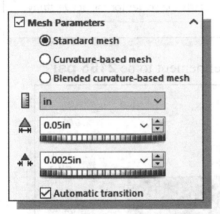

4. Click on the **OK** button to accept the *Mesh* settings and create a new mesh.

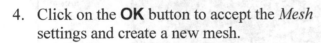

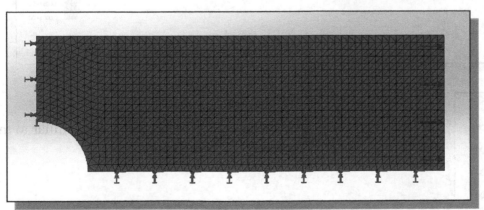

Mesh Details	
Study name	Static 1 (-Default-)
Mesh type	Planar 2D Mesh
Mesher Used	Standard mesh
Automatic Transition	On
Include Mesh Auto Loops	Off
Element size	0.05 in
Tolerance	0.0025 in
Mesh quality	High
Total nodes	4787
Total elements	2314
Time to complete mesh(hh:mm:ss)	00:00:01

5. On your own, examine the details of the refinement.

❖ The current coarse mesh consists of **4787** nodes and **2314** surface elements, which is about a 760% increase of elements.

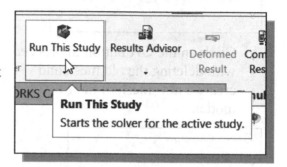

6. Click on the **Run This Study** button to start the *FEA Solver* to calculate the results.

Run This Study
Starts the solver for the active study.

7. The *FEA Solver* calculated the Max. Stress with the refinement to be **2165 psi**.

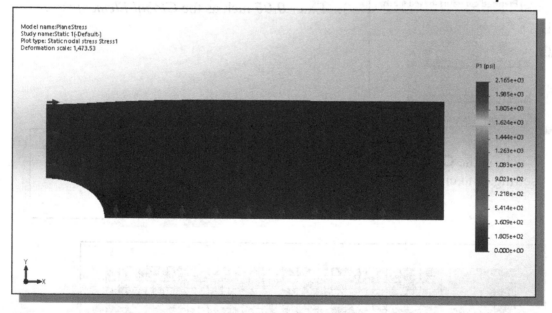

8. The Max. Displacement is **2.050e-4**, which is very similar to the result from the first mesh.

Refinement of the FEA Mesh – Global Element Size 0.03

We will next adjust the *Global Element size* to **0.03 inch** and observe the changes in the results.

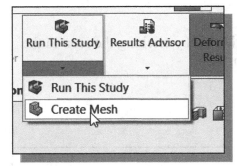

1. Choose **Create Mesh** by clicking the icon in the toolbar as shown.

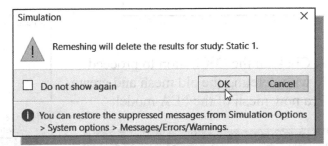

2. Click on the **OK** button to proceed with deleting the old mesh and create a new mesh of the FEA model.

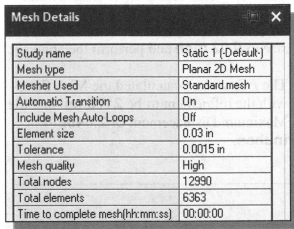

3. On your own, adjust the *Global Element size* to **0.03 inch** and perform the analysis.

4. The *FEA Solver* calculated the Maximum Stress with the refinement to be **2176 psi.** and the *Maximum Displacement* of **2.046e-4 inch**.

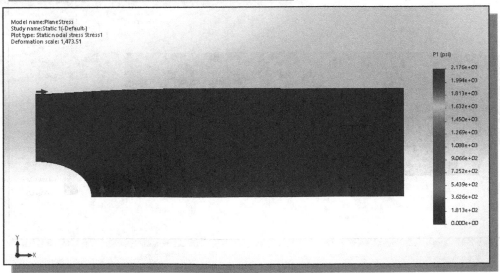

Refinement of the FEA Mesh – Global Element Size 0.02

We will next adjust the *Global Element size* to **0.02 inch** and observe the changes in the results.

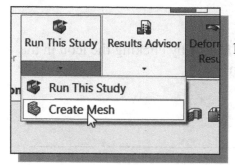

1. Choose **Create Mesh** by clicking the icon in the toolbar as shown.

2. Click on the **OK** button to proceed with deleting the old mesh and create a new mesh of the FEA model.

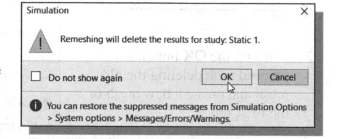

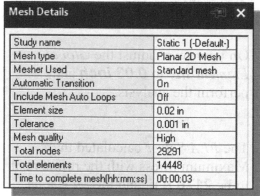

3. On your own, adjust the *Global Element size* to **0.02 inch** and perform the analysis.

4. The *FEA Solver* calculated the Max. Stress with the refinement to be **2179 psi** and the Maximum Displacement of **2.045e-4 inches**.

Comparison of Results

The accuracy of the SOLIDWORKS Simulation results for this problem can be checked by comparing them to the analytical results presented earlier. In the Preliminary Analysis section, the maximum stress was calculated using a stress concentration factor, and the value obtained was **2180 psi**. One should realize the analytical result is obtained through the use of charts from empirical data and therefore involves some degree of error. The maximum stress obtained by finite element analysis using SOLIDWORKS Simulation ranged from **2017** to **2178 psi**. In the *Preliminary Analysis* section, the maximum displacement was also estimated to be around **1.94E-4 inches**, measured from the center of the hole to one end of the plate. The maximum displacement obtained by finite element analysis using SOLIDWORKS Simulation was around **2.045E-4 inches**. The agreement between the analytical results and those from SOLIDWORKS Simulation demonstrates the potential of SOLIDWORKS Simulation as a very powerful design tool.

In FEA, the process of mesh refinement is called **Convergence Anal**ysis, or **H-convergence**. For our analysis, the refinement of the mesh does show the FEA results converging near the analytical results. The refinement to the size of **0.03~0.05 inches** is quite adequate for our analysis. Any further refinement does not provide any additional insight and is therefore not necessary.

Global Element size	Number of Elements	σ_{max} (psi)	D_{max} (in)
0.15	270	2017	2.045e-4
0.10	580	2108	2.045e-4
0.05	2314	2165	2.050e-4
0.03	6363	2176	2.046e-4
0.02	14448	2179	2.045e-4

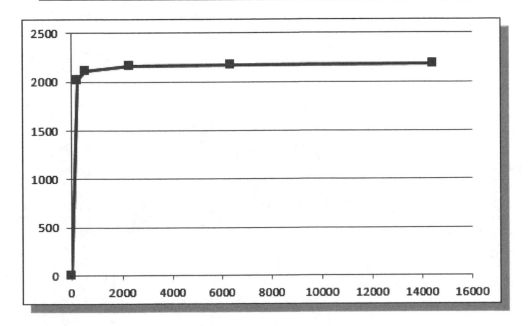

Questions:

1. What are the four basic types of two-dimensional solid elements?

2. What type of situation is more suitable to perform a *Plane Stress analysis* rather than a *Plane Strain analysis*?

3. For plane stress analysis using two-dimensional surface elements, how was the surface created?

4. Which type of two-dimensional solid elements is most suitable for members that are thin and whose loads act only in the plane?

5. Why is it important to recognize the symmetrical nature of design in performing FEA analysis?

6. Why is it important to not just concentrate on the stress results but also examine the displacements of FEA results?

7. In the tutorial, did the changes in the refinement of the elements help the stress results? By how much?

Exercises:

Determine the maximum stress produced by the loads.

1. Aluminum 6061-T6 plate.
   Dimensions are in inches, Plate thickness: ¼ in.

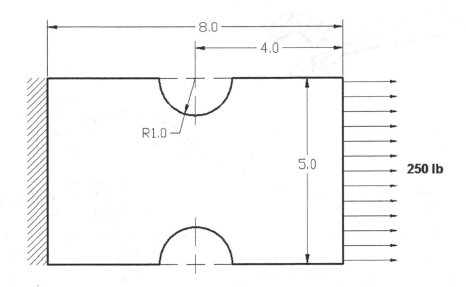

2. Material: Steel Plate
   Thickness: 25 mm

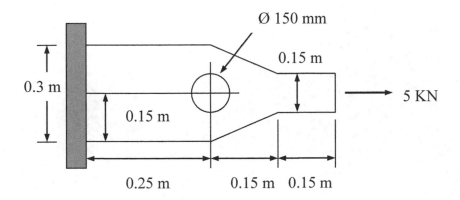

3. An upward force of 1000 N is applied to the handle of the wrench.

   Material: Steel
   Thickness: 8 mm

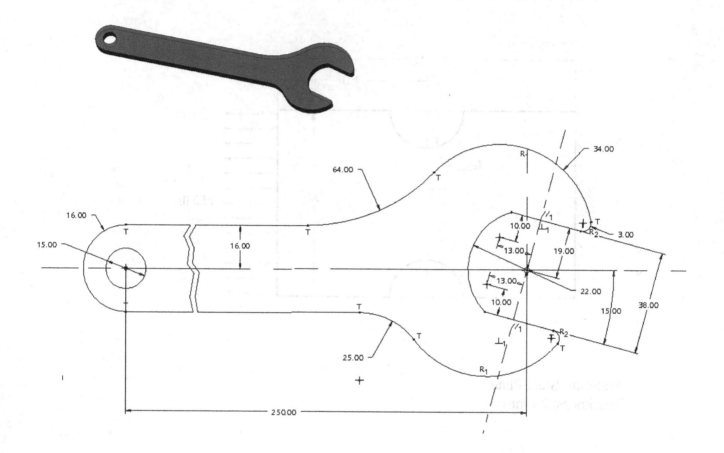

Chapter 11
Three-Dimensional Solid Elements

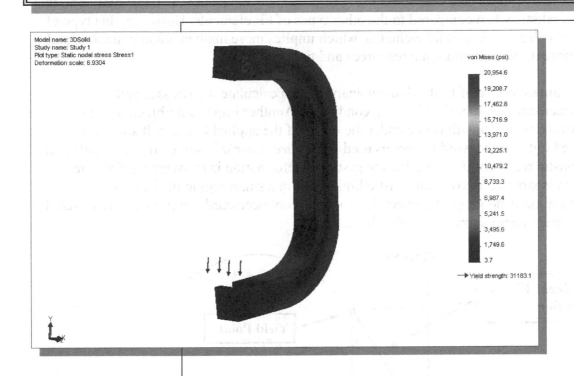

Model name: 3DSolid
Study name: Study 1
Plot type: Static nodal stress Stress1
Deformation scale: 6.9304

von Mises (psi)

20,954.6
19,208.7
17,462.8
15,716.9
13,971.0
12,225.1
10,479.2
8,733.3
6,987.4
5,241.5
3,495.6
1,749.6
3.7

→ Yield strength: 31183.1

Learning Objectives

- ◆ **Perform 3D Finite Element Analysis.**
- ◆ **Understand the concepts and theory of Failure Criteria.**
- ◆ **Create 3D Solid models and FE models.**
- ◆ **Use the SOLIDWORKS Simulation Mesh Control.**
- ◆ **Use the different Refinement options to examine Stress Results.**

Introduction

In this chapter, the general FEA procedure for using three-dimensional solid elements is illustrated. A finite element model using three-dimensional solid elements may look the most realistic when compared to the other types of FE elements. However, this type of analysis also requires more elements, which implies more mathematical equations and therefore more computational resources and time.

The main objective of finite element analysis is to calculate the stresses and displacements for specified loading conditions. Another important objective is to determine if *failure* will occur under the effect of the applied loading. It should be pointed out that the word *failure* as used for *failure criteria* is somewhat misleading. In the elastic region of the material, the system's deformation is recoverable. Once the system is stressed beyond the elastic limit, even in a small region of the system, deformation is no longer recoverable. This does not necessarily imply that the system has failed and cannot carry any further load.

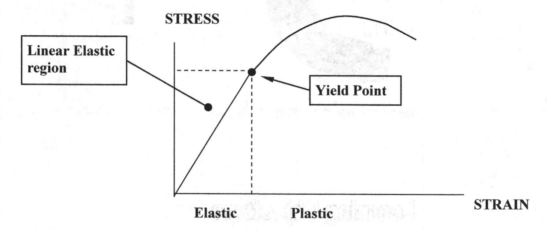

Stress-Strain diagram of typical ductile material

Several theories propose different failure criteria. In general, all these theories provide fairly similar results. The most widely used failure criteria are the **Von Mises yield criterion** and the **Tresca yield criterion**. Both the *Von Mises* and *Tresca* stresses are scalar quantities, and when compared with the yield stress of the material they indicate whether a portion of the system has exceeded the elastic state.

Von Mises Criterion: (Note that σ_1, σ_2, σ_3 are the three principal stresses.)

$$2\,\sigma_{yp}^{\,2} = (\sigma_1 - \sigma_2)^2 + (\sigma_2 - \sigma_3)^2 + (\sigma_3 - \sigma_1)^2$$

Tresca Yield Criterion:

$$\frac{1}{2}\,\sigma_{yp} = \frac{1}{2}\,(\sigma_1 - \sigma_3)$$

This chapter illustrates the general FEA procedure of using three-dimensional solid elements. The creation of a solid model is first illustrated, and *solid elements* are generated using the SOLIDWORKS Simulation mesh commands. In theory, all designs could be modeled with three-dimensional solid elements. The three-dimensional solid element is the most versatile type of element compared to the more restrictive one-dimensional or two-dimensional elements. The procedure involved in performing a three-dimensional solid FEA analysis is very similar to that of a two-dimensional solid FEA analysis, as was demonstrated in Chapter 10. As one might expect, the number of node-points involved in a typical three-dimensional solid FEA analysis is usually much greater than that of a two-dimensional solid FEA analysis. We will also use the SOLIDWORKS mesh-control tool to refine the mesh at the high-stress region, which provides a more efficient way of performing the convergence analysis.

Problem Statement

Determine the maximum normal stress in the AL6061 member shown; the c-link design is assembled to the frame at the upper hole (Ø 0.25) and a vertical load of 200 lbs. is applied at the top of the notch on the lower arm as shown.

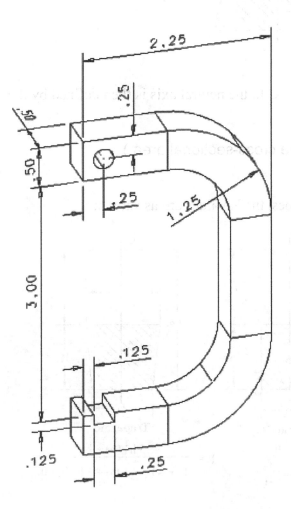
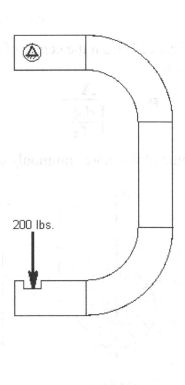

Preliminary Analysis

The analysis of stresses due to bending moment has been restricted to straight members as illustrated in chapters 7 through 9. Note that a good approximation may be obtained if the curvature of the member is relatively small in comparison to the cross section of the member. However, when the curvature is large, the stress distribution is no longer linear; it becomes hyperbolic. This is due to the fact that, in a curved member, the neutral axis of a transverse section does not pass through the centroid of that section.

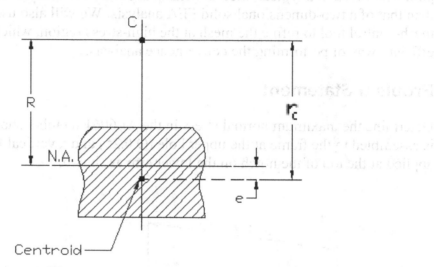

The distance R from the center of curvature C to the neutral axis is then defined by the equation:

$$R = \frac{A}{\int \frac{dA}{r_c}}$$ (A is the cross-sectional area.)

For some of the more commonly used shapes, the R values are as shown:

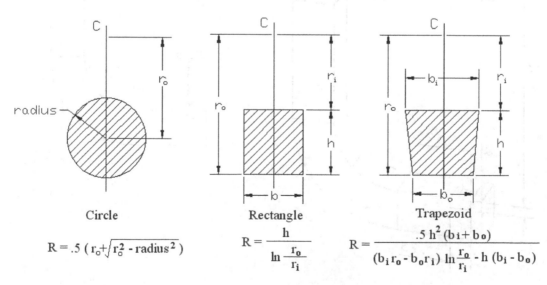

Circle

$$R = .5 \left(r_o + \sqrt{r_o^2 - radius^2} \right)$$

Rectangle

$$R = \frac{h}{\ln \frac{r_o}{r_i}}$$

Trapezoid

$$R = \frac{.5\,h^2\,(b_i + b_o)}{(b_i r_o - b_o r_i) \ln \frac{r_o}{r_i} - h\,(b_i - b_o)}$$

The bending stress in a **curved beam**, at location **r** measured from the center of curvature **C**, can be expressed as:

$$\sigma_r = \frac{M(R-r)}{Aer}$$

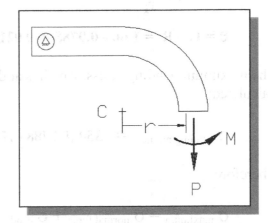

For our analysis, we can establish an equivalent load system in the upper portion of the member, through the centroid of the cross section, as shown.

P = 200 lbs.

M = 350 in-lb.

The normal stress at the cross section is a combination of the stresses from the normal force P and the bending moment M. The stress distributions are as shown in the figure below.

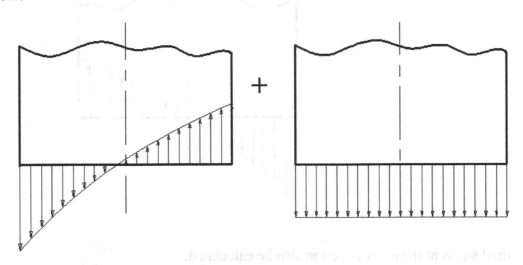

The maximum normal stress occurs on the inside edge, the edge that is closer to the center of the curvature C.

The normal stress from the normal force P:

$$\sigma_{normal\_force} = \frac{P}{A} = \frac{200 \text{ lb}}{.25 \text{ in}^2} = 800 \text{ psi}$$

The bending stress component:

$$\sigma_{\text{bending moment}} = \frac{M(R\text{-}r)}{Aer}$$

To calculate the bending stress, we will first calculate the **R** and **e** values:

$$R = \frac{h}{\ln \dfrac{r_o}{r_i}} = .5/ (\ln(1.25/.75)) = 0.9788$$

$$e = r_c - R = 1.00 - 0.9788 = 0.0212$$

The maximum bending stress, which is at the inside edge (r = 0.75), can now be calculated:

$$\sigma_{\text{bending moment}} = 350 \ (0.9788 - .75) / (0.25 \times .0212 \times .75) = 20146 \text{ psi}$$

Therefore

$$\sigma_{\text{maximum}} = \sigma_{\text{normal force}} + \sigma_{\text{bending moment}} = 800 + 20146 = 20946 \text{ psi}$$

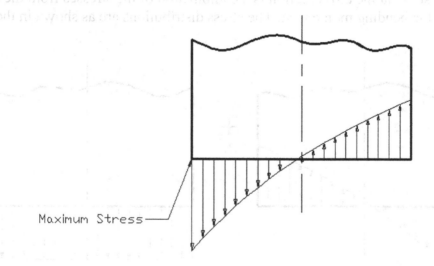

Maximum Stress

The normal stress at the outer edge can also be calculated:

$$\sigma_{\text{bending @ 1.25}} = 350 \ (0.9788 - 1.25) / (0.25 \times .0212 \times 1.25) = -14327 \text{ psi}$$

Therefore

$$\sigma_{\text{@1.25}} = \sigma_{\text{normal force}} + \sigma_{\text{bending moment}} = 800 - 14327 = -13527 \text{ psi}$$

Start SOLIDWORKS

1. Select the **SOLIDWORKS** option on the *Start* menu or select the **SOLIDWORKS** icon on the desktop to start SOLIDWORKS. The SOLIDWORKS main window will appear on the screen.

2. Select **Part** by clicking on the first icon in the *New SOLIDWORKS Document* dialog box as shown.

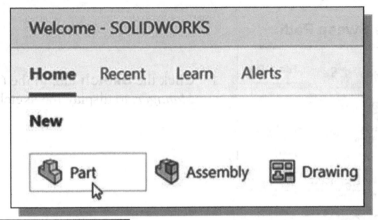

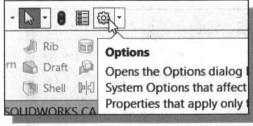

3. Select the **Options** icon from the *Menu* toolbar to open the *Options* dialog box.

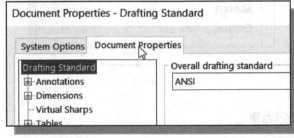

4. Switch to the **Document Properties** tab and reset the *Drafting Standard* to **ANSI** as shown in the figure.

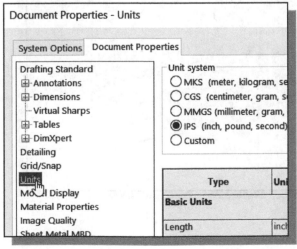

5. On your own, set the *Unit system* to **IPS (inch, pound, second)** as shown.

Create a CAD Model in SOLIDWORKS

To perform the surface FEA analysis, we will first create a solid model using the Sweep command. The **Sweep** operation is defined as moving a planar section through a planar (2D) or 3D path in space to form a three-dimensional solid object. In SOLIDWORKS, we create a *swept feature* by defining a 2D sketch of a cross section and a path. To create the path, we can use a sketch, existing model edges, or curves. The sketched profile is then swept along the planar path. The Sweep operation is used for objects that have uniform shapes along a trajectory.

➢ **Define the Sweep Path**

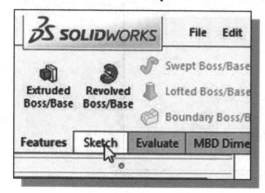

1. Click the **Sketch** tab, in the *Command Manager*, to display the sketch toolbar.

2. Click the **Sketch** icon, in the *Sketch* toolbar, to start a new 2D sketch.

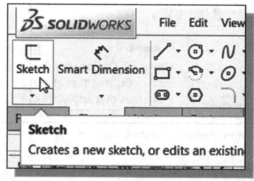

3. Click the **Front Plane,** in the graphics area, to align the sketching plane of our sketch as shown.

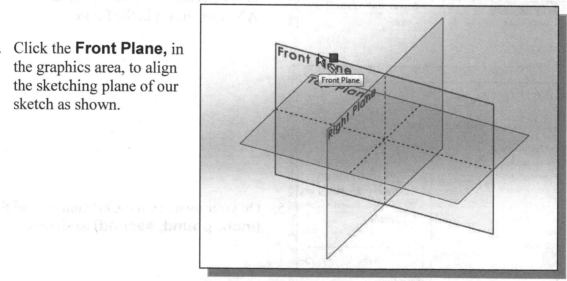

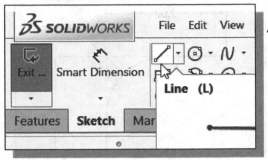

4. Click the **Line** icon in the *Sketch* toolbar as shown.

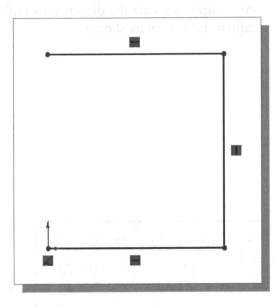

5. Start the first endpoint of the line series at the origin and create the three line segments as shown.

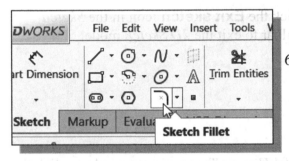

6. Click the **Fillet** icon in the *Sketch* toolbar as shown.

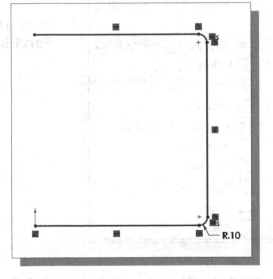

7. Create the two rounded corners as shown.

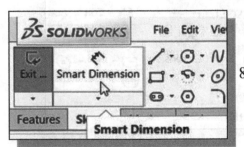

8. Click the **Smart Dimension** icon in the *Sketch* toolbar as shown.

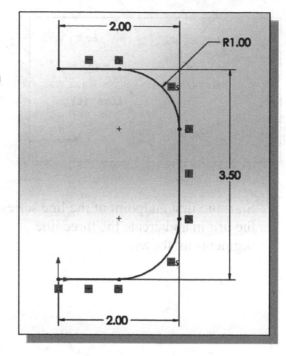

9. On your own, use the **Smart Dimension** command to create the dimensions and adjust the sketch as shown.

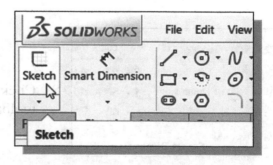

10. Click the **Exit sketch** icon in the *Sketch* toolbar to exit the *2D Sketch* mode.

> ➤ **Define the Sweep Section**

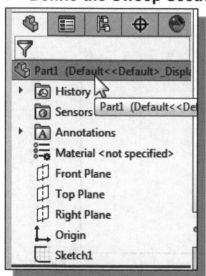

1. In the *Model History Tree* manager window, click the **Part1** item to deselect the *Sketch1* item.

2. Click the **Sketch** icon, in the *Sketch* toolbar, to create a new 2D sketch.

3. Inside the graphics area, expand the model history tree by clicking on the [+] icon in front of the Part1 item.

4. Click the **Right Plane**, in the *Model History Tree*, to align the sketching plane of our sketch as shown.

5. Click the **Normal To** icon to adjust the display to be perpendicular to the sketching plane.

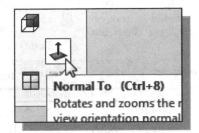

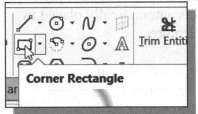

6. Click the **Rectangle** icon in the *Sketch* toolbar as shown.

7. Create a rectangle, roughly centered at the origin, as shown.

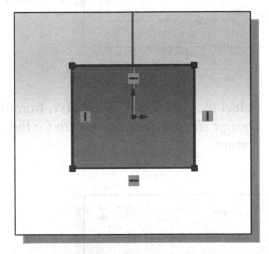

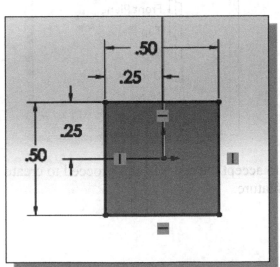

8. On your own, create and adjust the **dimensions** of the sketch as shown.

9. Click the **Exit Sketch** icon in the *Sketch* toolbar to exit the *2D Sketch* mode.

➢ **Create the Swept Feature**

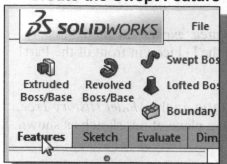

1. Select the **Features** tab in the *Command Manager* as shown.

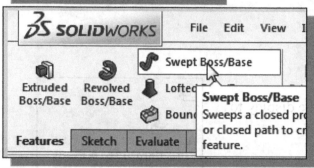

2. Click the **Swept Boss/Base** icon to execute the **Sweep** command.

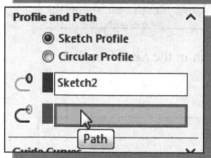

3. Note the last sketch, the profile sketch, was pre-selected and therefore it is automatically placed as the profile of the swept feature.

4. Click the **Path** selection box as shown.

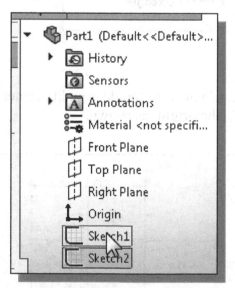

5. Select the first sketch, **Sketch1**, from the *Design History Tree* as the path for the swept feature.

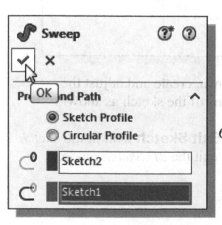

6. Click **OK** to accept the settings and proceed to create the swept feature.

➢ **Create a Cut Feature**

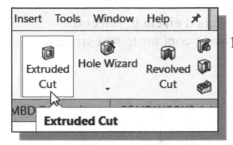

1. Click the **Extruded Cut** icon to execute the *Extrusion* command.

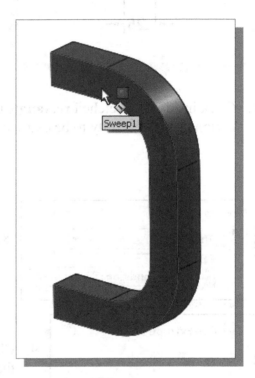

2. Click the **front surface** of the model, in the graphics area, to align the sketching plane of our sketch as shown.

3. On your own, create and modify the 2D sketch as shown.

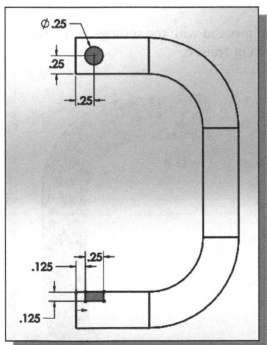

4. Click the **Exit Sketch** icon in the *Sketch* toolbar to exit the *2D Sketch* mode.

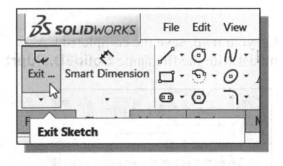

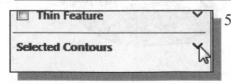

5. Click the **down arrow** next to the *Selected Contour* option to view the available options.

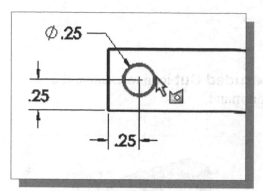

6. Click the sketched **circle** to select the contour as the 1st contour to be extruded.

7. Click inside the sketched **rectangle** to select the region as the 2nd entity to be extruded.

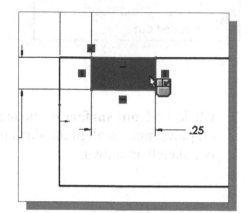

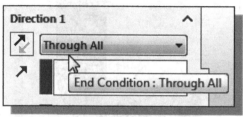

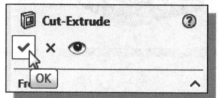

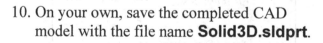

8. Set the extrusion *Direction 1* option to **Through All** as shown.

9. Click **OK** to proceed with the settings and create the extruded cut feature.

10. On your own, save the completed CAD model with the file name **Solid3D.sldprt**.

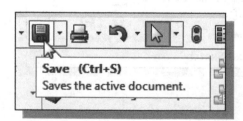

Activate the SOLIDWORKS Simulation Module

SOLIDWORKS Simulation is a multi-discipline Computer Aided Engineering (CAE) tool that enables users to simulate the physical behavior of a model, and therefore enables users to improve the design. SOLIDWORKS Simulation can be used to predict how a design will behave in the real world by calculating stresses, deflections, frequencies, heat transfer paths, etc.

The SOLIDWORKS Simulation product line features two areas of Finite Element Analysis: **Structure** and **Thermal**. *Structure* focuses on the structural integrity of the design, and *thermal* evaluates heat-transfer characteristics.

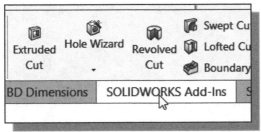

1. Start SOLIDWORKS Simulation by selecting the **SOLIDWORKS Add-Ins** tab in the *Command Manager* area as shown.

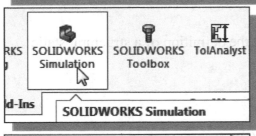

2. In the *SOLIDWORKS Office* list, choose **SOLIDWORKS Simulation** as shown.

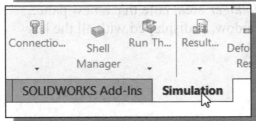

3. In the *Command Manager* area, choose **Simulation** as shown.

❖ Note that the SOLIDWORKS Simulation module is integrated as part of SOLIDWORKS. All of the SOLIDWORKS Simulation commands are accessible through the icon panel in the *Command Manager* area.

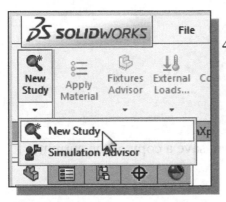

4. To start a new study, click the **New Study** item listed under the *Study Advisor* as shown.

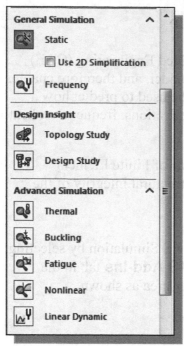

5. Select **Static** as the type of analysis to be performed with SOLIDWORKS Simulation.

❖   Note that different types of analyses are available, which include both structural static and dynamic analyses, as well as the thermal analysis.

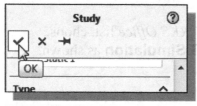

6. Click **OK** to start the definition of a structural static analysis.

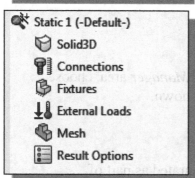

❖   In the *Feature Manager* area, note that a new panel, the *FEA Study* window, is displayed with all the key items listed.

❖   Also note that the **Static 1** tab is activated, which indicates the use of the FEA model.

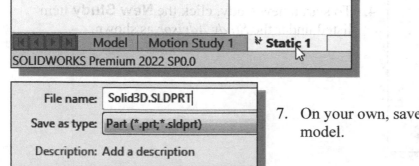

7. On your own, save a copy of the current model.

Assign the Element Material Property

Next, we will set up the *Material Property* for the elements. The *Material Property* contains the general material information, such as *Modulus of Elasticity, Poisson's Ratio*, etc. that is necessary for the FEA analysis.

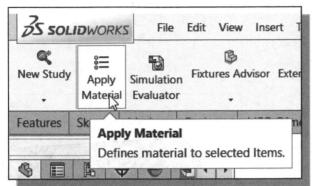

1. Choose the **Apply Material** option as shown.

❖ Note the default list of materials, which are available in the pre-defined SOLIDWORKS Simulation material library, is displayed.

2. Select **6061-T6 Aluminum** in the *Material* list as shown.

3. Set the **Units** option to display **English (IPS)** to make the selected material available for use in the current FEA model.

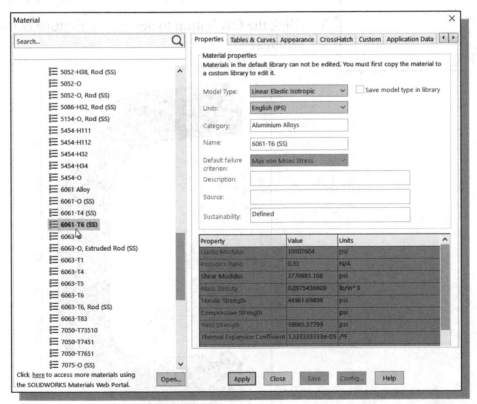

4. Click **Apply** to assign the material property then click **Close** to exit the material assignment command.

Apply Boundary Conditions – Constraints

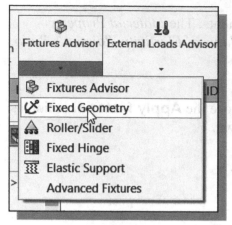

1. Choose **Fixed Geometry** from the Fixtures Advisor drop-down list as shown.

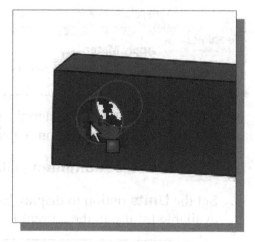

2. Select the **cylindrical surface** of the top portion of the model as the entity to apply the **Fixed** constraint.

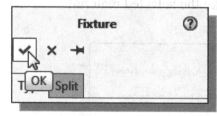

3. Click the **OK** button to accept the Fixture constraint settings.

Apply the External Load to the system

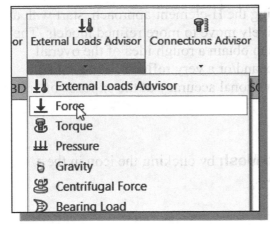

1. Choose **External Loads Advisor →
   Force** by clicking the icon in the toolbar as
   shown.

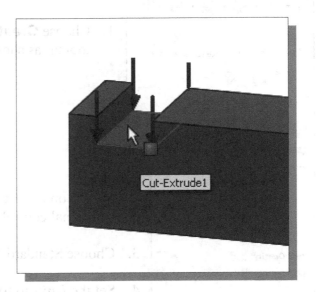

2. Select the **horizontal surface** of the
   notch of the lower section of the
   model as shown.

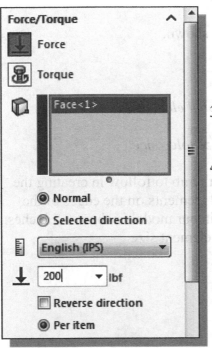

3. Set the Units to **English (IPS)**, to match with the
   systems units we are using.

4. Enter **200** lbf as the applied force.

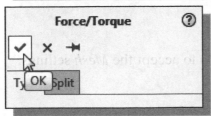

5. Click on the **OK** button to accept the *Force/Torque*
   settings and create the load.

Create the first FEA Mesh – Coarse Mesh

As a rule, when creating the first FEA mesh in using the H-element approach, start with a relatively small number of elements and progressively move to more refined models. The main objective of the first coarse mesh analysis is to obtain a rough idea of the overall stress distribution. In most cases, use of a complex and/or a very refined FEA model is not justifiable since it most likely provides computational accuracy at the expense of unnecessarily increased processing time.

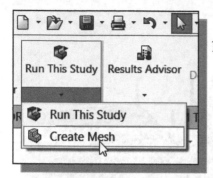

1. Choose **Create Mesh** by clicking the icon in the toolbar as shown.

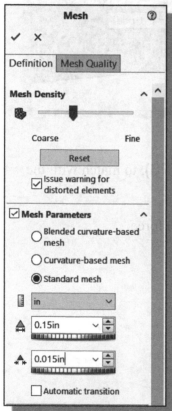

2. Switch on the **Mesh Parameters** options to show the additional control options.

3. Choose Standard Mesh as shown,

4. Set the *Units* to **inches**.

5. Enter **0.15** inch as the *Global element size*.

6. Enter **0.015** inch as the *Size tolerance*.

❖   In general, a good rule of thumb to follow in creating the first mesh is to have 3 to 4 elements on the edges of the model. The shortest edge in our model is about 0.5 inches, so we will use 0.15 as the element size.

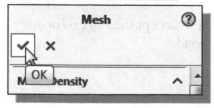

7. Click on the **OK** button to accept the *Mesh* settings.

❖ The *Automatic mesh* function in SOLIDWORKS Simulation has generated a fairly uniform and consistent mesh of elements.

8. In the *FEA Study* window, right-click on the *Mesh* item to display the option list and select **Details** as shown.

Mesh Details	
Study name	Static 1* (-Defau
DetailsMesh type	Solid Mesh
Mesher Used	Standard mesh
Automatic Transition	Off
Include Mesh Auto Loops	Off
Jacobian points for High quality mesh	16 points
Element size	0.15 in
Tolerance	0.015 in
Mesh quality	High
Total nodes	6291
Total elements	3679
Maximum Aspect Ratio	3.2089
Percentage of elements with Aspect Ratio < 3	99.8
Percentage of elements with Aspect Ratio > 10	0
Percentage of distorted elements	0
Number of distorted elements	0
Time to complete mesh(hh:mm:ss)	00:00:01
Computer name	

❖ The current coarse mesh consists of **6291** nodes and **3679** solid elements.

9. Click on the [**X**] icon to close the *Mesh Details* dialog box.

Run the Solver

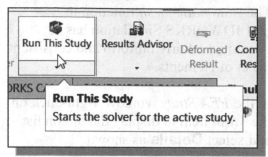

1. Click on the **Run This Study** button to start the *FEA Solver* to calculate the results.

❖ Once the *Solver* has completed the calculations, the display will switch to the stress distribution.

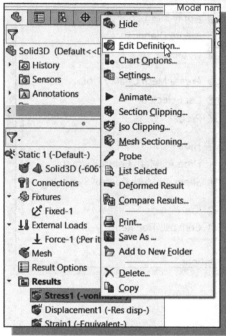

2. In the *FEA Study* window, right-click on the *Stress1* item to display the option list and select **Edit Definition** as shown.

3. Confirm the *Stress Component* is set to **VON: Von Mises Stress** as shown.

4. Set the display units to **psi**.

5. Set the *Deformed Shape* option to **Automatic**.

❖ The **Automatic** *Deformed Shape* option will enable the display of a scaled-up deformation. Note that the true deformation is relatively small, and the deformation will not be visible if the display option is set to **True scale**.

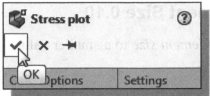

6. Click **OK** to display the results.

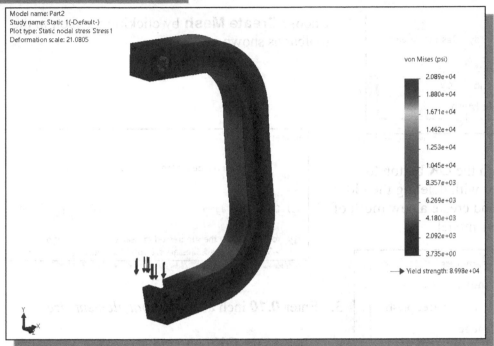

❖ Note the FEA calculated Max. Stress is very similar to the result from the preliminary analysis on page 11-6.

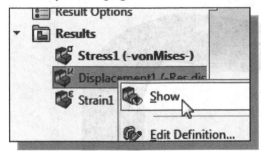

7. In the *FEA Study* window, right-click on the *Displacement1* item to display the option list and select **Show** to view the displacement.

❖ The max displacement occurred at the lower portion of the model, which matches with the applied constraint and loading situation.

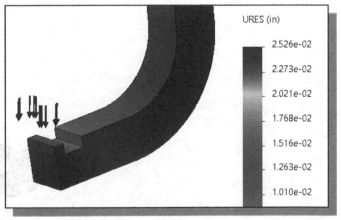

Refinement of the FEA Mesh – Global Element Size 0.10

One method of refinement is simply to adjust the *Global Element size* to a smaller value. We will need to delete the old mesh and create a new mesh.

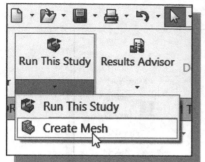

1. Choose **Create Mesh** by clicking the icon in the toolbar as shown.

2. Click on the **OK** button to proceed with deleting the old mesh and create a new mesh of the FEA model.

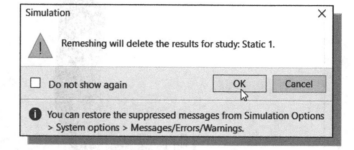

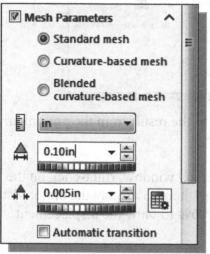

3. Enter *0.10* inch as the *Global element size*.

4. Click on the **OK** button to accept the *Mesh* settings and create a new mesh.

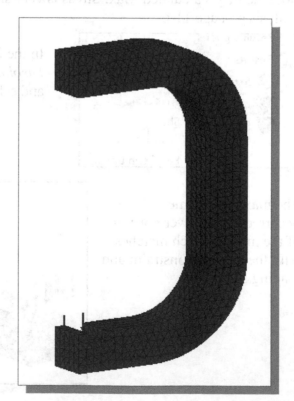

5. On your own, examine the details of the refinement.

❖ The current coarse mesh consists of **17660** nodes and **11078** solid elements, which is almost three times the number of elements compared to the first mesh.

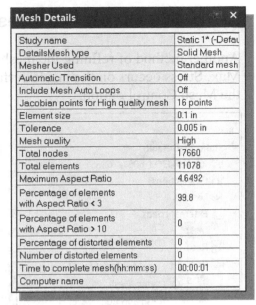

Mesh Details	
Study name	Static 1* (-Defau
DetailsMesh type	Solid Mesh
Mesher Used	Standard mesh
Automatic Transition	Off
Include Mesh Auto Loops	Off
Jacobian points for High quality mesh	16 points
Element size	0.1 in
Tolerance	0.005 in
Mesh quality	High
Total nodes	17660
Total elements	11078
Maximum Aspect Ratio	4.6492
Percentage of elements with Aspect Ratio < 3	99.8
Percentage of elements with Aspect Ratio > 10	0
Percentage of distorted elements	0
Number of distorted elements	0
Time to complete mesh(hh:mm:ss)	00:00:01
Computer name	

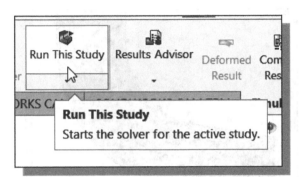

Run This Study   Results Advisor   Deformed Result   Com Res

Run This Study
Starts the solver for the active study.

6. Click on the **Run This Study** button to start the *FEA Solver* to calculate the results.

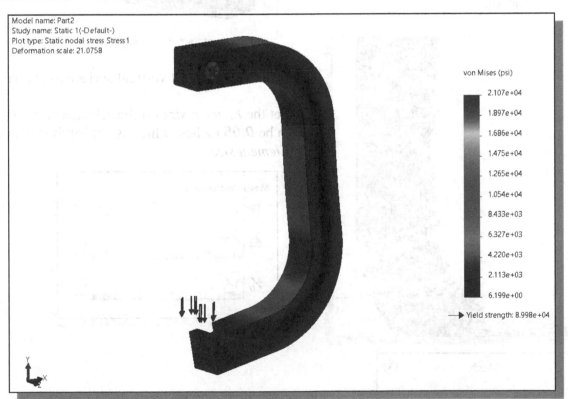

Model name: Part2
Study name: Static 1(-Default-)
Plot type: Static nodal stress Stress1
Deformation scale: 21.0758

von Mises (psi)

2.107e+04
1.897e+04
1.686e+04
1.475e+04
1.265e+04
1.054e+04
8.433e+03
6.327e+03
4.220e+03
2.113e+03
6.199e+00

→ Yield strength: 8.998e+04

❖ The *FEA Solver* calculated the Max. Von Mises Stress with the refinement to be **21070 psi**.

Refinement of the FEA Mesh – Mesh Control Option

Another method of refining the mesh is to only refine the high stress areas. Since the Max. Stress occurs on the inside surfaces, we will refine the mesh in those areas by using the Mesh Control option.

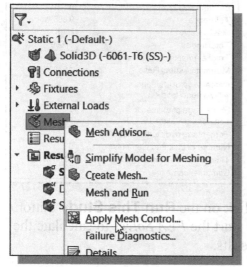

1. In the *FEA Study* window, right-click on the *Mesh* item to display the option list and select **Apply Mesh Control** as shown.

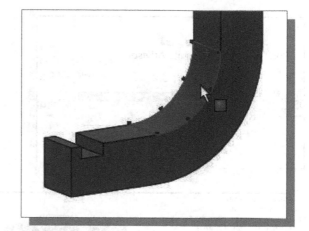

2. Select the lower inside curved surface as shown.

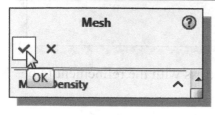

3. Select the adjacent vertical surface as shown.

4. Set the *Element size* on the selected surfaces to be **0.05** inches, which is half of the *Global Element size*.

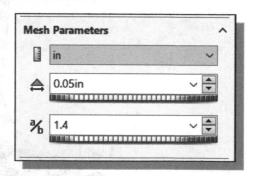

5. Click on the **OK** button to accept the *Mesh Control* settings.

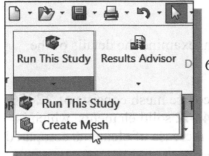

6. Choose **Create Mesh** by clicking the icon in the toolbar as shown.

7. Click on the **OK** button to proceed with deleting the old mesh and create a new mesh of the FEA model.

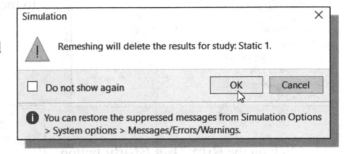

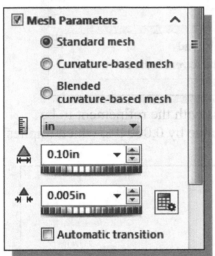

8. Confirm the *Global element size* is set to **0.10** inches as shown.

9. Click on the **OK** button to accept the *Mesh* settings and create a new mesh.

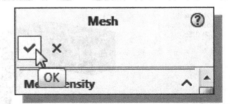

❖ The *Mesh Control* option allows us to refine the mesh in the high stress areas; this approach can be more effective than refining the *Global Element size*.

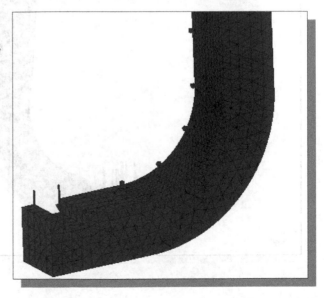

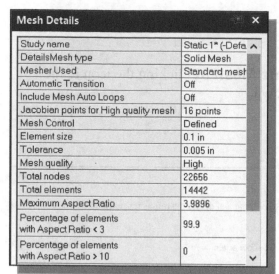

10. On your own, examine the details of the refinement.

❖ The current coarse mesh consists of **22656** nodes and **14442** solid elements, which is about a 30% increase of elements compared to the last mesh.

11. Click on the **Run This Study** button to start the *FEA Solver* to calculate the results.

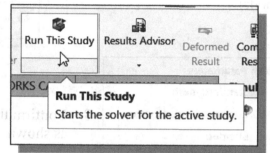

❖ The *FEA Solver* calculated the Max. Von Mises Stress with the refinement to be **21100 psi**. The refinement only changes the stress value by 0.0004%, which implies the current FEA mesh is quite adequate.

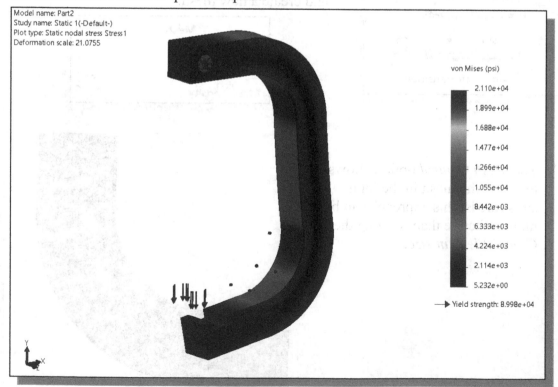

Refinement of the FEA Mesh – Automatic Transition

To confirm the FEA result has converged adequately, we will next adjust the mesh by using the *Automatic Transition* option.

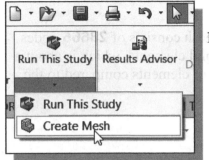

1. Choose **Create Mesh** by clicking the icon in the toolbar as shown.

2. Click on the **OK** button to proceed with deleting the old mesh and creating a new mesh of the FEA model.

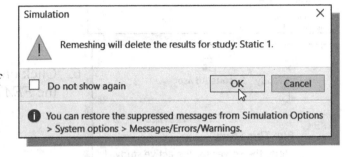

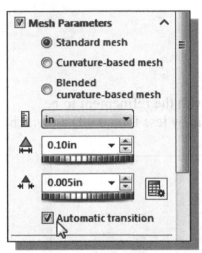

3. Switch on the **Automatic transition** option as shown.

❖ The *Automatic Transition* option will add additional elements to make the transition of different element sizes more gradual.

4. Click on the **OK** button to accept the *Mesh* settings and create a new mesh.

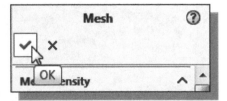

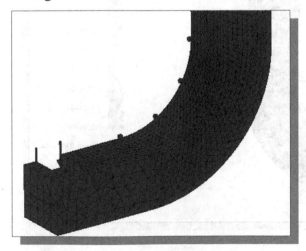

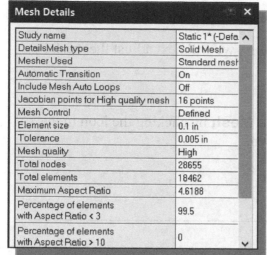

Mesh Details	
Study name	Static 1* (-Defa
DetailsMesh type	Solid Mesh
Mesher Used	Standard mesh
Automatic Transition	On
Include Mesh Auto Loops	Off
Jacobian points for High quality mesh	16 points
Mesh Control	Defined
Element size	0.1 in
Tolerance	0.005 in
Mesh quality	High
Total nodes	28655
Total elements	18462
Maximum Aspect Ratio	4.6188
Percentage of elements with Aspect Ratio < 3	99.5
Percentage of elements with Aspect Ratio > 10	0

5. On your own, examine the details of the refinement.

❖ The current mesh consists of **28665** nodes and **18462** solid elements, which is about a 28% increase of elements compared to the last mesh.

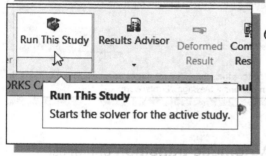

6. Click on the **Run This Study** button to start the *FEA Solver* to calculate the results.

Run This Study
Starts the solver for the active study.

❖ The *FEA Solver* calculated the Max. Von Mises Stress with the refinement to be **21080 psi**. The refinement only changes the stress value by less than 0.001%, which implies the previous FEA mesh is quite adequate.

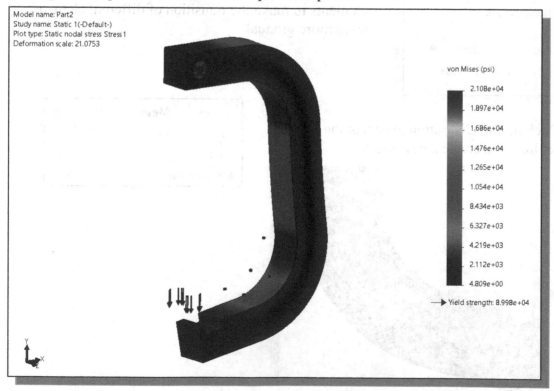

Comparison of Results

In FEA, the process of mesh refinement is called convergence analysis. The convergence study is necessary for both surface and solid elements. For our analysis, the refinement of the mesh shows the FEA results converging near the analytical result. The refinement to the size of **0.10 inches** is quite adequate for our analysis. Note the use of the *Mesh Control* and the *Automatic Transition* options allow for more efficient and effective refinements of the mesh in the high stress areas.

The accuracy of the SOLIDWORKS Simulation results for this problem can be checked by comparing them to the analytical results presented earlier. In the *Preliminary Analysis* section, the maximum stress was calculated using the curved beam theory and the value obtained was **20946 psi**. The maximum Von Mises stress obtained by finite element analysis using SOLIDWORKS Simulation ranged from **20890** to **21170 psi**. The agreement between the analytical result and those from SOLIDWORKS Simulation is quite good.

Global Element size	Mesh Control	Number of Elements	σ_{max} (psi)
0.15		3679	20890
0.10		11078	21170
0.10	0.05	14442	21100
0.10	0.05 -Auto Transition	18462	21080

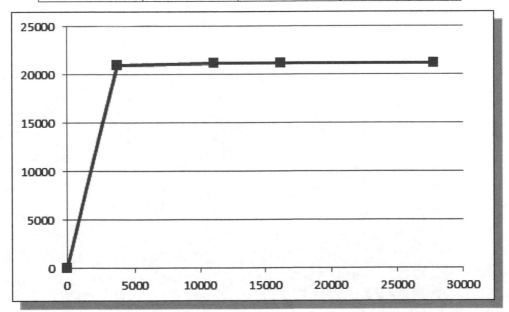

> You are also encouraged to create and compare FEA analyses of the same problem using one-dimensional beam elements and 2D surface elements. What are the advantages/limitations of performing multiple analyses using different types of FEA elements?

Questions:

1. What are the most widely used failure criteria?

2. What are the main objectives of finite element analysis?

3. What will happen to the system if it is stressed beyond the elastic limit?

4. What are the necessary items to create a swept feature in SOLIDWORKS?

5. What is the purpose of doing a convergence study?

6. Under what condition is the bending stress developed in a curved beam no longer linear?

7. Can a curved beam problem be analyzed with the 1D beam element? What is the main limitation of performing such an analysis?

8. What are the advantages of using 3D elements over using the 1D or 2D elements?

Exercises:

1. For the steel U-shape design (diameter: ¾″), determine the maximum stress developed under a loading of P=200 lb.

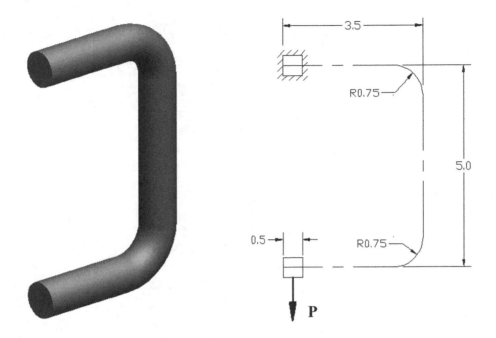

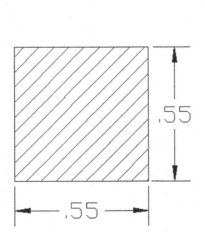

2. For the above problem, perform additional FEA analyses using the two cross-sections shown below and compare the results.

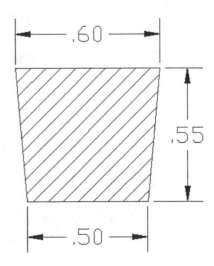

Notes:

Chapter 12
3D Thin Shell Analysis

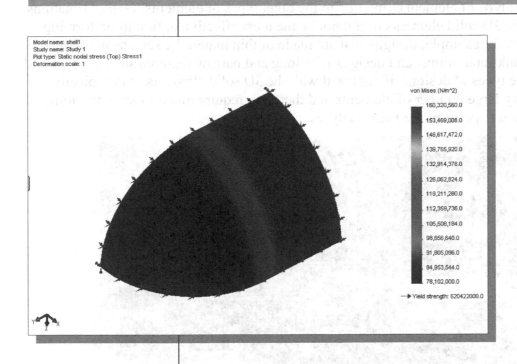

Model name: shell1
Study name: Study 1
Plot type: Static nodal stress (Top) Stress1
Deformation scale: 1

von Mises (N/m^2)

160,320,560.0
153,469,008.0
146,617,472.0
139,765,920.0
132,914,376.0
126,062,824.0
119,211,280.0
112,359,736.0
105,508,184.0
98,656,640.0
91,805,096.0
84,953,544.0
78,102,000.0

→ Yield strength: 620422000.0

Learning Objectives

- ♦ **Understand the procedures to create Solid and Shell FEA models.**
- ♦ **Create a surface Model and perform an FEA Analysis using Shell elements.**
- ♦ **Apply proper constraints to the different types of elements.**
- ♦ **Understand the Basic differences of the different 3D FEA elements.**

Introduction

In the previous chapter, we examined the use of the three-dimensional solid element to perform FEA analyses. Although the three-dimensional (3D) solid element is perhaps the most versatile type of element compared to the other types of elements, certain situations do exist where 3D solid elements might not be the most effective option in performing FEA analyses; for example, designs that are made of thin materials, such as pressure containers, tanks and drums, and designs with long and narrow features such as pipes and frames. These types of designs, if analyzed with the 3D solid elements, will typically result in a very large number of elements and therefore require much more computing power and time to perform the FEA analyses.

The three-dimensional **thin shell** element is another type of 3D element that is particularly designed to aid the FEA analyses of thin shell designs. In theory, all designs could be modeled with three-dimensional solid elements. A thin shell design can also be analyzed using 3D solid elements. Two of the main considerations in selecting the element type to use are (1) the amount of time it takes to perform the analysis and (2) the accuracy of the results.

With the advancements in computer technology in the last few decades, it is now quite feasible to perform multiple FEA analyses using different types of elements. In many instances, performing FEA analyses may be faster than hand calculations; and one should also realize that hand calculations might not be possible in many situations. In this chapter, the procedures to perform FEA analyses using the above two types of elements are illustrated. To assure the accuracy of FEA results, performing a second and/or a third analysis can be very practical and effective.

In FEA analysis, the 3D solid element is the most versatile type of element, as all designs are three-dimensional objects.

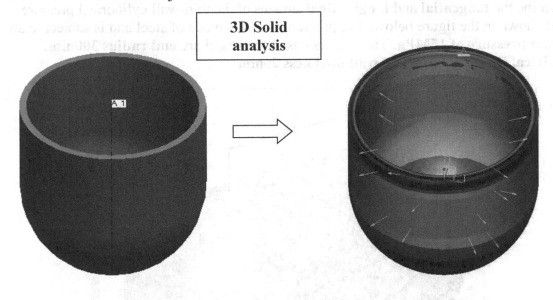

In FEA analysis, if a part is relatively thin compared to its length and width, using *shell elements* would be more efficient than using 3D solid elements. In SOLIDWORKS Simulation, we can also create a surface model, which is an idealization of the design.

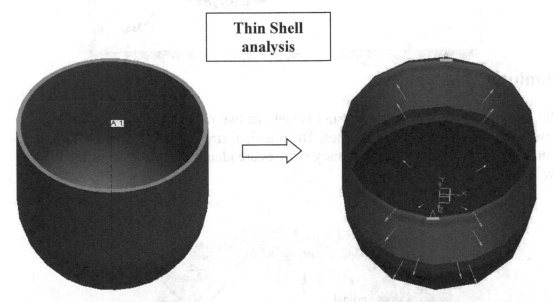

Besides selecting the proper type of elements to be used in the FEA analyses, options to simplify the FEA analyses, such as symmetry, should also be considered. Symmetry is an important characteristic that is often seen in designs. Symmetrical designs are generally more pleasing to the eye and also provide the desirable functionalities in designs. As it was illustrated in Chapter 11, symmetry can greatly reduce the computing time in FEA analyses.

Problem Statement

Determine the **tangential** and **longitudinal** stresses of the thin-wall cylindrical pressure vessel shown in the figure below. The pressure vessel is made of steel and is subject to an **internal pressure of 15MPa**. The dimensions of the vessel are **end radius 300mm**, **cylindrical length 500mm** and **wall thickness 25mm**.

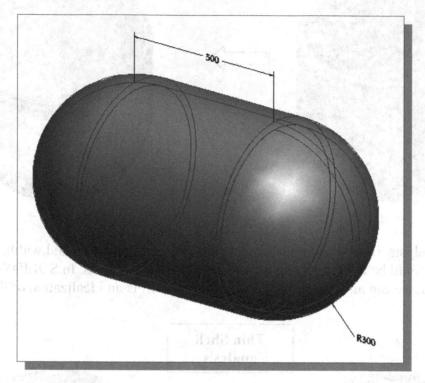

Preliminary Analysis

On the *cylindrical* section of the pressure vessel, the two principal stresses are (1) the **tangential** and (2) **longitudinal** stresses. The principal stresses on the *hemispherical* ends are **tangential stress**. These primary stresses are identified as shown in the figure below.

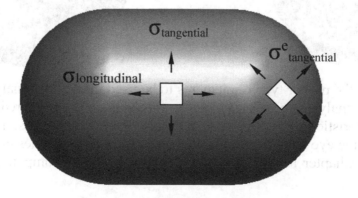

From *Strength of Materials*, the principal stresses on the **cylindrical section** of the pressure vessel walls, which are the tangential and longitudinal stresses, can be determined by

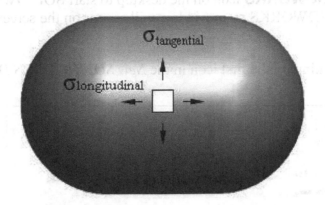

$$\sigma_{tangential} = \frac{Pd_i}{2t} = 15.0 \times 10^6 \times (2 \times 0.275)/(2 \times 0.025)$$

$$= 165 \text{ MPa}$$

$$\sigma_{longitudinal} = \frac{Pd_i}{4t} = 15.0 \times 10^6 \times (2 \times 0.275)/(4 \times 0.025)$$

$$= 82.5 \text{ MPa}$$

The principal stresses on the **hemispherical section** are tangential stresses, which can be determined by

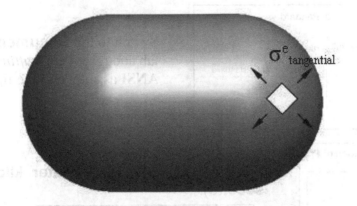

$$\sigma^e_{tangential} = \frac{Pd_i}{4t} = 15.0 \times 10^6 \times (2 \times 0.275)/(4 \times 0.025)$$

$$= 82.5 \text{ MPa}$$

In the following sections, we will perform three FEA analyses using two different types of FEA elements: **Shell** and **3D Solid** elements.

Start SOLIDWORKS

1. Select the **SOLIDWORKS** option on the *Start* menu or select the **SOLIDWORKS** icon on the desktop to start SOLIDWORKS. The SOLIDWORKS main window will appear on the screen.

2. Select **Part** by clicking on the first icon in the *New SOLIDWORKS Document* dialog box as shown.

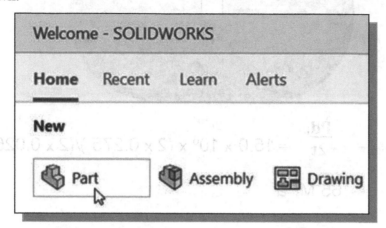

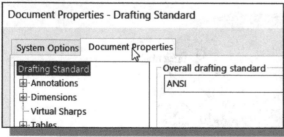

3. Select the **Options** icon from the *Menu* toolbar to open the *Options* dialog box.

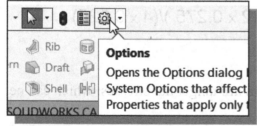

4. Switch to the **Document Properties** tab and reset the *Drafting Standard* to **ANSI** as shown in the figure.

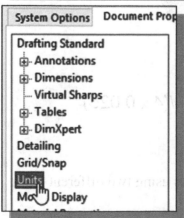

5. Set the *Unit system* to **MKS (meter, kilogram, second)**.

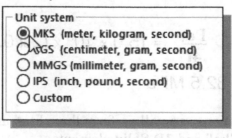

Create a 3D Solid Model in SOLIDWORKS

To perform the 3D Solids FEA analysis, we will first create a solid model using the **Revolve** command. The **Revolve** operation is defined as creating a solid by taking a planar section rotating about an axis of rotation.

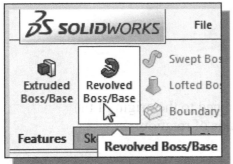

1. Click the **Revolved Boss/Base** icon, in the *Features* toolbar, to create a new revolved feature.

2. Click the **Right Plane**, in the graphics area, to align the sketching plane of our sketch as shown.

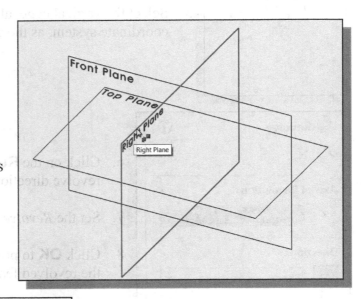

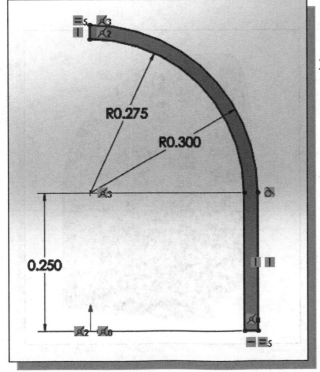

3. On your own, create the 2D sketch as shown. Note the top vertical edge, the bottom horizontal edge and the centers of the arcs are aligned to the origin of the coordinate system.

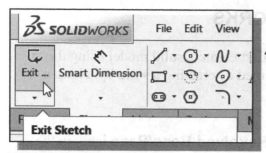

4. Click the **Exit sketch** icon in the *Sketch* toolbar to exit the *2D Sketch* mode.

5. Select the vertical edge, aligned to the origin of the coordinate system, as the axis of rotation.

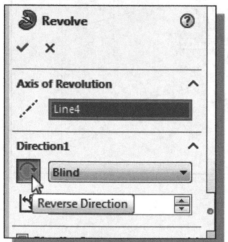

6. Click on the **Reverse Direction** icon to set the revolve direction into the screen if necessary.

7. Set the *Revolve Angle* to **90.00 deg**.

8. Click **OK** to proceed with the settings and create the revolved feature.

❖ Note that the symmetrical nature of the design makes it more effective to perform the FEA analysis with respect to the symmetry planes.

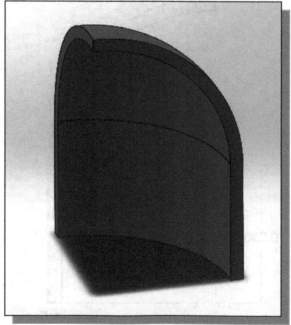

Activate the SOLIDWORKS Simulation Module

SOLIDWORKS Simulation is a multi-discipline Computer Aided Engineering (CAE) tool that enables users to simulate the physical behavior of a model, and therefore enables users to improve the design. SOLIDWORKS Simulation can be used to predict how a design will behave in the real world by calculating stresses, deflections, frequencies, heat transfer paths, etc.

The SOLIDWORKS Simulation product line features two areas of Finite Element Analysis: **Structure** and **Thermal**. *Structure* focuses on the structural integrity of the design, and *thermal* evaluates heat-transfer characteristics.

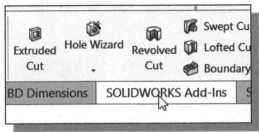

1. Start SOLIDWORKS Simulation by selecting the **SOLIDWORKS Add-Ins** tab in the *Command Manager* area as shown.

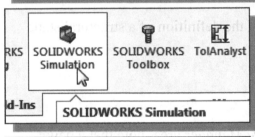

2. In the *SOLIDWORKS Office* list, choose **SOLIDWORKS Simulation** as shown.

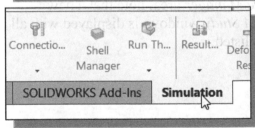

3. In the *Command Manager* area, choose **Simulation** as shown.

❖ Note that the SOLIDWORKS Simulation module is integrated as part of SOLIDWORKS. All of the SOLIDWORKS Simulation commands are accessible through the icon panel in the *Command Manager* area.

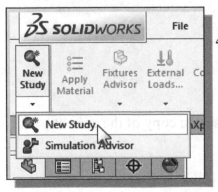

4. To start a new study, click the **New Study** item listed under the *Study Advisor* as shown.

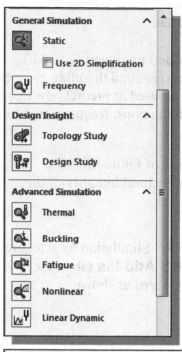

5. Select **Static** as the type of analysis to be performed with SOLIDWORKS Simulation.

❖ Note that different types of analyses are available, which include both structural static and dynamic analyses, as well as the thermal analysis.

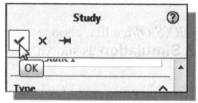

6. Click **OK** to start the definition of a structural static analysis.

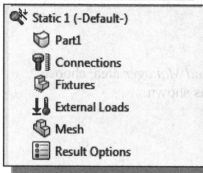

❖ In the *Feature Manager* area, note that a new panel, the *FEA Study* window, is displayed with all the key items listed.

❖ Also note the **Static 1** tab is activated, which indicates the use of the FEA model.

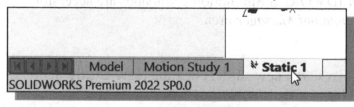

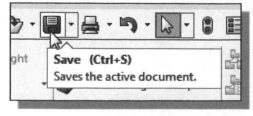

7. On your own, save a copy of the current model.

Assign the Element Material Property

Next, we will set up the *Material Property* for the elements. The *Material Property* contains the general material information, such as *Modulus of Elasticity*, *Poisson's Ratio*, etc. that is necessary for the FEA analysis.

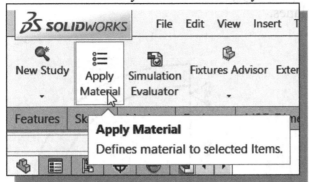

1. Choose **Apply Material** option as shown.

❖ Note the default list of materials, which are available in the pre-defined SOLIDWORKS Simulation material library, is displayed.

2. Select **Alloy Steel** in the *Material* list as shown.

3. Confirm the **Units** option to display **SI – N/m^2 (Pa)**.

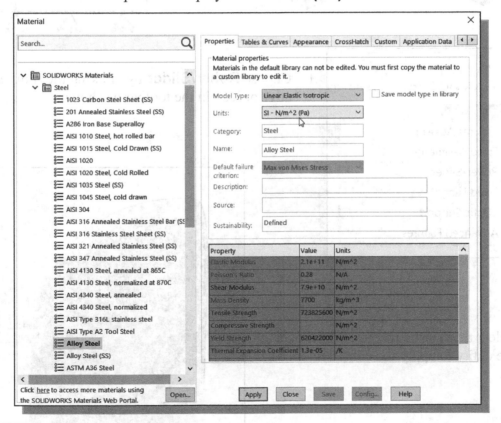

4. Click **Apply** to assign the material property, then click **Close** to exit the Material Assignment command.

Apply Boundary Conditions – Constraints

For our solid analysis, the model we created is a ⅛ solid model of the pressure container which is symmetrical about the horizontal mid-plane and two vertical planes. We will need to apply roller constraints to the three symmetry planes, motions not allowed in the perpendicular directions of the symmetry planes.

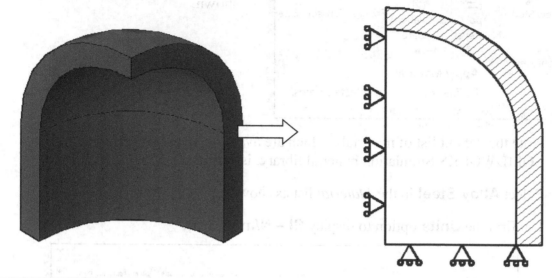

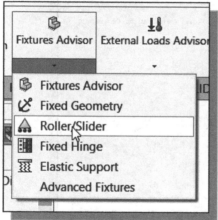

1. Choose **Roller/Slider** by clicking the fixtures advisor icon in the toolbar as shown.

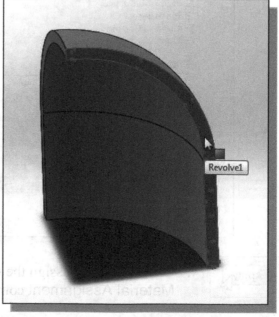

2. Select the **front surface** of the model as the entity to apply the constraint to.

3. On your own, rotate the 3D model to view the back plane as shown.

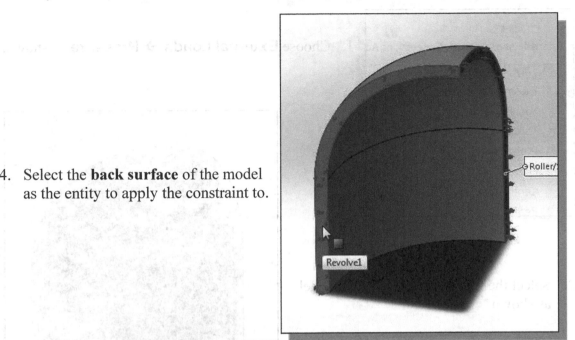

4. Select the **back surface** of the model as the entity to apply the constraint to.

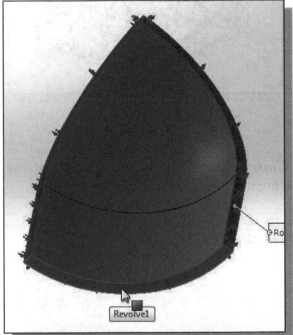

5. On your own, rotate and select the bottom surface as shown.

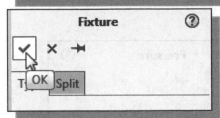

6. Click on the **OK** button to accept the Fixture constraint settings.

Apply the Pressure to the System

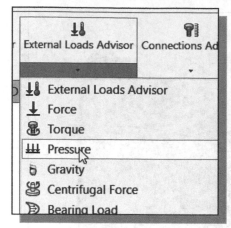

1. Choose **External Loads → Pressure** as shown.

2. Select the **inside surfaces** of the model as shown.

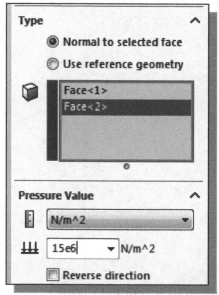

3. Set the *Units* to **N/m^2**, to match with the systems units we are using.

4. Enter **15e6** N/m^2 as the applied force.

5. Click on the **OK** button to accept the *Pressure* settings and create the load.

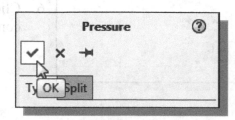

Create the first FEA Mesh – Coarse Mesh

As a rule, in creating the first FEA mesh in using the H-element approach, start with a relatively small number of elements and progressively move to more refined models. The main objective of the first coarse mesh analysis is to obtain a rough idea of the overall stress distribution.

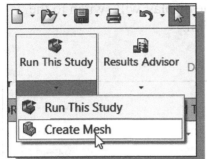

1. Choose **Create Mesh** by clicking the icon in the toolbar as shown.

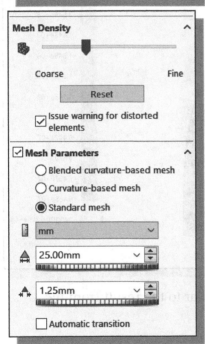

2. Switch on the **Mesh Parameters** options, to show the additional control options.

3. Set the mesh option to **Standard mesh**.

4. Set the *Units* to **millimeters** as shown.

5. Enter **25mm** as the *Global Element size*.

❖ In general, a good rule of thumb to follow in creating the first mesh is to have about 3 to 4 elements on the edges of the model. Since our model has a fairly small cross section, we will use the wall thickness as the element size.

6. Click on the **OK** button to accept the *Mesh* settings.

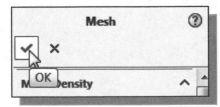

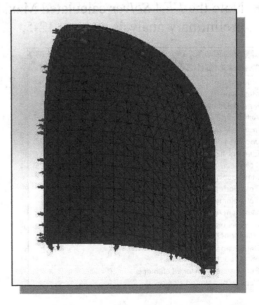

Run the Solver and view the results

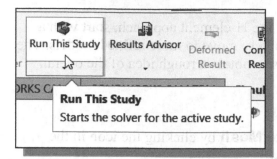

1. Click on the **Run This Study** button to start the *FEA Solver* to calculate the results.

❖ Once the solver has completed the calculations, the display will switch to the stress distribution.

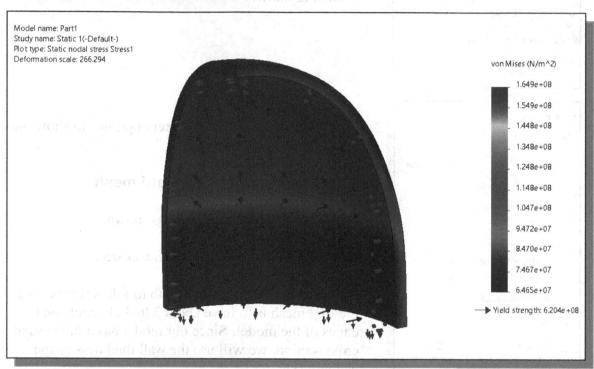

Model name: Part1
Study name: Static 1(-Default-)
Plot type: Static nodal stress Stress1
Deformation scale: 266.294

von Mises (N/m^2)

1.649e+08
1.549e+08
1.448e+08
1.348e+08
1.248e+08
1.148e+08
1.047e+08
9.472e+07
8.470e+07
7.467e+07
6.465e+07

➤ Yield strength: 6.204e+08

❖ Note the *FEA Solver* calculated Max. Stress is very similar to the result from the preliminary analysis on page 12-5.

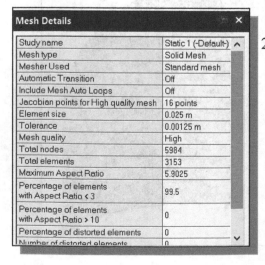

Mesh Details	✕
Study name	Static 1 (-Default-)
Mesh type	Solid Mesh
Mesher Used	Standard mesh
Automatic Transition	Off
Include Mesh Auto Loops	Off
Jacobian points for High quality mesh	16 points
Element size	0.025 m
Tolerance	0.00125 m
Mesh quality	High
Total nodes	5984
Total elements	3153
Maximum Aspect Ratio	5.9025
Percentage of elements with Aspect Ratio < 3	99.5
Percentage of elements with Aspect Ratio > 10	0
Percentage of distorted elements	0
Number of distorted elements	0

2. On your own, examine the mesh details. Current mesh consists of **5984** nodes and **3153** elements.

Refinement of the FEA Mesh – Global Element Size 12.5

To confirm the FEA results are valid, we will adjust the mesh by using a *Global Element size* of 12.5mm.

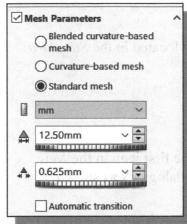

1. On your own, adjust the mesh size to **12.5mm** as shown.

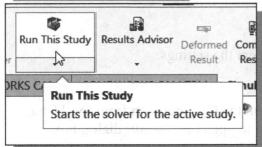

2. Click on the **Run This Study** button to start the *FEA Solver* to calculate the results.

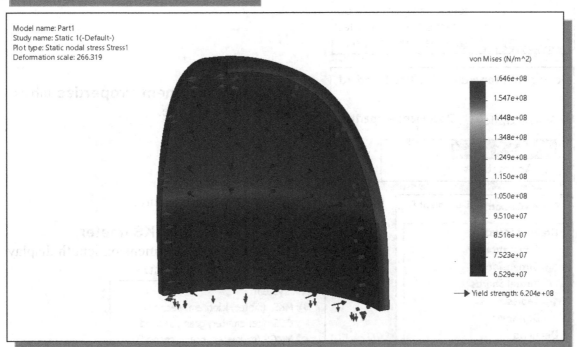

❖ The *FEA Solver* calculated the Max. Von Mises Stress with the refinement to be **164.6MPa**, which is almost the same as the previous mesh. This confirms the FEA is adequate for the model.

Start a New 3D Surface Model

To perform the 3D FEA thin shell analysis, we will first create a new surface model.

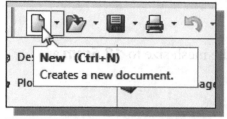

1. Click on the **New** icon, located in the *Standard* toolbar as shown.

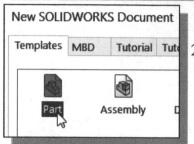

2. Select **Part** by clicking on the first icon in the *New* SOLIDWORKS *Document* dialog box as shown.

3. Click on the **OK** button to accept the settings.

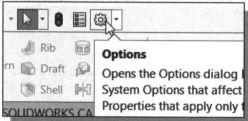

4. Click on the **Options** icon from the *Menu* toolbar to open the *Options* dialog box.

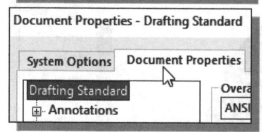

5. Select the **Document Properties** tab as shown in the figure.

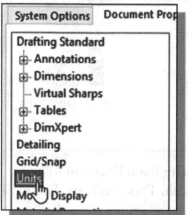

6. Click **Units** as shown in the figure.

7. On your own, set the *Units* to **MKS (meter, kilogram, second)** and the dimension length display to 4 digits after the decimal point.

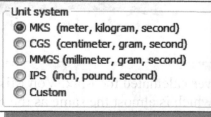

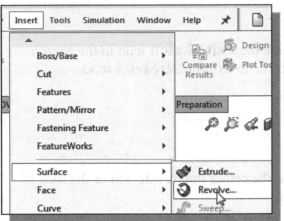

8. In the pull-down menu, select **Insert→ Surface → Revolve**.

9. Click the **Front Plane**, in the graphics area, to align the sketching plane of our sketch as shown.

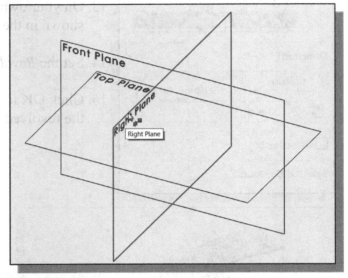

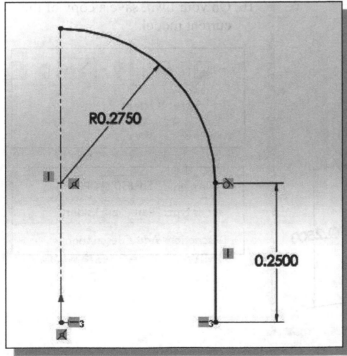

10. On your own, create the 2D sketch as shown. Note the sketch consists of an **arc** connected to a **vertical line** and also a **vertical center line**. Also, note the alignment between the top endpoint, the center point, and the bottom endpoint to the origin.

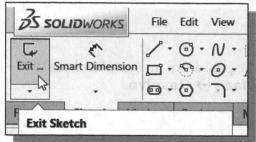

11. Click the **Exit Sketch** icon in the *Sketch* toolbar to exit the *2D Sketch* mode.

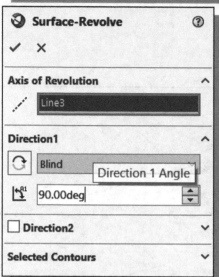

12. Confirm the center line is selected as the axis of rotation.

13. On your own, adjust the rotation direction as shown in the below figure.

14. Set the *Revolve Angle* to **90.00 deg** as shown.

15. Click **OK** to proceed with the settings and create the revolved feature.

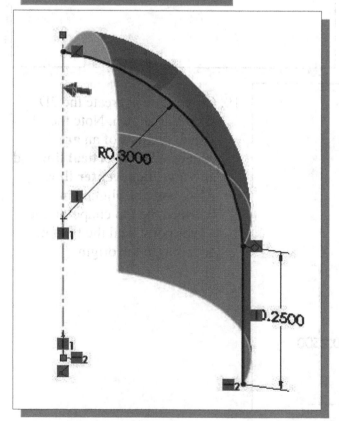

16. On your own, save a copy of the current model.

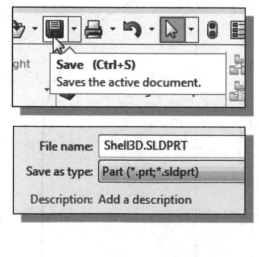

Start a New FEA Study

1. To start a new study, click the **New Study** item listed under the *Study Advisor* as shown.

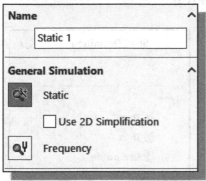

2. Select **Static** as the type of analysis to be performed with SOLIDWORKS Simulation.

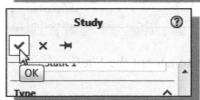

3. Click **OK** to start the definition of a structural static analysis.

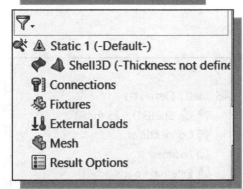

❖ In the *Feature Manager* area, note that a new panel, the *FEA Study* window, is displayed with all the key items listed.

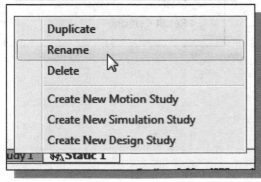

4. Right-click once on the **Static 1** tab to display the option list and select **Rename** as shown.

5. On your own, rename the FEA study to **Shell**.

Completing the Definition of the Surface Model

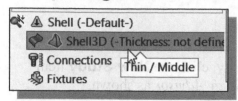

1. Click on the model name in the study window and notice the thickness of the model has not been defined.

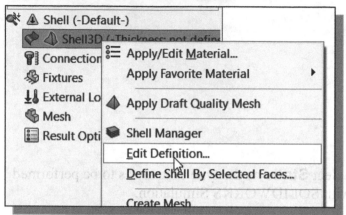

2. Right-click once on the model name to bring up the option list and select **Edit Definition** as shown.

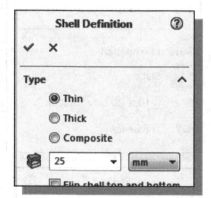

3. Confirm the *Type* option is set to **Thin** as shown.

4. Enter **25 mm** as the surface thickness as shown.

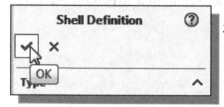

5. Click on the **OK** button to accept the *Shell Definition* settings.

6. Notice the warning icon is removed in the study window, indicating the surface model is fully defined.

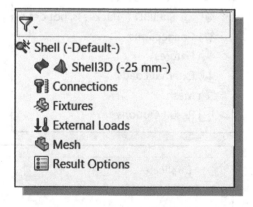

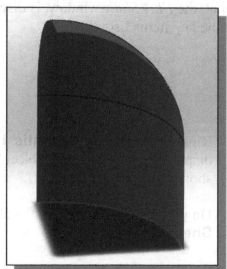

Assign the Element Material Property

Next, we will set up the *Material Property* for the elements. The *Material Property* contains the general material information, such as *Modulus of Elasticity*, *Poisson's Ratio*, etc. that is necessary for the FEA analysis.

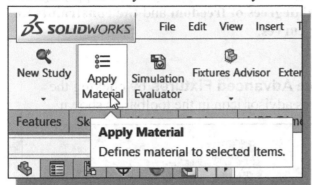

1. Choose the **Apply Material** option as shown.

❖ Note the default list of materials, which are available in the pre-defined SOLIDWORKS Simulation material library, is displayed.

2. Select **Alloy Steel** in the *Material* list as shown.

3. Set the **Units** option to display **SI – N/m^2 (Pa)**.

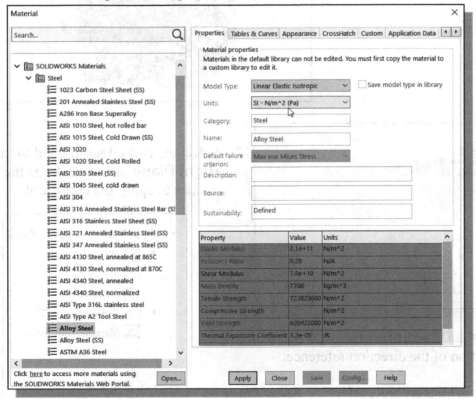

4. Click **Apply** to assign the material property then click **Close** to exit the Material Assignment command.

Apply Boundary Conditions – Constraints

For our solid analysis, the model we created is a ⅛ solid model of the pressure container, which is symmetrical about the horizontal mid-plane and two vertical planes. The most important differences between 3D shell analysis and 3D solid analysis is in applying the constraints; **shell elements have rotational degrees of freedom** and **the constraints for shell models must be applied to edges or curves.**

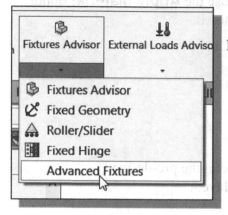

1. Choose **Advanced Fixtures** by clicking the fixtures advisor icon in the toolbar as shown.

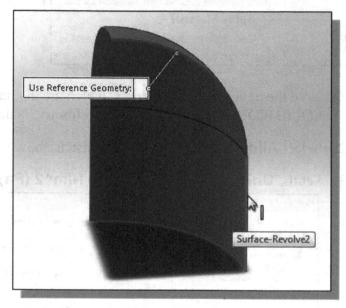

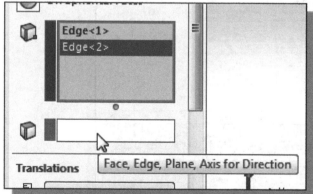

2. Select the **two edges** aligned to the **Right Plane** of the model as the entities to apply the constraint.

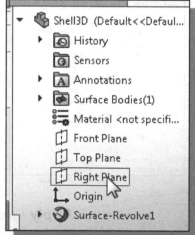

3. Click in the **Direction reference** box to activate the selection of the direction reference.

4. In the model history tree, select the *Front Plane* of the CAD model as the direction reference.

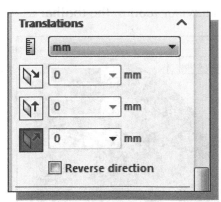

5. Set the distance measurement to **millimeters**, to match with the systems units we are using.

6. In the *Translation constraints* list, click on the **Normal to Plane** icon to activate the constraint.

7. Set the *Normal to Plane* distance to **0** as shown.

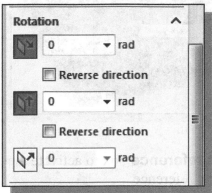

8. In the *Rotation constraints* list, click on the **Along Plane Dir 1** and **Along Plane Dir 2** icons to activate the constraint.

9. Set the *Along Plane Dir 1* and *Along Plane Dir 2* angles to **0** as shown.

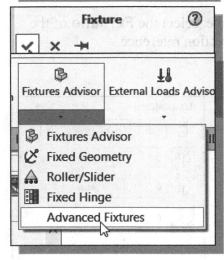

10. Click on the **OK** button to accept the Fixture constraint settings.

11. Choose **Advanced Fixtures** by clicking the fixtures advisor icon in the toolbar as shown.

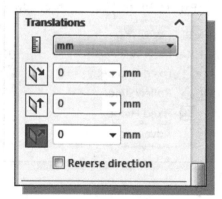

12. Select the **two edges** on the *left side* of the model as the entities to apply the constraint to.

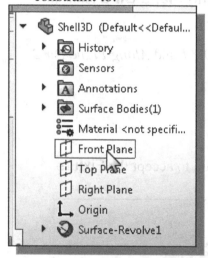

13. Click in the **Direction reference** box to activate the selection of the direction reference.

14. In the model history tree, select the **Ft Plane** of the CAD model as the direction reference.

15. Set the distance measurement to **millimeters**, to match with the systems units we are using.

16. In the *Translations constraints* list, click on the **Normal to Plane** icon to activate the constraint.

17. Set the *Normal to Plane* distance to **0** as shown.

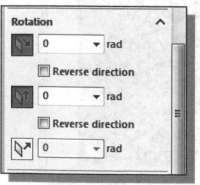

18. In the *Rotation constraints* list, click on the **Along Plane Dir 1** and **Along Plane Dir 2** icons to activate the constraint.

19. Set the *Along Plane Dir 1* and *Along Plane Dir 2 angles* to **0** as shown.

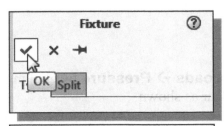

20. Click on the **OK** button to accept the Fixture constraint settings.

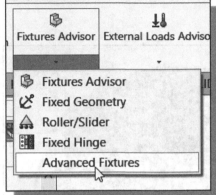

21. Choose **Advanced Fixtures** by clicking the fixture icon in the toolbar as shown.

22. Select the **inside edge** on the *bottom surface* of the model as the entity to apply the constraint to.

23. On your own, set the **Top Plane** of the CAD model as the direction reference.

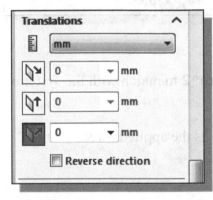

24. Set the distance measurement to **millimeters** to match with the systems units we are using.

25. In the *Translations constraints* list, click on the **Normal to Plane** icon to activate the constraint.

26. Set the *Normal to Plane* distance to **0** as shown.

27. In the *Rotation constraints* list, click on the **Along Plane Dir 1** and **Along Plane Dir 2** icons to activate the constraint.

28. Set the *Along Plane Dir 1* and *Along Plane Dir 2* angles to **0** as shown.

29. Click on the **OK** button to accept the settings.

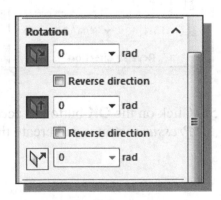

Apply the Pressure to the system

1. Choose **External Loads → Pressure** by clicking the icon in the toolbar as shown.

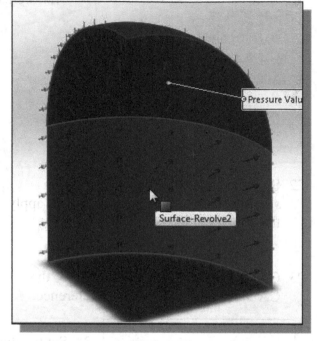

2. Select the **inside faces** of the model as shown.

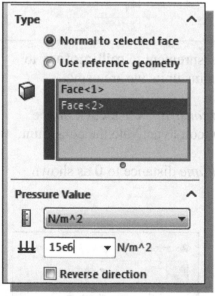

3. Set the *Units* to **N/m^2** to match with the systems units we are using.

4. Enter *15e6* N/m^2 as the applied force.

5. Click on the **OK** button to accept the *Pressure* settings and create the load.

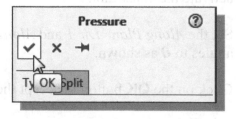

Create the first FEA Mesh – Coarse Mesh

As a rule, in creating the first FEA mesh in using the H-element approach, start with a relatively small number of elements and progressively move to more refined models. The main objective of the first coarse mesh analysis is to obtain a rough idea of the overall stress distribution.

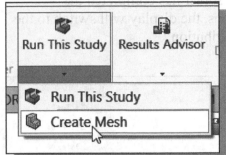

1. Choose **Create Mesh** by clicking the icon in the toolbar as shown.

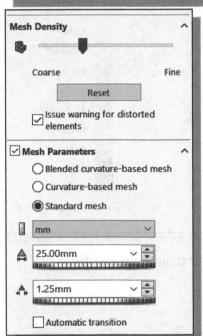

2. Switch on the **Mesh Parameters** option, to show the additional control options.

3. Set the *mesh* to **Standard mesh** and *Units* to **millimeters** as shown.

4. Enter **25** millimeters as the *Global element size*.

❖ We are using the same element size as the previous 3D solid analysis; this will allow us to compare the two analyses.

5. Click on the **OK** button to accept the *Mesh* settings.

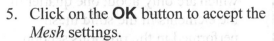

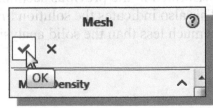

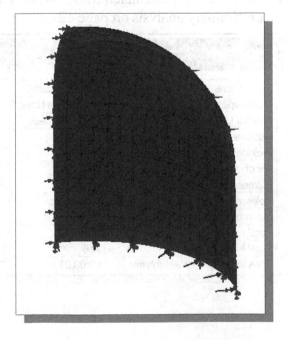

Run the Solver and view the results

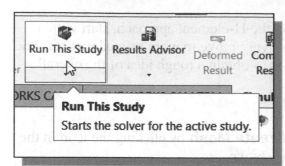

1. Click on the **Run This Study** button to start the *FEA Solver* to calculate the results.

❖ Once the solver has completed the calculations, the display will switch to the stress distribution.

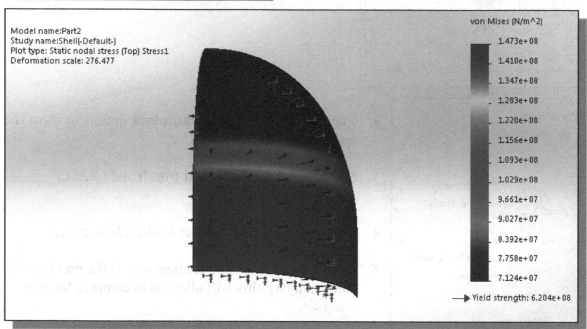

Model name:Part2
Study name:Shell(-Default-)
Plot type: Static nodal stress (Top) Stress1
Deformation scale: 276.477

von Mises (N/m^2)

1.473e+08
1.410e+08
1.347e+08
1.283e+08
1.220e+08
1.156e+08
1.093e+08
1.029e+08
9.661e+07
9.027e+07
8.392e+07
7.758e+07
7.124e+07

Yield strength: 6.204e+08

❖ Note the FEA calculated Max. Stress is about 11% lower than the result from the preliminary analysis on page 12-5.

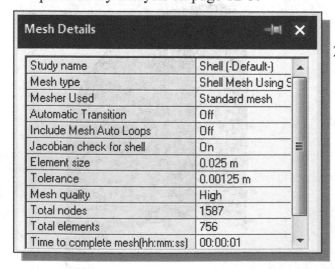

Mesh Details	
Study name	Shell (-Default-)
Mesh type	Shell Mesh Using S
Mesher Used	Standard mesh
Automatic Transition	Off
Include Mesh Auto Loops	Off
Jacobian check for shell	On
Element size	0.025 m
Tolerance	0.00125 m
Mesh quality	High
Total nodes	1587
Total elements	756
Time to complete mesh(hh:mm:ss)	00:00:01

2. On your own, examine the mesh details. Current mesh consists of **1587** nodes and **756** elements, which are only about one quarter of the elements in the solid mesh performed in the previous section. This also indicates the solution time is much less than the solid analysis.

Refinement of the FEA Mesh – Global Element Size 10.0

The shell model, without the small cross section, allows us to create a more uniform mesh; we will adjust the mesh to a global element size of 10mm.

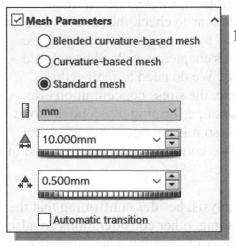

1. On your own, adjust the mesh size to **10mm** as shown.

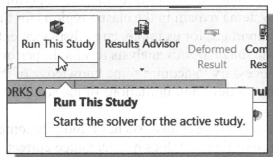

2. Click on the **Run This Study** button to start the *FEA Solver* to calculate the results.

Run This Study
Starts the solver for the active study.

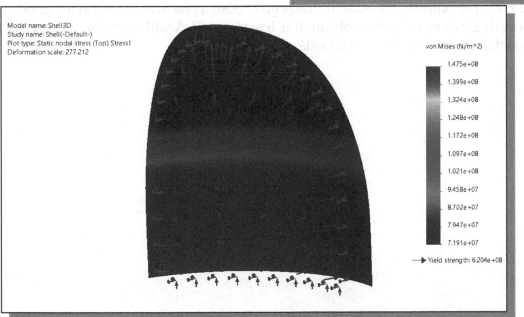

Model name: Shell3D
Study name: Shell(-Default-)
Plot type: Static nodal stress (Top) Stress1
Deformation scale: 277.212

von Mises (N/m^2)

1.475e+08
1.399e+08
1.324e+08
1.248e+08
1.172e+08
1.097e+08
1.021e+08
9.458e+07
8.702e+07
7.947e+07
7.191e+07

→ Yield strength: 6.204e+08

❖ The *FEA Solver* calculated the Max. Von Mises Stress with the refinement to be **147.5MPa**, which is about the same as the previous mesh. Note the refinement generated 4624 elements, which is about 6 times more elements than the previous mesh. No more refinement is needed as the results showed a good convergence.

Notes on FEA Linear Static Analyses

In examining the FEA results, one should first examine the deformed shape to check for proper placement of boundary conditions and to determine if the calculated deformation of the model is reasonable. It is always important to perform a *convergence study* to obtain more accurate results. Besides using hand calculations to check the results of the FEA analyses, it is also quite feasible to check the results by using other element types. For example, the curved-beam problem, illustrated in this chapter, can also be analyzed using the 1D beam elements and/or 2D surface elements. We do need to realize that different elements have different limitations; for example, the stress concentration effects are not present with the beam elements. But the purpose of performing a second and/or a third analysis is to assure the results of the first analysis, so it is not necessary to expect all element types to produce exactly the same results. This concept is further illustrated in the next chapter.

It should be emphasized that, when performing FEA analysis, besides confirming that the systems remain in the elastic regions for the applied loading, other considerations are also important; for example, large displacements and buckling of beams can also invalidate the *linear statics* analysis results. In performing finite element analysis, it is also necessary to acquire some knowledge of the theory behind the method and understand the restrictions and limitations of the software. There is no substitution for experience.

Finite element analysis has rapidly become a vital tool for design engineers. However, use of this tool does not guarantee correct results. The design engineer is still responsible for doing approximate calculations, using good design practice, and applying good engineering judgment to the problem. It is hoped that FEA will supplement these skills to ensure that the best design is obtained.

Questions:

1. For designs that are thin and symmetrical about an axis, what are the different FEA analyses available in SOLIDWORKS Simulation?

2. What are the advantages and disadvantages of performing a shell analysis over a 3D solid analysis?

3. In SOLIDWORKS Simulation, how do you set up the FEA model to be a surface model?

4. What are the requirements to apply constraints on thin shell models?

5. How do you specify the thickness of a thin shell model?

6. What are the differences in applying constraints on thin shell models vs. solid models?

Exercises:

1. Determine the **maximum Von Mises stress** of the thin-wall round-bottom cylindrical pressure vessel shown in the figure below; dimensions are in mm. The pressure vessel is made of steel and is subject to an **internal pressure of 10MPa**. For the FEA analyses, set all degrees of freedom to **Fixed** at the top edge of the opening. (The Radius 225 and Radius 200 arcs share the center point, which is measured 200 mm along the vertical axis. The Radius 265 and Radius 235 arcs also share the same center point at the origin.)

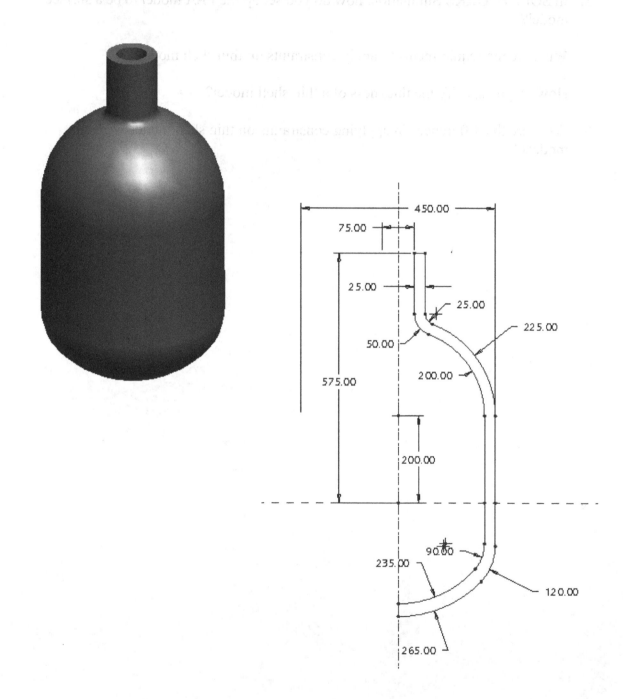

2. Determine the **maximum Von Mises stress** of the thin-wall flat-bottom cylindrical pressure vessel shown in the figure below; dimensions are in inches. The pressure vessel is made of steel and is subject to an **internal pressure of 45Psi**. For the FEA analyses, set all degrees of freedom to **Fixed** at the top edge of the opening.

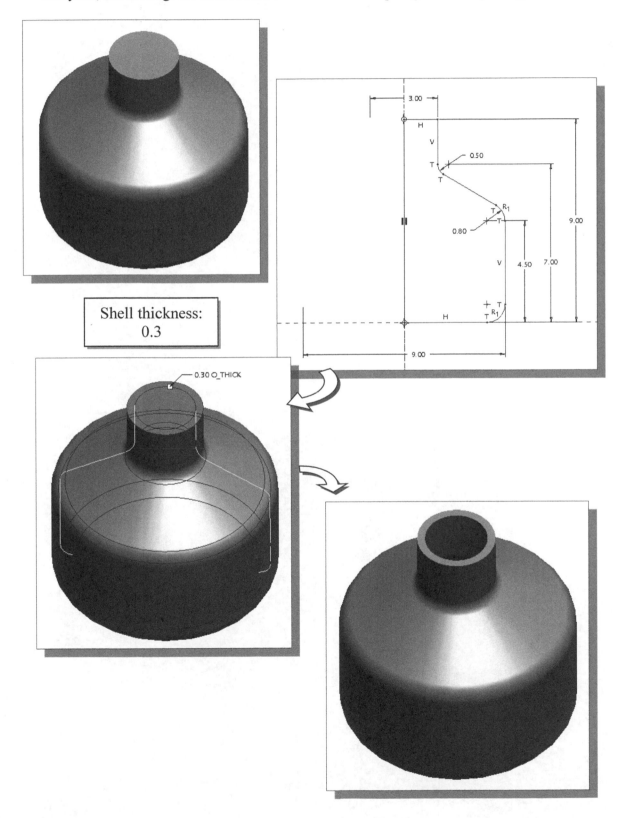

Shell thickness:
0.3

Notes:

Chapter 13
FEA Static Contact Analysis

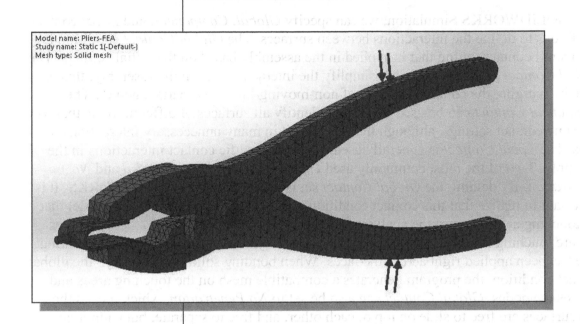

Model name: Pliers-FEA
Study name: Static 1(-Default-)
Mesh type: Solid mesh

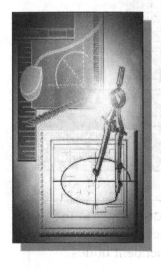

Learning Objectives

- **Understand the Basic concepts and procedures of Contact Analysis.**
- **Perform FEA Contact Analysis on Assembly Model.**
- **Set up the Contact sets with Global Contact and Local Contact Settings.**
- **Use the Animate option to view the FEA results.**
- **Refine the mesh by using Apply Mesh Control option.**

Introduction

The most critical aspect in performing a stress analysis on an assembly is to accurately represent how the parts interact with each other. In finite element analysis, this is typically referred to as *Contact Analysis*.

With SOLIDWORKS Simulation, we can specify *Global*, *Component*, and *Local* contact conditions to define the interactions between surfaces. The *Global Contact* condition is the overall contact setting that is applied in the assembly based on the initial settings. The *Global Contact* is typically used to simplify the interaction within the assembly; this is done by merging the contacting nodes of non-moving/low stress parts together. The *Component Contact* can be used to quickly identify all surfaces of different parts that use the same contact settings, although this may result in many unnecessary interactions of parts. The *Local Contact* is generally used to define specific contact interactions in the assembly. Two of the most commonly used *Contact* conditions are *Bonded* and *No Penetration*. By default, the *Global Contact* set is set to *Bonded* in SOLIDWORKS. It is important to realize that this contact condition only applies to surfaces in the model that are touching at the start of the analysis. If the contact setting is set to *Bonded*, all faces that are touching are assumed to be rigidly locked together, almost as if perfectly strong glue has been applied right across the faces. When bonding solid faces through the global contact condition, the program generates a compatible mesh on the touching areas and merges the nodes. *Global Contact* can also be set to *No Penetration*, which means the two surfaces are free to slide on top of each other, and free to separate, but cannot go through each other. We can also use the *Shrink Fit* option of *local contact* to define a shrink fit contact condition between initially interfering components.

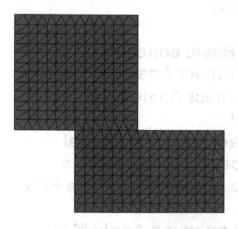

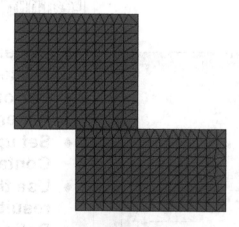

Bonded: Nodes at the touching surfaces of the two parts are merged.

No Penetration: Nodes at the touching surfaces of the two parts are independent nodes.

Contact is a common source of nonlinearity. Although nonlinear studies are typically used to solve the majority of contact problems, SOLIDWORKS allows us to perform **static** studies to solve contact problems with small and large displacements. If nonlinearity other than that caused by contact is present, static studies should not be used. Common sources of nonlinearity are material properties and changing loads and restraints. The properties dialogs of **static** and **nonlinear** studies provide an option to use large displacements. Use the small displacement formulation only if the expected motions are small and the parts are independently stable in directions other than the primary contact direction. One should also realize that *contact analysis* is calculation intensive. The more contact surface sets defined in an assembly, the more calculations are required.

When a small gap exists between two faces in an assembly, no matter how small a gap, the global contact doesn't apply any contact condition to those faces. Therefore, if a small tolerance gap between two objects exists, no contact will be recognized, and if a load is applied to one of the objects, this object will pass right through the other object. However, it is good engineering practice to design with appropriate tolerance gaps just like parts fit together in real life.

The first concern of performing a contact analysis is to identify where the global contact condition is being applied. To do this, the SOLIDWORKS **interference detection** tool can be used. This tool will show us all the interfaces in the model where faces are touching. If there's an interface between two parts that doesn't show up in this interference check, it won't be recognized by the *Global Contact*.

With SOLIDWORKS, multiple approaches are feasible to perform FEA analyses on assemblies. However, it is important to go through the entire assembly carefully, considering the interactions between each part, and make sure they have been accurately represented.

Problem Statement

Determine the effects of the **pliers** on the **Cast Iron fork**.

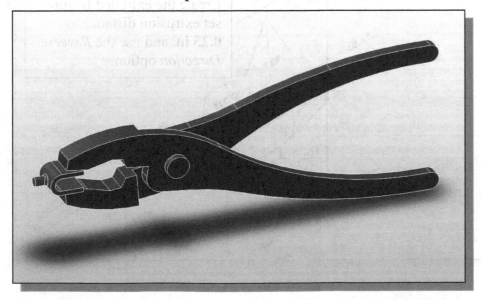

Parts

- Three parts are required for the assembly: (1) **Pliers-Jaw** (Alloy Steel SS), (2) **Pin** (Alloy Steel SS), and (3) **Fork** (Malleable Cast iron). Create the three parts as shown below, then save the models as separate part files: *Pliers-Jaw*, *Pin*, and *Fork*. (Close all part files or exit SOLIDWORKS after you have created the parts.)

(1) Pliers-Jaw (Alloy Steel SS) (Note that the teeth in the jaw surface are removed to simplify the analysis.)

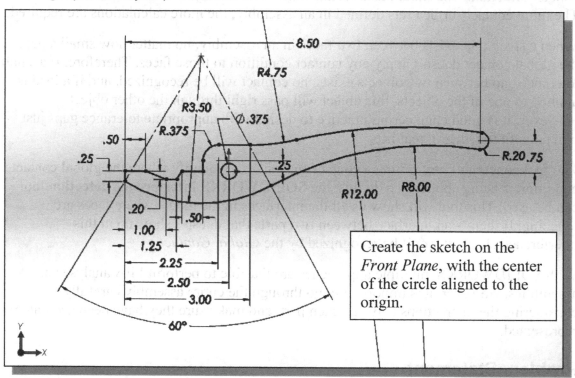

Create the sketch on the *Front Plane*, with the center of the circle aligned to the origin.

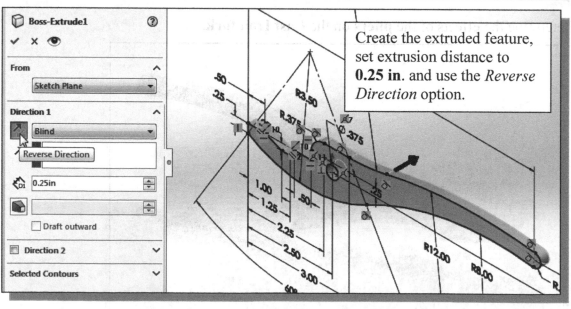

Create the extruded feature, set extrusion distance to **0.25 in**. and use the *Reverse Direction* option.

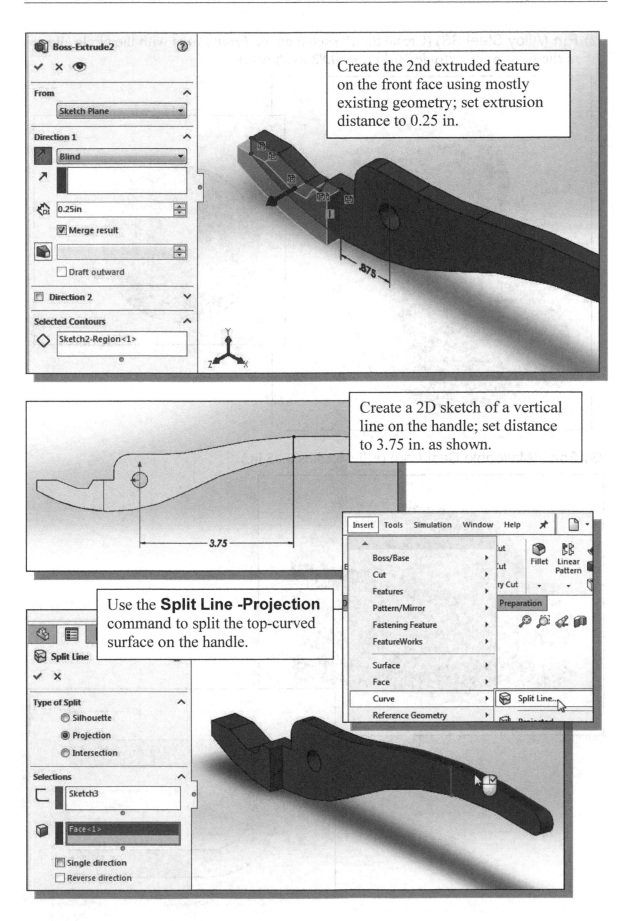

Create the 2nd extruded feature on the front face using mostly existing geometry; set extrusion distance to 0.25 in.

Create a 2D sketch of a vertical line on the handle; set distance to 3.75 in. as shown.

Use the **Split Line -Projection** command to split the top-curved surface on the handle.

(2) Pin (Alloy Steel SS) (Create the 1st sketch on the *Front Plane* with the circle aligned to the *Origin* and extrude *Mid-Plane* **0.83** as shown.)

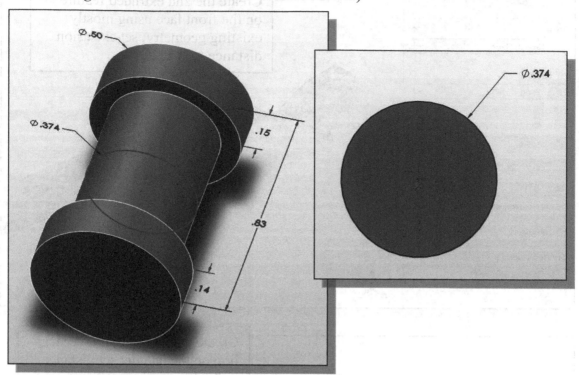

(3) Fork (Malleable Cast Iron) (Fillet Radius 0.05 in.)

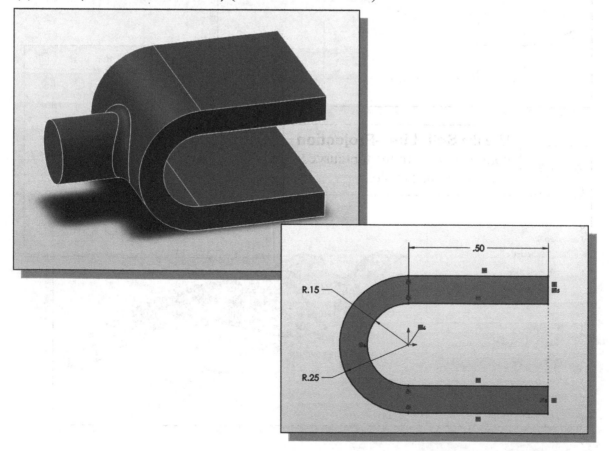

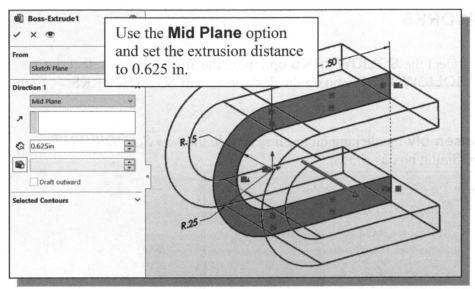

Use the **Mid Plane** option and set the extrusion distance to 0.625 in.

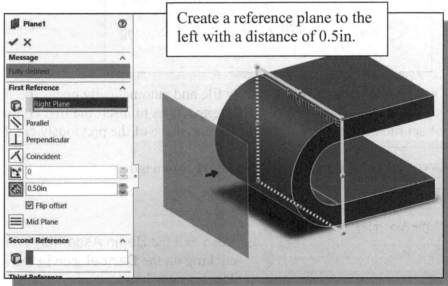

Create a reference plane to the left with a distance of 0.5in.

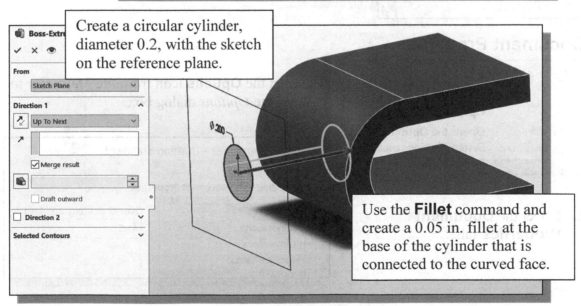

Create a circular cylinder, diameter 0.2, with the sketch on the reference plane.

Use the **Fillet** command and create a 0.05 in. fillet at the base of the cylinder that is connected to the curved face.

Start SOLIDWORKS

1. Select the **SOLIDWORKS** option on the *Start* menu or select the **SOLIDWORKS** icon on the desktop to start SOLIDWORKS.

2. Select **Assembly** by clicking on the first icon in the *New SOLIDWORKS Document* dialog box as shown.

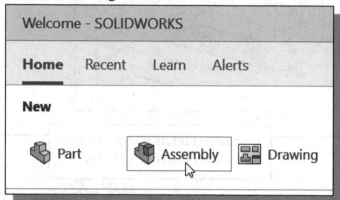

❖ SOLIDWORKS opens an assembly file and automatically opens the *Begin Assembly Property Manager*. SOLIDWORKS expects us to insert the first component. We will first set the document properties to match those of the previously created parts.

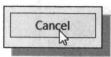

3. Cancel the **Open Part** command.

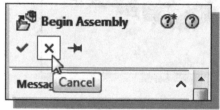

4. Also exit the **Begin Assembly** command by clicking on the **Cancel** icon in the *Property Manager* as shown.

Document Properties

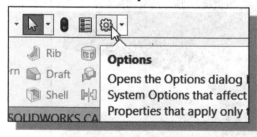

1. Select the **Options** icon from the *Menu Bar* to open the *Options* dialog box.

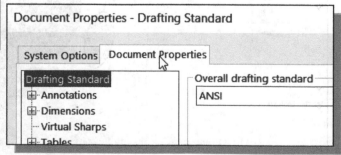

2. Select the **Document Properties** tab.

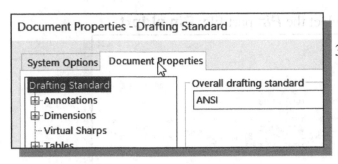

3. Select **ANSI** in the pull-down selection window under the *Overall drafting standard* panel to reset to the default settings.

4. Click **Units**, select the **IPS (inch, pound, second)** unit system, and select **.123** in the *Decimals* spin box for the *Length* units as shown.

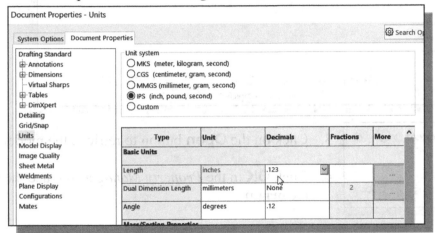

5. Click **OK** in the *Document Properties* window to accept the settings.

Insert the First Component

The first component inserted in an assembly should be a fundamental part or subassembly. The first component in an assembly file sets the orientation of all subsequent parts and subassemblies. The origin of the first component is aligned to the origin of the assembly coordinates and the part is grounded (all degrees of freedom are removed). The rest of the assembly is built on the first component, the ***base component***. In most cases, this *base component* should be one that is **not likely to be removed** and **preferably a non-moving part** in the design. Note that there is no distinction in an assembly between components; the first component we place is usually considered the *base component* because it is usually a fundamental component to which others are constrained. We can change the base component if desired. For our project, we will use the ***Pin*** as the base component in the assembly.

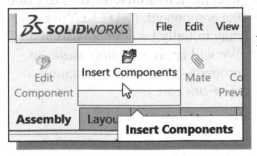

1. In the *Assembly* toolbar, select the **Insert Components** command by left-clicking the icon.

2. Use the browser to locate and select the **Pin** part file: **Pin.sldprt**.

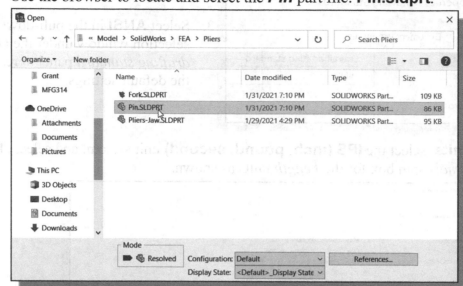

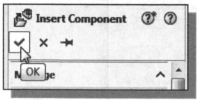

3. Click on the **Open** button to retrieve the model.

4. Click **OK** in the *Property Manager* to place the *Pin* at the **origin**.

Insert the Second Component

We will retrieve the *Fork* part as the second component of the assembly model.

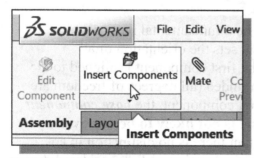

1. In the *Assembly* toolbar, select the **Insert Components** command by left-clicking the icon.

2. Select the **Fork** design (part file: **Fork.sldprt**) in the browser. And click on the **Open** button to retrieve the model.

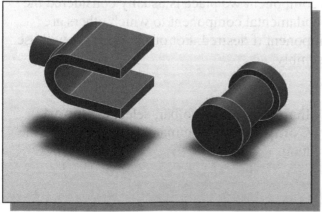

3. Place the **Fork** to the left of the Pin, as shown in the figure. Click once with the **left-mouse-button** to place the component. (**NOTE:** Your *part* may initially be oriented differently. We will apply *assembly mates* to properly control the orientation of the *Fork* part.)

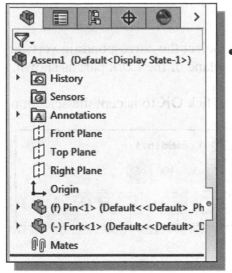

- Inside the *Feature Manager Design Tree*, the retrieved parts are listed in their corresponding order. The **(f)** in front of the *Pin* part signifies the part is **fixed** and all *six degrees of freedom* are restricted. The number behind the filename is used to identify the number of copies of the same component in the assembly model.

Assembly Mates

To assemble components into an assembly, we need to establish the assembly relationships between components. **Assembly Mates** create a parent/child relationship that allows us to capture the design intent of the assembly. Because the component that we are placing actually becomes a child to the already assembled components, we must use caution when choosing mate types and references to make sure they reflect the intent.

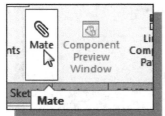

1. In the *Assembly* toolbar, select the **Mate** command by left-clicking once on the icon.

2. Inside the graphics window, click on the arrow symbol in front of *Assem1* and select the **Front Plane** of the *Pin* part as shown.

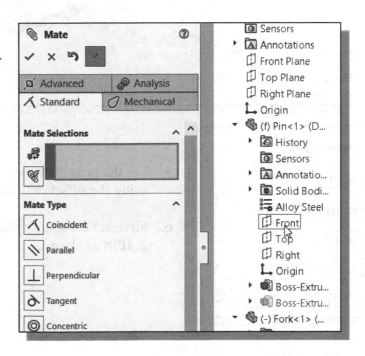

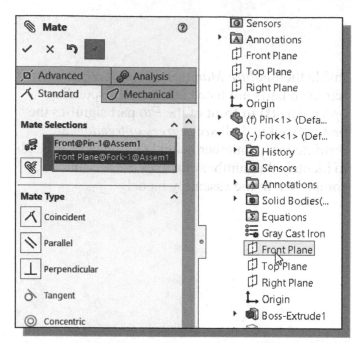

3. Select the corresponding vertical plane of the *Fork* part as shown.

4. Click **OK** to accept the selection.

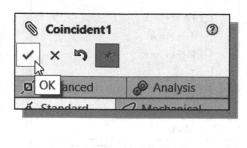

5. On your own, repeat the above steps and align the horizontal datum planes as shown.

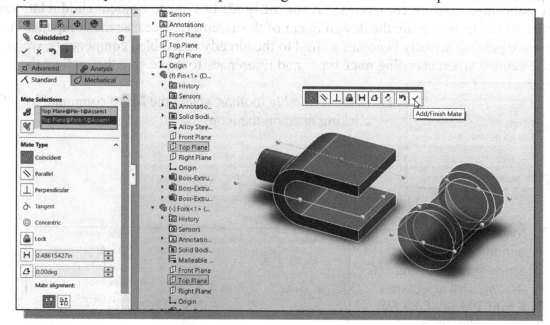

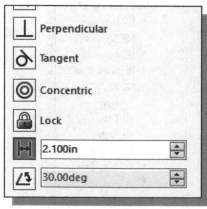

- For the next mate, we will align the Right Planes using the offset option.

6. First, set the *Mate* option to **Distance** and enter **2.10in** as shown.

7. Select the **Right Plane** of the *Pin* part and one of the rectangular faces of the *Fork* part as shown. (Use the *Flip Mate* option to reverse the direction if necessary.)

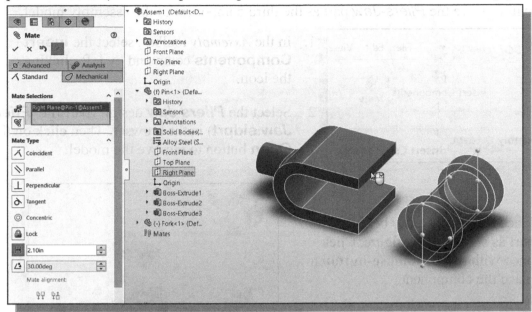

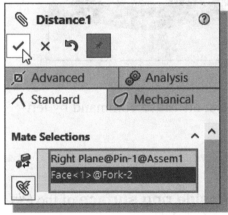

8. Click **OK** to position the *Fork* part to the left at the distance specified relative to the *Pin* as shown.
- Note that both parts are fully constrained; these two pieces are the two non-moving parts in the assembly.

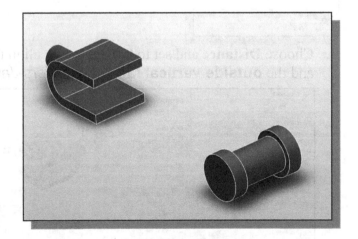

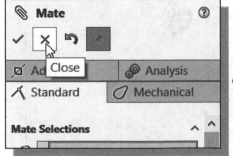

9. Click **Close** to end the **Mate** command as shown.

Insert the Third Component

We will retrieve the *Pliers-Jaw* part as the third component of the assembly model.

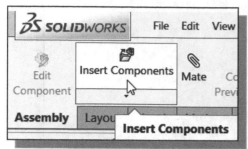

1. In the *Assembly* toolbar, select the **Insert Components** command by left-clicking on the icon.

2. Select the **Pliers-Jaw** design (part file: **Pliers-Jaw.sldprt**) in the browser. Then click on the **Open** button to retrieve the model.

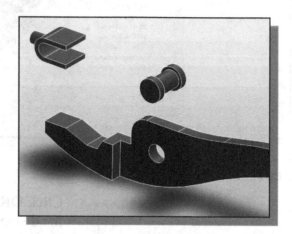

3. Place the **Pliers-Jaw** below the *Pin* part as shown in the figure. Click once with the **left-mouse-button** to place the component.

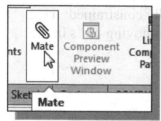

4. In the *Assembly* toolbar, select the **Mate** command by left-clicking once on the icon.

5. Choose **Distance** and set to **0.01inch** then align the **inside ring surface** of the *Pin* and the **outside vertical face** of the *Pliers-Jaw* part as shown.

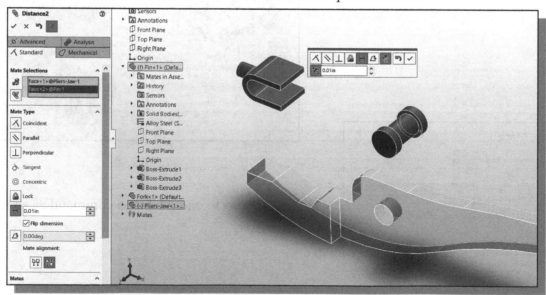

6. On your own, apply a **Tangent** constraint to the cylindrical surface of the *Pin* part and the hole of the *Pliers-Jaw* part. (Use the *Flip Mate* option if necessary.)

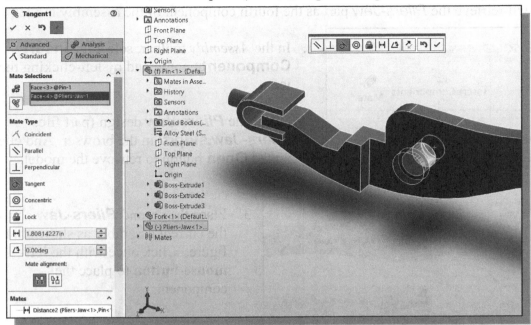

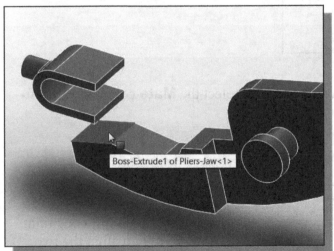

7. On your own, exit the *Mate* command and use the *drag & drop* approach to reposition the Jaw piece as shown.

8. On your own, add another **Coincident** constraint; first select the ***Jaw* surface** as shown.

9. Select the **bottom edge** of the *Fork* part as shown. The lower jaw is now fully constrained.

10. Click **OK** to accept the constraint then exit the **Mate** command.

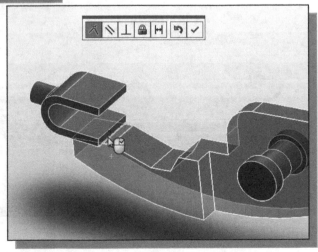

Insert the Upper Jaw Component

We will retrieve the *Pliers-Jaw* part as the fourth component of the assembly model.

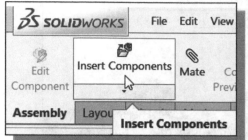

1. In the *Assembly* toolbar, select the **Insert Components** command by left-clicking the icon.

2. Select the *Pliers-Jaw* design (part file: *Pliers-Jaw*.**sldprt**) in the browser. And click on the **Open** button to retrieve the model.

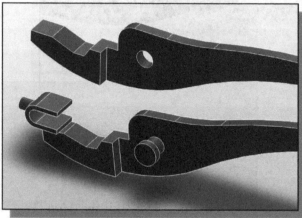

3. Place the second *Pliers-Jaw* above the current assembly as shown in the figure. Click once with the **left-mouse-button** to place the component.

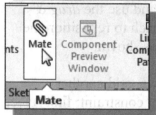

4. In the *Assembly* toolbar, select the **Mate** command by left-clicking once on the icon.

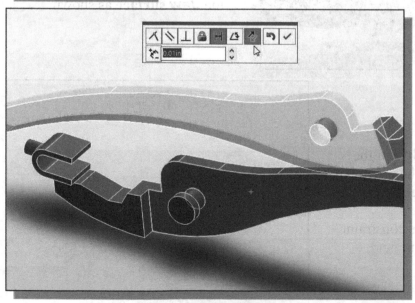

5. Apply a **Distance** (0.01in) constraint and align the two inside handle surfaces of the two pliers-jaw parts as shown. (Hint: Use the **Flip Mate Alignment** icon to flip the part.)

6. Click **OK** to accept the constraint then exit the **Mate** command.

7. Apply a **Tangent** constraint to the cylindrical surface of the *Pin* part and the hole of the second *Pliers-Jaw* part. (Hint: Use the **Flip Mate Alignment** icon to flip the part if necessary.)

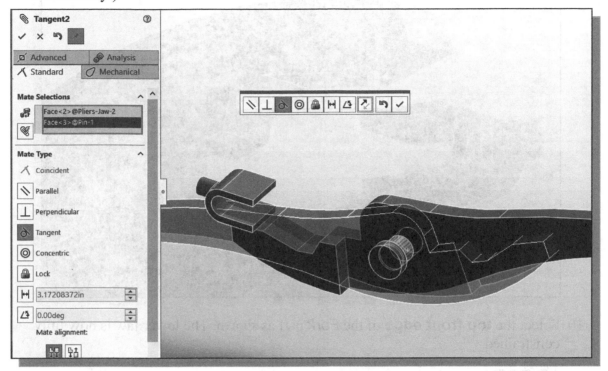

8. Reposition the upper jaw part roughly as shown; this can be done by using the **drag and drop** option available in SOLIDWORKS.

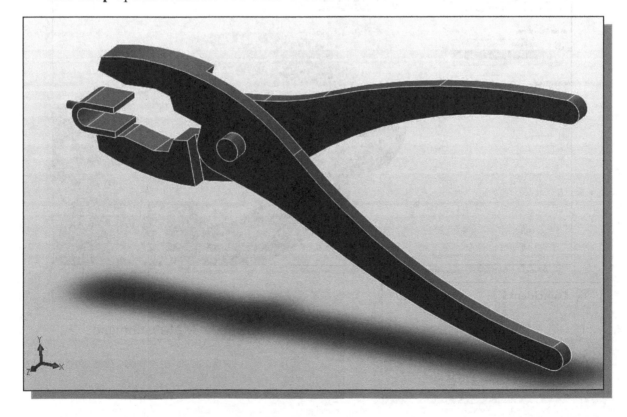

9. For the next **Coincident** constraint, first select the ***Jaw* surface** as shown.

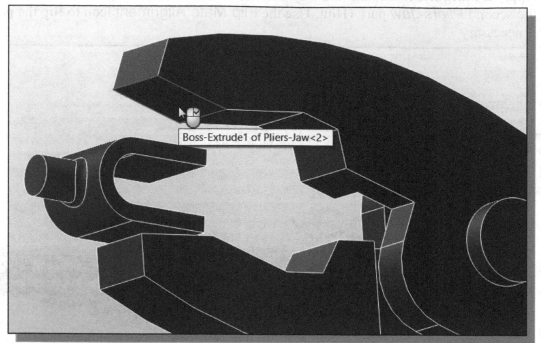

10. Select the **top front edge** of the *Fork* part as shown. The lower jaw is now fully constrained.

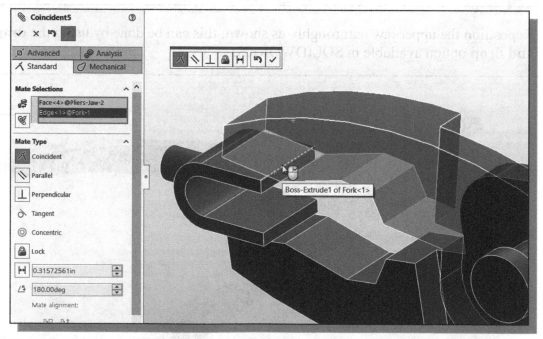

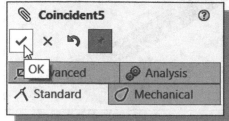

11. Click **OK** to accept the **Mate** command.

Identifying Coincident Surfaces in the Model

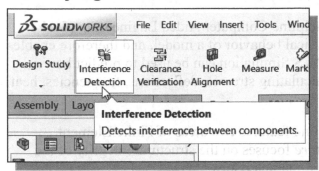

1. In the *Evaluate* toolbar, select the **Interference Detection** command as shown.

2. In the *Properties Manager*, switch **on** the **Treat coincidence as interference** option as shown.

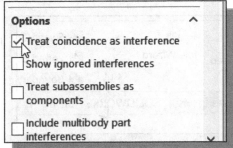

3. Click on the **Calculate** button to detect any interference existing in the current model.

4. Confirm the four **Coincident** locations, as listed in the *Results* window, are the ones we have established.

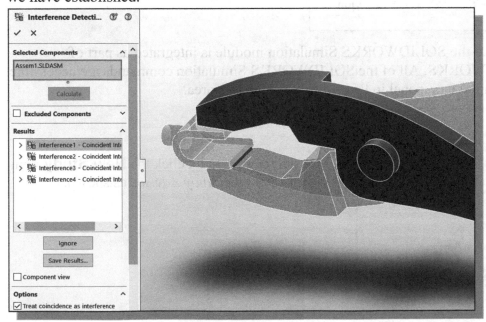

Activate the SOLIDWORKS Simulation Module

SOLIDWORKS Simulation is a multi-discipline Computer Aided Engineering (CAE) tool that enables users to simulate the physical behavior of a model, and therefore enables users to improve the design. SOLIDWORKS Simulation can be used to predict how a design will behave in the real world by calculating stresses, deflections, frequencies, heat transfer paths, etc.

The SOLIDWORKS Simulation product line features two areas of Finite Element Analysis: **Structure** and **Thermal**. *Structure* focuses on the structural integrity of the design, and *thermal* evaluates heat-transfer characteristics.

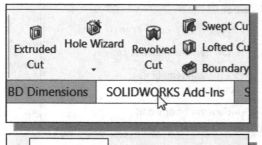

1. Start SOLIDWORKS Simulation by selecting the **SOLIDWORKS Add-Ins** tab in the *Command Manager* area as shown.

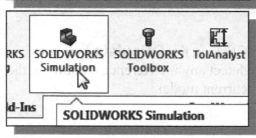

2. In the *Command Manager* toolbar, choose **SOLIDWORKS Simulation** as shown.

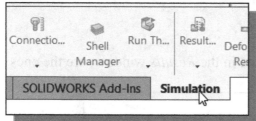

3. In the *Command Manager* area, choose **Simulation** as shown.

❖ Note that the SOLIDWORKS Simulation module is integrated as part of SOLIDWORKS. All of the SOLIDWORKS Simulation commands are accessible through the icon panel in the *Command Manager* area.

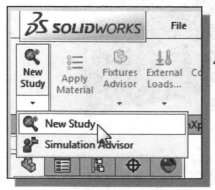

4. To start a new study, click the **New Study** item listed under the *Study Advisor* as shown.

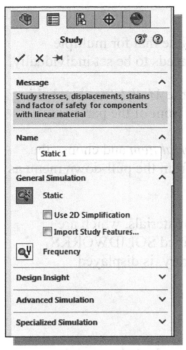

5. Select **Static** as the type of analysis to be performed with SOLIDWORKS Simulation.

❖ Note that different types of analyses are available, which include both structural static and dynamic analyses, as well as the thermal analysis.

6. Click **OK** to start the definition of a structural static analysis.

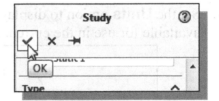

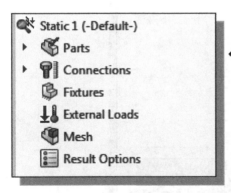

❖ In the *Feature Manager* area, note that a new panel, the *FEA Study* window, is displayed with all the key items listed.

❖ Also note the **Static 1** tab is activated, which indicates the use of the FEA model.

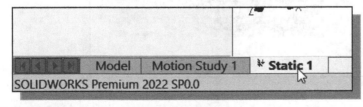

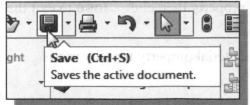

7. On your own, save a copy of the current assembly model.

Assign the Element Material Property

Next, we will set up the *Material Property* for the elements. Note that for multiple components in the model, the material property for each part needs to be set individually.

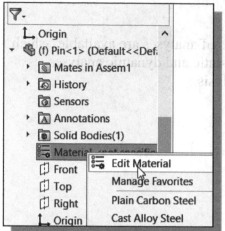

1. Expand the *Pin* part in the *Model Tree* by clicking on the arrow symbol in front of the part name.

2. Right-click on the *material item* and choose the **Edit Material** option from the pull-down menu as shown.

❖ Note the default list of materials, which are available in the pre-defined SOLIDWORKS Simulation material library, is displayed.

3. Select **Alloy Steel (SS)** in the *Material* list as shown.

4. Set the **Units** option to display **English (IPS)** to make the selected material available for use in the current FEA model.

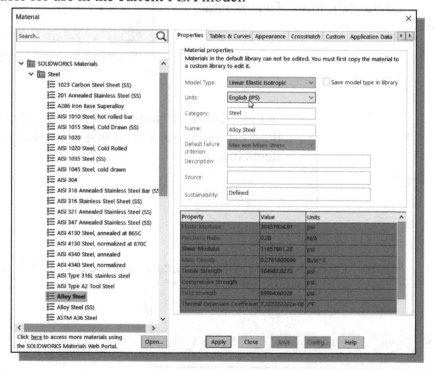

5. Click **Apply** to assign the material property then click **Close** to exit the Material Assignment command.

6. On your own, repeat the above steps and set the material property for all parts as shown on pages 13-4 and 13-5.

Apply Boundary Conditions – Constraints

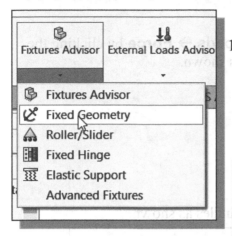

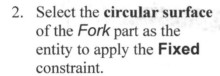

1. Choose **Fixed Geometry** by clicking the triangle icon below the Fixtures Advisor icon as shown.

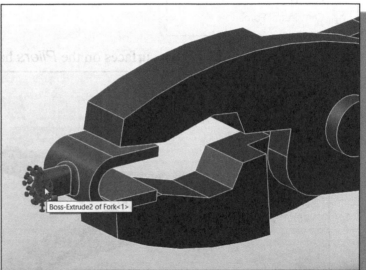

2. Select the **circular surface** of the *Fork* part as the entity to apply the **Fixed** constraint.

3. Also select the two end faces of the *Pin* part to apply the **Fixed** constraint as shown.

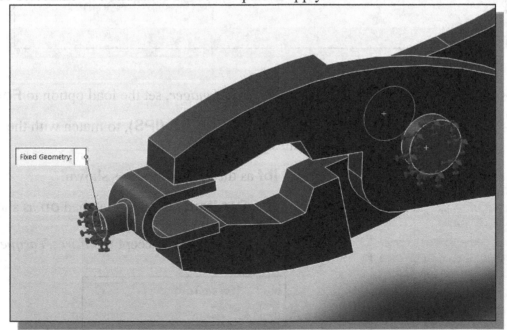

- Note the applied fixture constraints will allow the two *Pliers-Jaws* to pivot about the *Pin*, which will result in the clamping of the *Fork* part.

Apply the External Load on the Handles

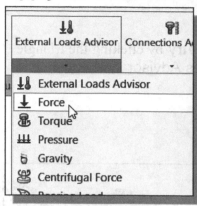

1. Choose **External Loads → Force** by clicking the icon in the toolbar as shown.

2. Select the **top and bottom** surfaces on the *Pliers* handles as shown.

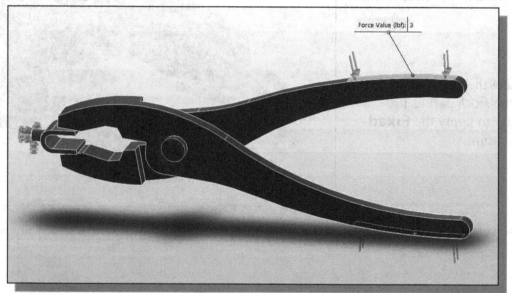

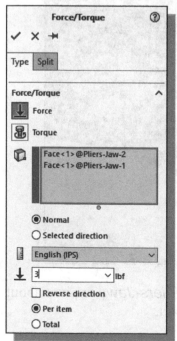

3. In the *Property Manager*, set the load option to **Force**.

4. Set the *Units* to **English (IPS)**, to match with the systems units we are using.

5. Enter **3 lbf** as the applied force as shown.

6. Confirm the **Per Item** option is switched **on** as shown.

7. Click on the **OK** button to accept the *Force/Torque* settings

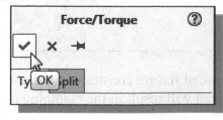

Global Contact Settings

By default, the *Global Contact* condition is set to **Bonded**. Note that three options, **Bonded**, **No Penetration** and **Allow Penetration**, are available.

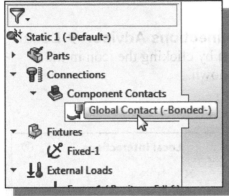

1. In the *Property Manager*, expand the connections list by clicking on the triangle in front of the items as shown.

- Note the default *Global Contact* condition is set to **Bonded**, as shown in the expanded list.

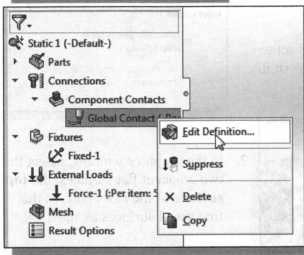

2. In the *Property Manager*, right-click once on the *Global Contact* item to bring up the option menu and select the **Edit Definition** option as shown.

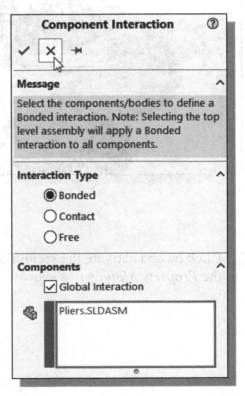

- Note the three contact types, **Bonded**, **Contact** and **Free**, are available for selection.

3. Click **Cancel** to exit the *Component Contact* option dialog box without making any changes.

Set up Specific Local Surfaces Interaction

Local Interaction sets can be created by selecting surfaces that are currently in contact or will be in contact as the loads are applied.

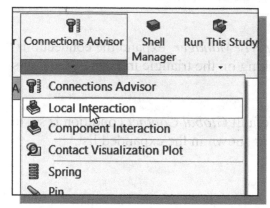

1. Choose **Connections Advisor → Local Interaction** by clicking the icon in the toolbar as shown.

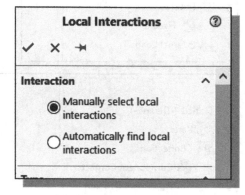

- Note the default *Interaction* condition is set to *Manually select local interactions* as shown in the option list.

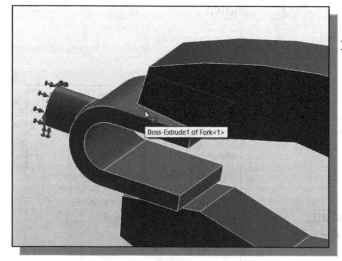

2. In the graphics window, select the two adjacent flat surfaces, the **top surface** of the *Fork* part, as the first set of surfaces as shown.

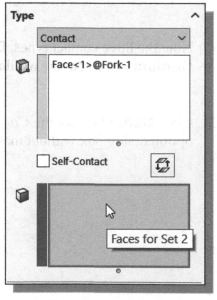

3. Click on and activate the **second selection box** in the *Property Manager* as shown.

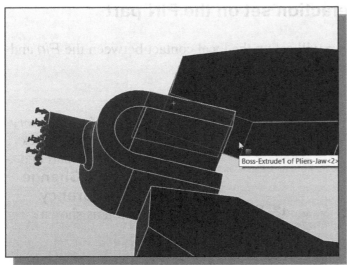

4. Select the related flat face on the upper jaw of the *Pliers-Jaw* part as shown. (Hint: Use the **Dynamic Rotate** option to adjust the display first.)

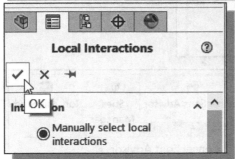

5. In the *Local Interactions* dialog box, click **OK** to accept the setting.

6. On your own, repeat the above steps and create another contact set with the bottom surface of the *Fork* part and the lower *Pliers-Jaw* as shown.

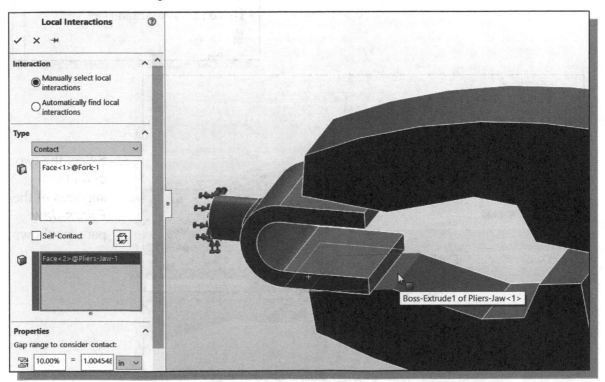

Set up another Surface Interaction set on the PIN part

For the next *Surface Interaction Set*, we will set up the local contact between the *Pin* and the two *Pliers-Jaw* parts.

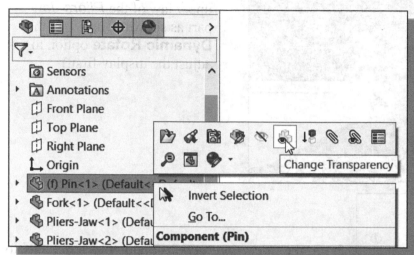

1. In the *Model History Tree* window, click on the **Pin** part and select the **Change Transparency** option as shown.

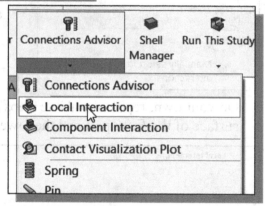

2. Choose **Connections Advisor → Local Interaction** by clicking the icon in the toolbar as shown.

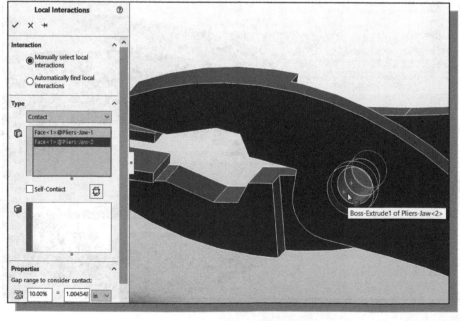

3. Select the two cylindrical surfaces of the *Pliers-Jaw* parts as shown.

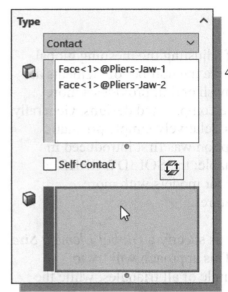

4. Click on and activate the **second selection box** in the *Property Manager* as shown.

5. Rotate and zoom-in the assembly and select the inside cylindrical surface of the *Pin* part as shown. (Hint: You can also use the right-click **Select Other** option.)

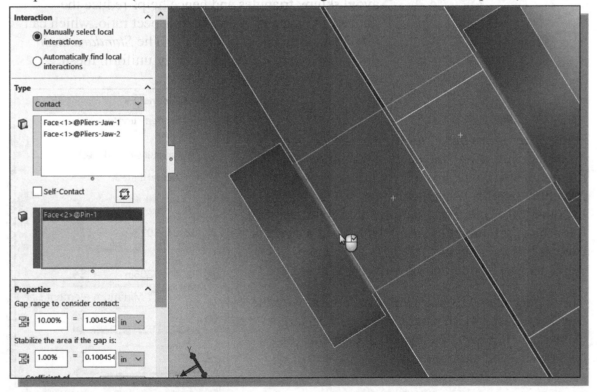

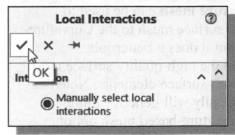

6. Click **OK** to accept the settings and create the contact set.

SOLIDWORKS Curvature-Based Mesh

In the previous chapter, we have examined the effects of adjusting the meshing global size and also using the local mesh control option to increase the number of elements at high stress area. The SOLIDWORKS Curvature-based mesh option provides a more advanced option to control the sizes of the mesh for more complicated designs. Generally speaking, the standard mesh is best used for flat surfaces, relatively simple prismatic geometry. The SOLIDWORKS curvature-based mesh option was first introduced in 2008; and in 2016, another meshing option became available: the SOLIDWORKS Blended-Curvature mesh. This technique is better suited for models with more complicate geometry containing 3D complex surface features.

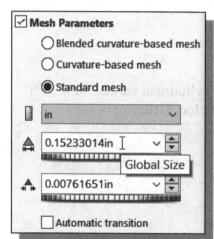

The **Standard mesh** lets us specify a *Global Element Size* and a *Tolerance value*. This approach will try to maximize the smallest angle of all triangles, while the geometry is being split into element triangles or tetrahedrons. In essence, the *Standard mesh* tends to avoid skinny triangles and hence helps reduce the number of elements with a high aspect ratio, which can affect the accuracy of our results. The *Standard mesh* does a great job of creating a fairly uniform mesh.

With a **Curvature based mesh**, the surface element size is determined by the entered Maximum and minimum sizes; which is also adjusted by the minimum number of elements that fit in a hypothetical circle. For example, the default value of 8 for the minimum number of elements in a circle is used; if this value is increased 16, the number of elements will be doubled on curved edges.

The surface elements sizes are less uniform with the Curvature based mesh; for simple surfaces, larger elements will be created and small elements for complex curved surfaces.

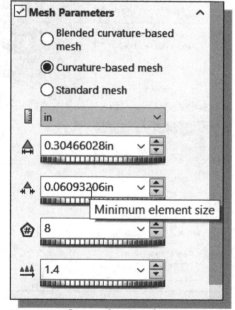

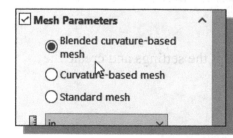

The **Blended-Curvature mesh** can be used to generate very similar surface mesh to the Curvature-based mesh option, but it does a better job transitioning between the high-quality surface mesh and the less refined sub-surface elements. Note that this mesh option generally will generate more elements than the Curvature-based mesh option.

Create the FEA Mesh

For the frequency analysis, it is still necessary to confirm the convergence by using a coarse mesh as the first FEA mesh for the analysis.

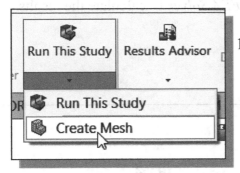

1. Choose **Create Mesh** by clicking the icon in the toolbar as shown.

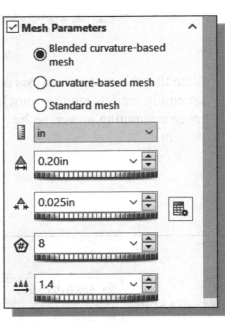

2. Switch **on** the **Mesh Parameters** option to show the additional control options.

3. Set the *Mesh* to **Blended curvature-based mesh** and *Units* to **inches** as shown.

4. Enter *0.20* inch as the *Global element size*.

5. Enter *0.025* inch as the *Size tolerance*.

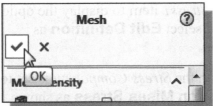

6. Click on the **OK** button to accept the *Mesh* settings. (The mesh option generated 11481 nodes & 6108 elements.)

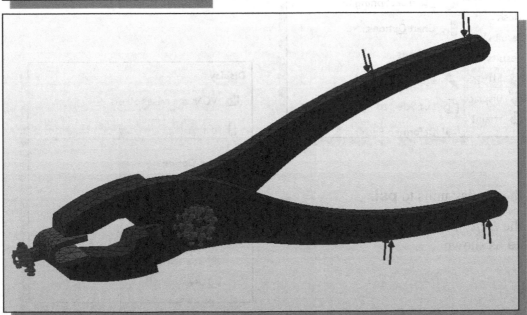

Run the Solver and View the Results

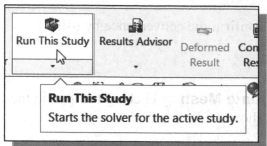

1. Click on the **Run This Study** button to start the *FEA Solver* to calculate the results.

- Note that the contact analysis on assembly models requires much more computing power, so be patient.

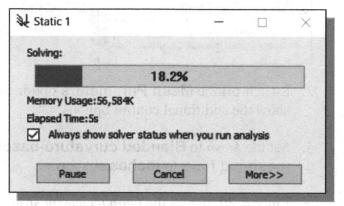

2. In the *FEA Study* window, right-click on the *Stress1* item to display the option list and select **Edit Definition** as shown.

3. Confirm the *Stress Component* is set to **VON: Von Mises Stress** as shown.

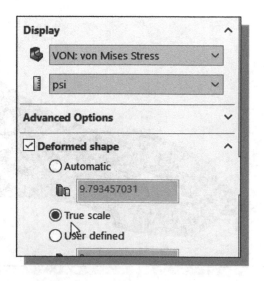

4. Set the display units to **psi**.

5. Set the *Deformed Shape* option to **True scale** as shown.

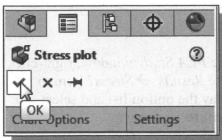

6. Click **OK** to display the results.

7. Right click on the *Stress1 (-vonMises-)* and choose **Show** to show the **Stress result**.

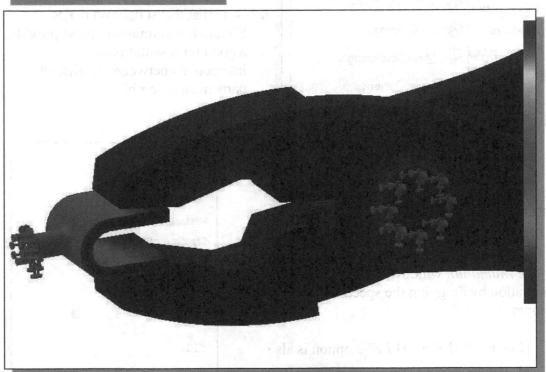

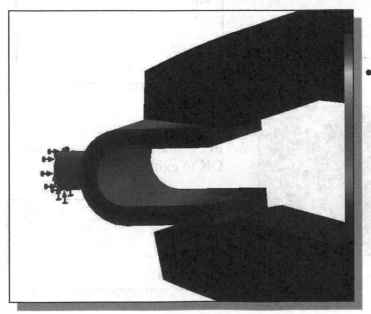

- Note that most of the higher stresses occurred on the curved surfaces of the *Fork* part and near the Pin part.

Use the Animate Option

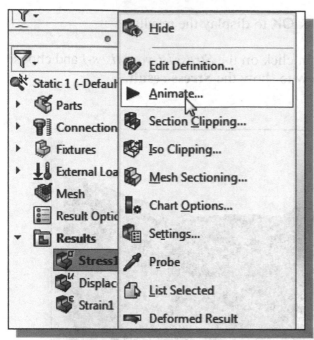

1. In the *FEA Study* window, right-click on the *Results* → *Stress1* item to display the option list and select **Animate** as shown.

❖ Note that the SOLIDWORKS Simulation animation option provides a good representation of the interactions between the different parts in an assembly.

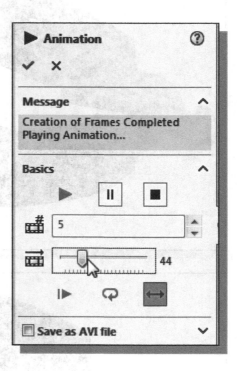

2. In the *Animation* window, adjust the speed of the animation by dragging the speed slider control as shown.

   • Note the *SAVE as AVI File* option is also available.

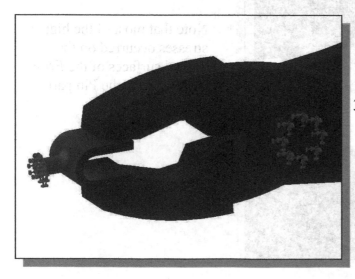

3. In the *Animation* window, click **OK** to exit the option.

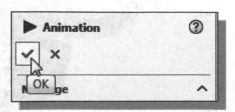

Refinement of the FEA Mesh – Apply Mesh Control

One way of refinement is to adjust the *Element size* to a smaller value in the area showing higher stresses. We will use the **Apply Mesh Control** option and refine the mesh on the *Fork* part.

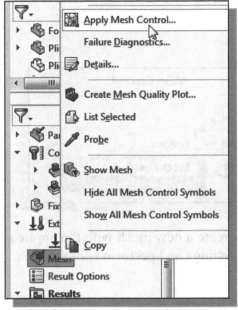

1. In the *FEA Study* window, right-click on the *Mesh* item to display the option list and select **Apply Mesh Control** as shown.

2. Select the **three curved surfaces** as shown.

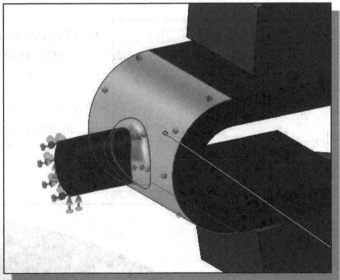

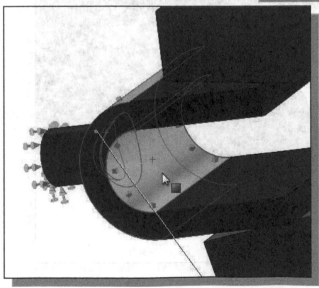

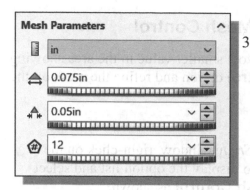

3. In the *Mesh Parameters* box, enter **0.075** inch as the *Maximum element size*, 0.05 inch as the minimum *element size*, 12 as the minimum number of elements in a circle as shown.

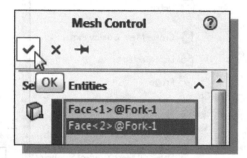

4. Click on the **OK** button to accept the *Mesh* settings.

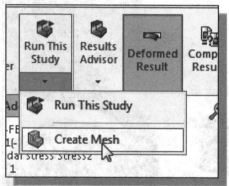

5. On your own, create a new mesh with the applied *Mesh Control* settings and perform another FEA analysis.

6. On your own, repeat the mesh refinement process on the other concerned regions of the assembly and perform an FEA convergence study.

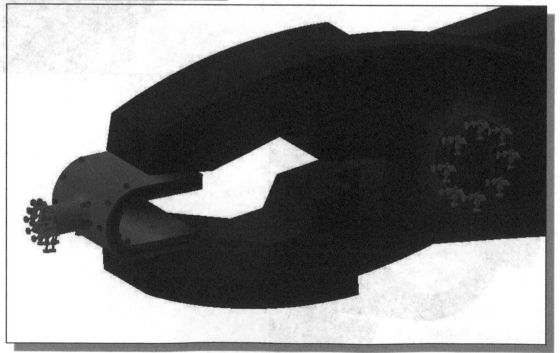

Use the Section Clipping Option

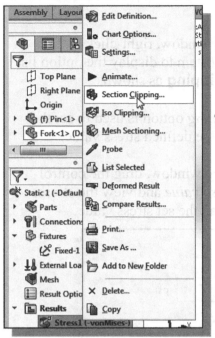

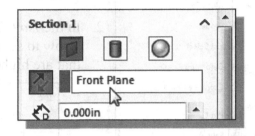

1. In the *FEA Study* window, right-click on the *Results* → *Stress1* item to display the option list and select **Section Clipping** as shown.

- Note the default section plane is set to the *Front Plane*; other planes can also be used.

2. In the graphics window, drag the control arrow to move the section plane and view the stresses on the inside of the plane.

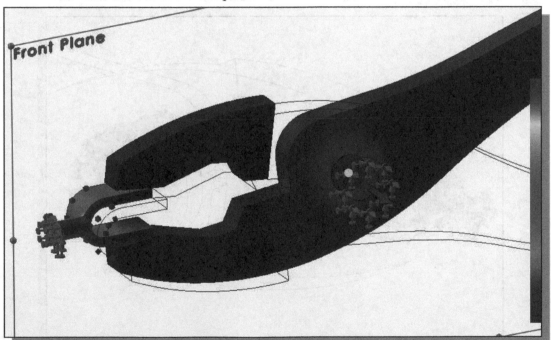

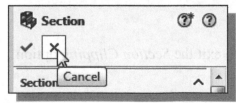

3. Click **Cancel** to exit the *Section Clipping* option.

Use the Iso Clipping Option

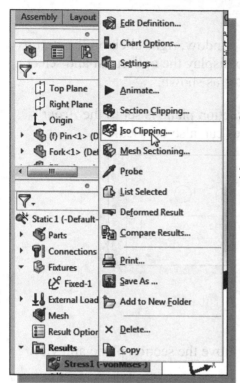

1. In the *FEA Study* window, right-click on the *Results* → *Stress1* item to display the option list and select **Iso Clipping** as shown.

- Note the *Iso Clipping* option is used to set the display based on the defined stress value.

2. In the *Iso Clipping* window, drag the control arrow to set the *Iso value* and view the stresses that are higher than the set stress value.

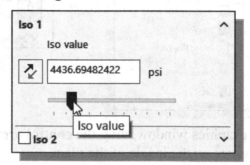

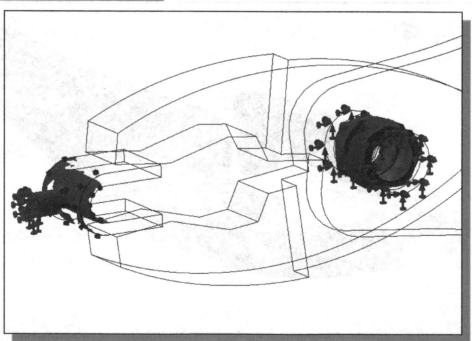

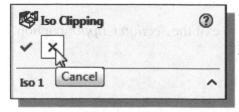

3. Click **Cancel** to exit the *Section Clipping* option.

Set up a Contact Pressure Plot

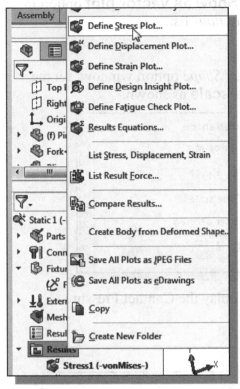

1. In the *FEA Study* window, right-click on the *Results* item to display the option list and select **Define Stress Plot** as shown.

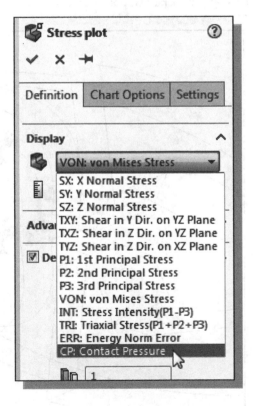

2. In the *Display* window, select **CP: Contact Pressure** as the stress plot type.

*   Note the *Iso Clipping* option is used to set the display based on the defined stress value.

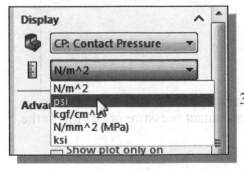

3. Set the display plot units to **psi** as shown.

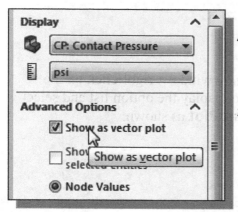

4. Switch on the **Show as vector plot** option in the *Advanced options* list.

5. In the *Deformed Shape* option window, set the option to **True scale** as shown.

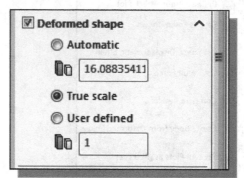

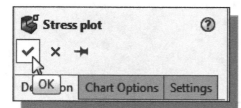

6. Click **OK** to display the Contact Pressure.

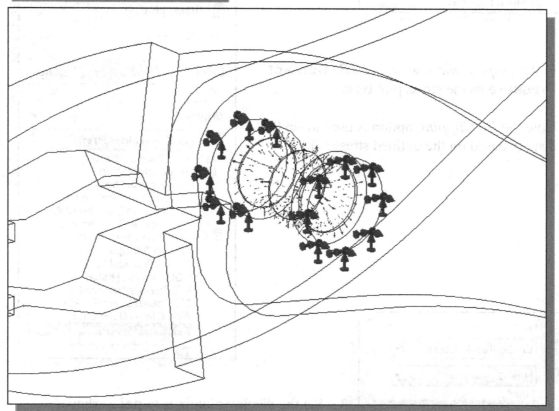

- Note the contact pressure plot identifies the higher contact pressure occurring in the *Pin* and the two *Pliers-Jaws*.

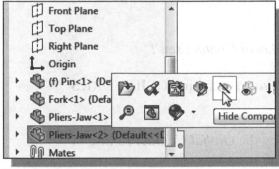

7. On your own, hide the upper *Pliers-Jaw* component as shown.

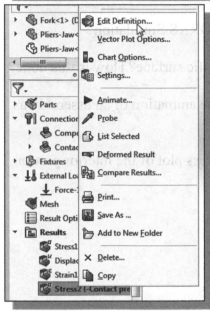

8. In the *FEA Study* window, right-click on the *Results → Stress2* item to display the option list and select **Edit Definition** as shown.

9. In the *Advanced Options* list, turn off the **Show as vector plot** option as shown.

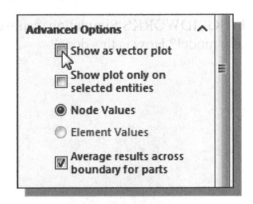

10. Click **OK** to view the contact pressure plot.

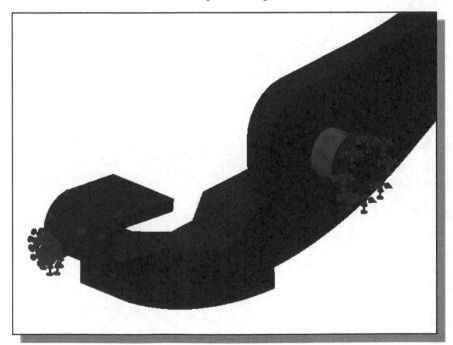

Questions:

1. What is the difference between *Global* and *Local Contact* sets?

2. What is the default setting of the *Global Contact* setting in SOLIDWORKS Simulation?

3. Describe the **Bonded** contact condition set by the *Global Contact* setting in SOLIDWORKS Simulation.

4. How do we create the *Local Contact* sets in SOLIDWORKS Simulation?

5. In performing a mesh refinement, can we refine a specific surface? How is this done?

6. In SOLIDWORKS Simulation, is it possible to save the animation of an assembly as an AVI movie file? How is this done?

7. In SOLIDWORKS Simulation, can we examine the stress plot of the interior portion of a model? How is this done?

Exercises:

Perform FEA static contact analysis on other types of pliers that you own; use the same *Fork* part to compare the results.

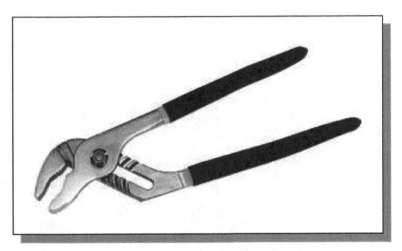

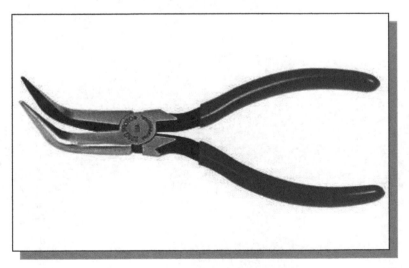

Notes:

Chapter 14
Dynamic Modal Analysis

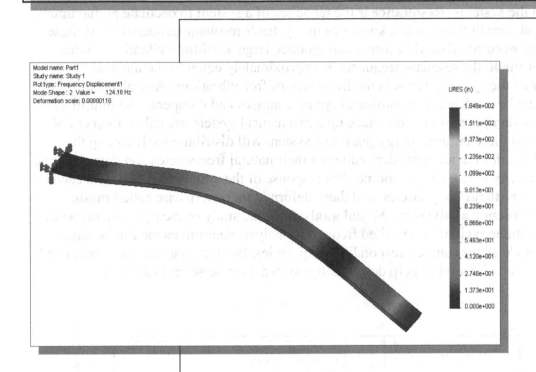

Model name: Part1
Study name: Study 1
Plot type: Frequency Displacement1
Mode Shape : 2 Value = 124.18 Hz
Deformation scale: 0.00880116

URES (in)

1.648e+002
1.511e+002
1.373e+002
1.236e+002
1.099e+002
9.613e+001
8.239e+001
6.866e+001
5.493e+001
4.120e+001
2.746e+001
1.373e+001
0.000e+000

Learning Objectives

- ♦ **Understand the Basic concepts of Natural Frequencies.**
- ♦ **Perform FEA Frequency Analysis on simple Systems.**
- ♦ **Use the Equivalent Mass Spring System approach in performing Modal Analysis.**

Introduction

All objects will vibrate when subjected to impact, noise or vibration. And many systems can **resonate**, where small forces can result in large deformation, and damage can be induced in the systems. **Resonance** is the tendency of a system to oscillate at maximum amplitude at certain frequencies, known as the system's resonant frequencies. At these frequencies, even small driving forces can produce large amplitude vibrations. When damping is small, the resonant frequency is approximately equal to the **natural frequency** of the system, which is the frequency of free vibrations. Any physical structure can be modeled as a number of springs, masses and dampers. The multitude of spring-mass-damper systems that make up a mechanical system are called **degrees of freedom**, and the vibration energy put into a system will distribute itself among the degrees of freedom in amounts depending on their natural frequencies and damping, and on the frequency of the energy source. The response of the system is different at each of the different **natural frequencies**, and these deformation patterns are called **mode shapes**. **Frequency analysis** (or **Modal analysis**) is the study of the dynamic properties of systems under excitation. Detailed frequency analysis determines the fundamental vibration mode shapes and corresponding frequencies. Both the natural frequencies and mode shapes can be used to help design better systems for noise and vibration applications.

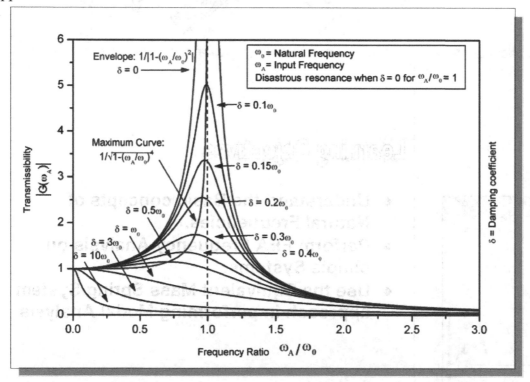

Frequency analysis can be relatively simple for basic components of a simple system and extremely complicated for a complex system, such as a structure exposed to periodic wind loading or during seismic activities. With the advancements of computers, the accurate determination of natural frequencies and mode shapes are best suited to using special techniques such as Finite Element Analysis.

Problem Statement

1) Determine the **natural frequencies** and **mode shapes** of a cantilever beam. The beam is made of steel and has the dimensions of 14.5" x 1/8" x 1".

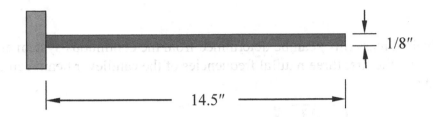

2) Determine the first **natural frequency** and **mode shapes** of a cantilever beam with a small half pound object attached to the free end of the cantilever beam.

Preliminary Analysis

The cantilever beam is an example of a system, which can be modeled as a simple spring-mass system. In order to model the vibration of the cantilever beam, the end of the beam is chosen as a reference point at which the characteristics and response of the beam are measured. An equivalent system is then built so that the natural frequency of the system can be determined. The equivalent spring constant can be calculated using beam deflection formulae. Calculation of an equivalent mass is necessary because all points along the beam's length do not have the same response as the end of the beam. This means that the equivalent mass, m_e, cannot be determined simply by using the masses of the beam but must be found by equating the energy of the system as it vibrates.

From *Strength of Materials*, the deflection at the tip of the cantilever beam can be determined by

$$y = \frac{PL^3}{3EI}$$

The deflection equation of the mass spring system is

$$F = ky$$

So the equivalent spring constant can be expressed as

$$k = \frac{3EI}{L^3}$$

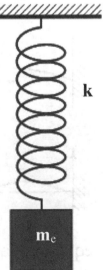

1) For the simple mass-spring system, the natural frequency can be expressed as

$$\omega_n = \sqrt{\frac{k}{m_e}} \quad \text{(in rad/sec)}$$

The equivalent mass (m_e) can be determined from the continuous system analytical approach, and the first three natural frequencies of the cantilever beam can then be expressed as

$$\omega_n = \alpha_n^2 \sqrt{\frac{EI}{m_b L^3}} \quad \text{where} \quad \alpha_n = 1.875, 4.694, 7.855$$

The first three natural frequencies and mode shapes of a cantilever beam are as shown in the figure below.

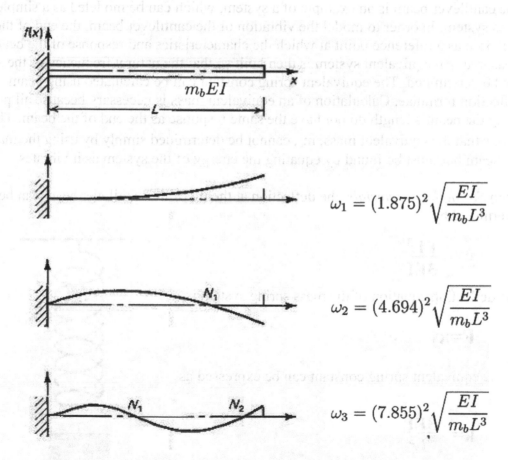

$$\omega_1 = (1.875)^2 \sqrt{\frac{EI}{m_b L^3}}$$

$$\omega_2 = (4.694)^2 \sqrt{\frac{EI}{m_b L^3}}$$

$$\omega_3 = (7.855)^2 \sqrt{\frac{EI}{m_b L^3}}$$

Frequencies and Mode Shapes for Cantilever Beam

For our preliminary analysis, the first three natural frequencies calculations are as follows

$$\omega_n = \alpha_n^2 \sqrt{\frac{EI}{m_b L^3}} \quad \text{where} \quad \alpha_n = 1.875, 4.694, 7.855$$

$E = 3.0 \times 10^7$ psi **(Modulus of Elasticity for Steel)**

$I = bh^3/12 = $ **1** $\times$ **(0.125)**^{3} **/12** $= 1.627 \times 10^{-4}$ in^4

$L = 14.5$ in

$m_b = $ Volume $\times$ Density $=$ **1** $\times$ **0.125** $\times$ **14.5** $\times$ **0.0007324** $= 0.001326 \frac{\text{lbsec}^2}{\text{in}}$

Therefore,

$\omega_1 = (1.875)^2 \times (3.0 \times 10^7 \times 1.627 \times 10^{-4} / (0.001326 \times (14.5)^3))^{1/2}$
$\quad = 122.16$ (rad/sec)
$\quad = 19.71$ (Hz)

$\omega_2 = (4.694)^2 \times (3.0 \times 10^7 \times 1.627 \times 10^{-4} / (0.001326 \times (14.5)^3))^{1/2}$
$\quad = 765.62$ (rad/sec)
$\quad = 121.85$ (Hz)

$\omega_3 = (7.855)^2 \times (3.0 \times 10^7 \times 1.627 \times 10^{-4} / (0.001326 \times (14.5)^3))^{1/2}$
$\quad = 2143.99$ (rad/sec)
$\quad = 341.22$ (Hz)

2) The equivalent mass (m_e) for the cantilever beam with a mass attached at the end has a similar format so that the first natural frequency can be expressed as

$$\omega_1 = \sqrt{\frac{3EI}{(m_a + 0.236 m_b) L^3}}$$

$\omega_1 = (3 \times 3.0 \times 10^7 \times 1.627 \times 10^{-4} / ((0.5/(32.2 \times 12) + 0.236 \times 0.001326) \times (14.5)^3)^{1/2}$
$\quad = 54.67$ (rad/sec)
$\quad = 8.70$ (Hz)

The Cantilever Beam Modal Analysis program

The *Cantilever Beam Modal Analysis* program is a custom-built MS Windows based computer program that can be used to perform simple Frequency analysis on uniform cross section cantilever systems; rectangular and circular cross section calculations are built-in, but other cross sections can also be used. The program is based on the analytical methods described in the previous section. Besides calculating the first six natural frequencies of a cantilever beam, the associated Frequency shapes are also constructed and displayed. The program is very compact, 75 Kbytes in size, and it will run in any Microsoft Windows based computer system.

❖ First download the **CBeamModal.zip** file, which contains the *Cantilever Beam Frequency Analysis* program, from the SDC Publications website.

1. Launch your internet browser, such as the *MS Internet Explorer* or *Mozilla Firefox* web browsers.

2. In the *URL address* box, enter:
   www.SDCPublications.com/downloads/978-1-63057-484-0

3. Click the **Download File** button to download the *CBeamModal.zip* file to your computer.

4. On your own, extract the content of the ZIP file to a folder on the desktop.

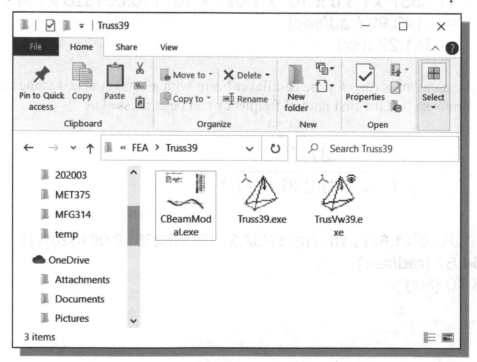

5. Start the ***Cantilever Beam Frequency Analysis*** program by double-clicking on the **CBeamModal.exe** icon.

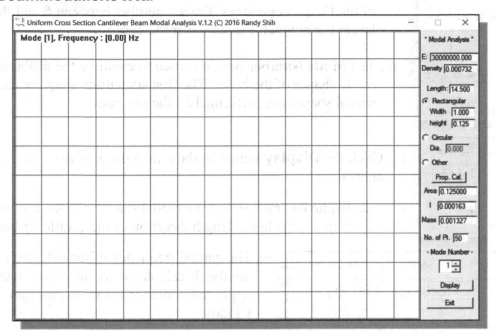

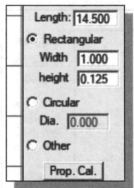

❖ The controls to the inputs and the display options are listed toward the right edge of the program's main window.

6. The first two items in the control panel are the required material information: the **Modulus of Elasticity** (E) and the **Density** of the material. Note that it is critical to use the same units for all values entered.

7. Enter the **physical dimensions** of the cantilever beam in the mid-section of the control panel:

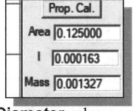

- For rectangular shapes, select the **Rectangular** option and enter the **Width** and **Height** dimensions of the cross section.
- For circular shapes, select the **Circular** option and enter the **Diameter** value.
- For any other cross section shapes, select the **Other** option and enter the **Area** and **Area Moment Of Inertia** information in the next section.

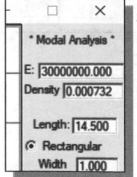

8. Press the **Property Calculation** button to calculate the associated **Area, Area Moment of Inertia** and **Mass** information for the rectangular or circular cross sections. For any other shapes, enter the information in the edit boxes. Note that the mass information is always calculated based on the density value.

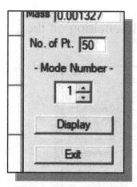

9. The **Number of Point** option is used to control the smoothness of the Frequency shapes. Enter a number between **5** and **80**; the larger the number, the smoother the curve.

10. The **Mode Number** option is used to examine the different mode shapes of the beam. The first six natural frequencies modal shapes are performed by the program.

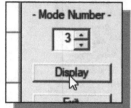

11. Click the **Display** button to show the results of the Frequency analysis.

➤ The displayed graph shows the modal shape with the horizontal axis corresponds to the length direction of the cantilever beam.

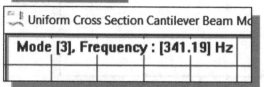

➤ The natural frequency of the selected mode number is calculated and displayed near the upper left corner of the main program window.

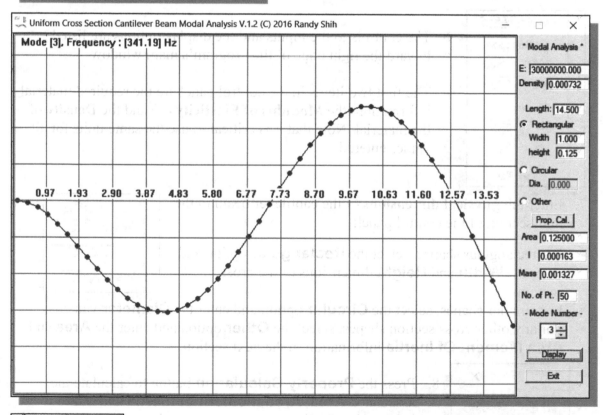

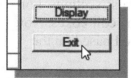

12. Click the **Exit** button to end the program.

Start SOLIDWORKS

1. Select the **SOLIDWORKS** option on the *Start* menu or select the **SOLIDWORKS** icon on the desktop to start SOLIDWORKS. The SOLIDWORKS main window will appear on the screen.

2. Select **Part** by clicking on the first icon in the *New SOLIDWORKS Document* dialog box as shown.

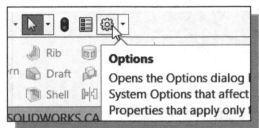

3. Select the **Options** icon from the *Menu* toolbar to open the *Options* dialog box.

4. Switch to the **Document Properties** tab and reset the *Drafting Standard* to **ANSI** as shown in the figure.

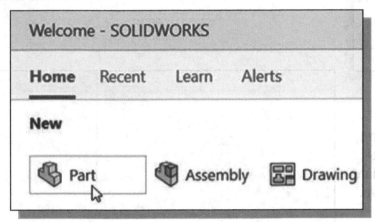

5. On your own, set the *Unit system* to **IPS (inch, pound, second)** as shown.

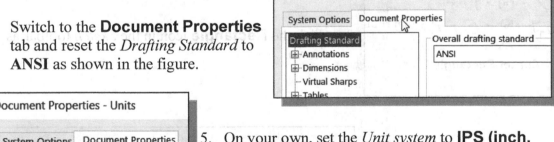

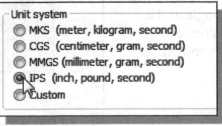

Create the CAD Model

To perform the surface FEA analysis, we will first create a solid model using the **Extrude** command.

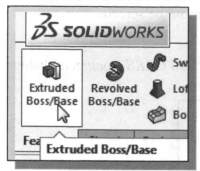

1. Click the **Extruded Boss/Base** icon, in the *Features* toolbar, to create a new extruded feature.

2. Click the **Right Plane**, in the graphics area, to align the sketching plane of our sketch as shown.

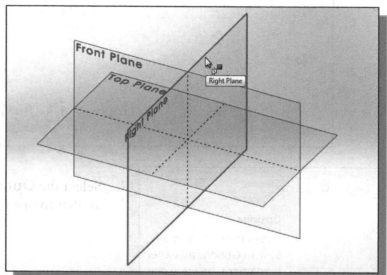

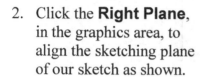
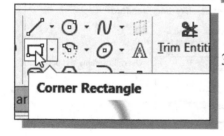

3. Click the **Rectangle** icon in the *Sketch* toolbar as shown.

4. Start the lower left corner of the rectangle at the origin and create a rectangle as shown.

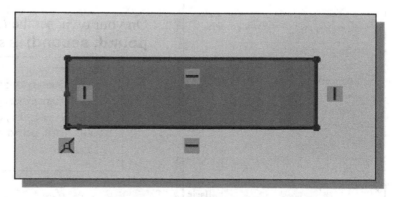

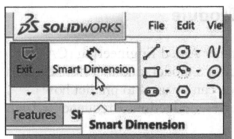

5. Click the **Smart Dimension** icon in the *Sketch* toolbar as shown.

6. On your own, create the dimensions and adjust the sketch as shown.

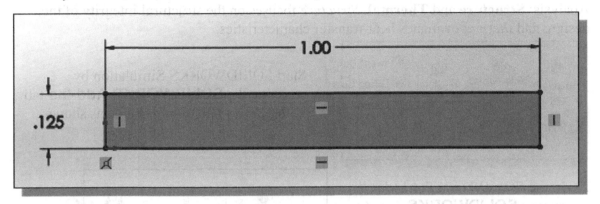

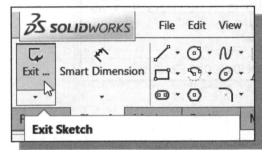

7. Click the **Exit Sketch** icon in the *Sketch* toolbar to exit the *2D Sketch* mode.

8. On your own, using the extrusion distance of **14.5** in, create the solid feature as shown.

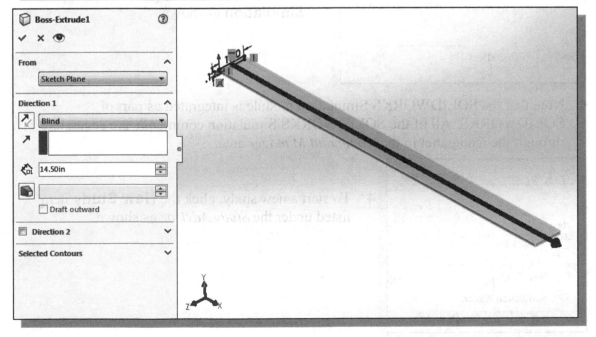

Activate the SOLIDWORKS Simulation Module

SOLIDWORKS Simulation is a multi-discipline Computer Aided Engineering (CAE) tool that enables users to simulate the physical behavior of a model, and therefore enables users to improve the design. SOLIDWORKS Simulation can be used to predict how a design will behave in the real world by calculating stresses, deflections, frequencies, heat transfer paths, etc.

The SOLIDWORKS Simulation product line features two areas of Finite Element Analysis: **Structure** and **Thermal**. *Structure* focuses on the structural integrity of the design, and *thermal* evaluates heat-transfer characteristics.

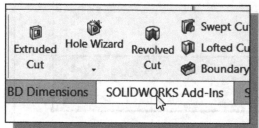

1. Start SOLIDWORKS Simulation by selecting the **SOLIDWORKS Add-Ins** tab in the *Command Manager* area as shown.

2. In the *SOLIDWORKS Add-Ins* list, choose **SOLIDWORKS Simulation** as shown.

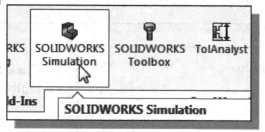

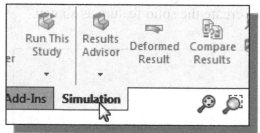

3. In the *Command Manager* area, choose **Simulation** as shown.

❖ Note that the SOLIDWORKS Simulation module is integrated as part of SOLIDWORKS. All of the SOLIDWORKS Simulation commands are accessible through the icon panel in the *Command Manager* area.

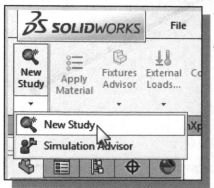

4. To start a new study, click the **New Study** item listed under the *Study Advisor* as shown.

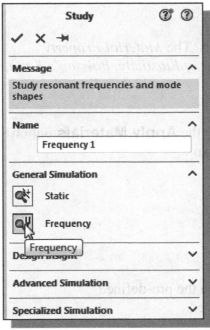

5. Select **Frequency** as the type of analysis to be performed with SOLIDWORKS Simulation.

❖ Note that different types of analyses are available, which include both structural static and dynamic analyses, as well as thermal analysis.

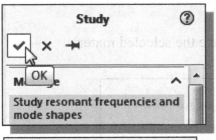

6. Click **OK** to start the definition of a frequency analysis.

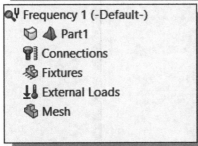

❖ In the *Feature Manager* area, note that a new panel, the *FEA Study* window, is displayed with all the key items listed. The icon in front of the Study name identifies the selected type of analysis.

❖ Also note that **Frequency 1** tab is activated as the current FEA model.

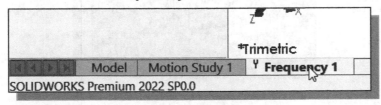

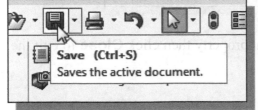

7. On your own, save a copy of the current model.

Assign the Element Material Property

Next, we will set up the *Material Property* for the elements. The *Material Property* contains the general material information, such as *Modulus of Elasticity*, *Poisson's Ratio*, etc. that is necessary for the FEA analysis.

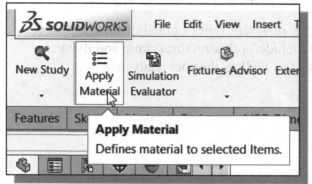

1. Choose the **Apply Materials** option as shown.

❖ Note the default list of materials, which are available in the pre-defined SOLIDWORKS Simulation material library, is displayed.

2. Select **Alloy Steel** in the *Material* list as shown.

3. Set the **Units** option to display **English (IPS)** to make the selected material available for use in the current FEA model.

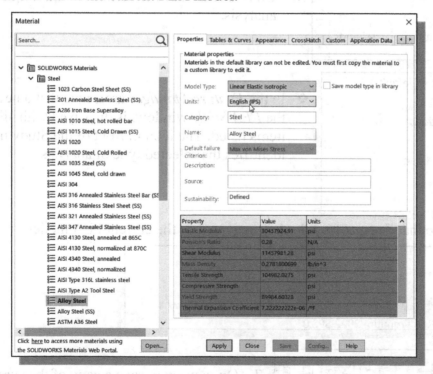

4. Click **Apply** to assign the material property then click **Close** to exit the Material Assignment command.

Apply Boundary Conditions – Constraints

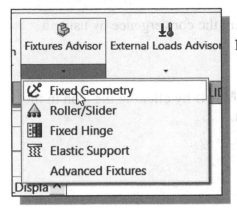

1. Choose **Fixed Geometry** by clicking the fixture icon in the toolbar as shown.

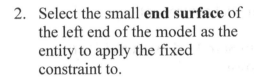

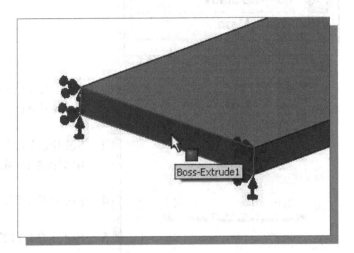

2. Select the small **end surface** of the left end of the model as the entity to apply the fixed constraint to.

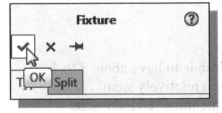

3. Click on the **OK** button to accept the *Fixture constraint* settings.

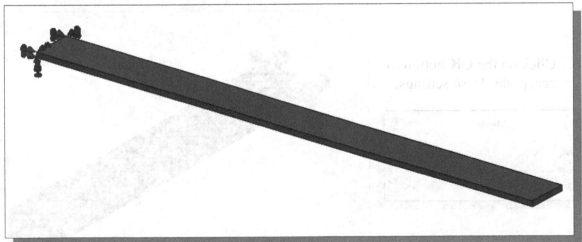

➤ For *frequency analysis*, we do not need to apply any external load to the system. The natural frequencies of physical systems are dynamic properties, which can be determined through the equivalent mass-spring system.

Create the first FEA Mesh

For the frequency analysis, it is still necessary to confirm the convergence by using a coarse mesh as the first FEA mesh for the analysis.

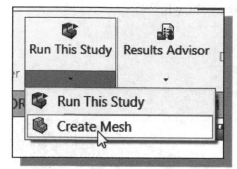

1. Choose **Create Mesh** by clicking the icon in the toolbar as shown.

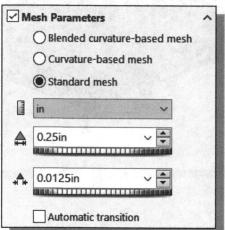

2. Switch on the **Mesh Parameters** option, to show the additional control options.

3. Set the *Mesh* to **Standard mesh** and *Units* to **inches** as shown.

4. Enter **0.25** inch as the *Global element size*.

5. Enter **0.0125** inch as the *Size tolerance*.

❖ A good rule of thumb to follow in creating the first mesh is to have about 3 to 4 elements on the edges of the model, but since we have a relatively small cross section, we will use the height of the cross section as the initial element size.

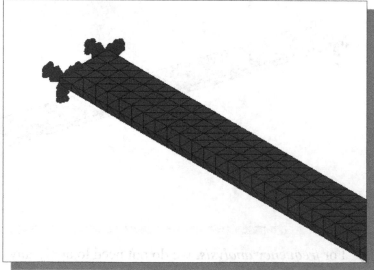

6. Click on the **OK** button to accept the *Mesh* settings.

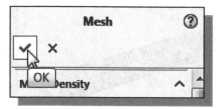

7. In the *FEA Study* window, right-click on the *Mesh* item to display the option list and select **Details** as shown.

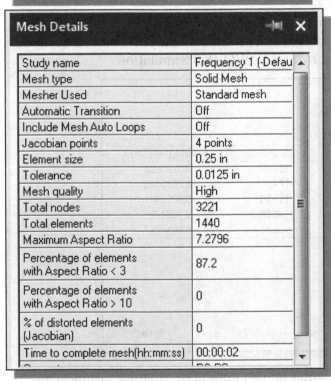

❖ The current mesh consists of **3221** nodes and **1440** solid elements.

8. Click on the [**X**] icon to close the *Mesh Details* dialog box.

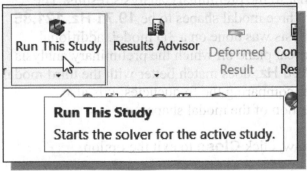

9. Click on the **Run This Study** button to start the *FEA Solver* to calculate the results.

Run This Study
Starts the solver for the active study.

Viewing the results

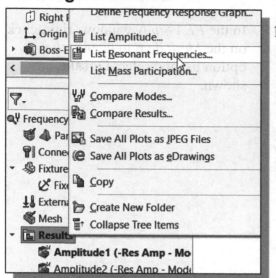

1. In the *FEA Study* window, right-click on the *Results* item to display the option list and select **List Resonant Frequencies** as shown.

❖ Note that the first five natural frequencies of the FEA solutions are shown in the *List Modes* dialog box. The three natural frequencies from our analytical calculations are very similar to the frequencies found by SOLIDWORKS Simulation.

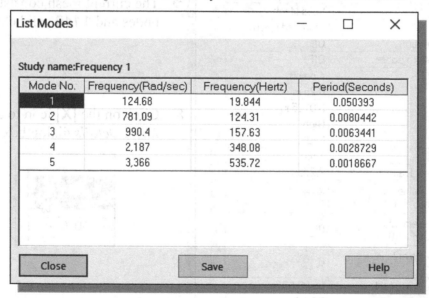

Mode No.	Frequency(Rad/sec)	Frequency(Hertz)	Period(Seconds)
1	124.68	19.844	0.050393
2	781.09	124.31	0.0080442
3	990.4	157.63	0.0063441
4	2,187	348.08	0.0028729
5	3,366	535.72	0.0018667

- Note that the FEA results are very similar to the preliminary analysis results. The preliminary analysis calculated the first three modal shapes to be **19.71 Hz**, **121.85 Hz** and **341.22 Hz**. Since the FEA analysis was done on a 3D model, additional modal shapes may exist beyond the vertical plane on which the preliminary analysis was based. The fourth FEA result, **348.08 Hz**, does match better with the third modal hand calculation of **341.22 Hz**. Besides comparing the frequencies, it is also necessary to view and examine the direction of the modal shapes.

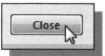

2. In the *List Modes* window, click **Close** to exit the option.

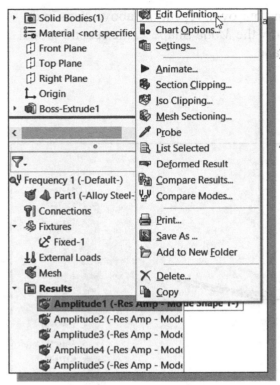

3. In the *FEA Study* window, right-click on the first item **(Mode shape 1)** under *Results* to display the option list and select **Edit Definition** as shown.

4. Confirm the *Plot Step – Mode* is set to **1**; this will display the first modal shape.

5. Confirm the *Deformed Shape* option is set to **Automatic**.

6. Click **OK** to display the results.

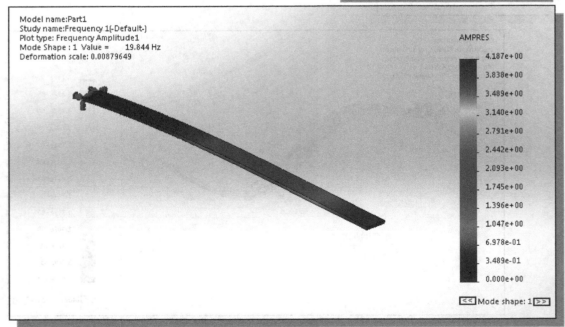

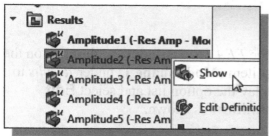

7. On your own, repeat the above steps and show the *Mode* Shape **2**, the second modal shape.

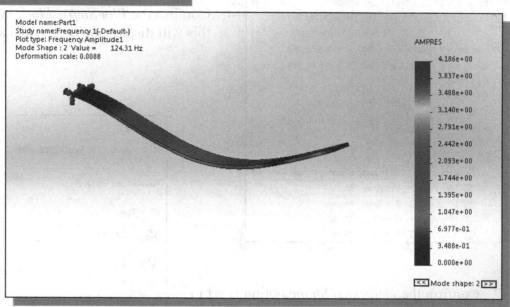

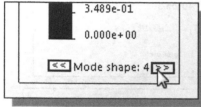

8. On your own, use the Mode Shape selection icon and set the *Mode Shape* to **4**, and display the fourth modal shape.

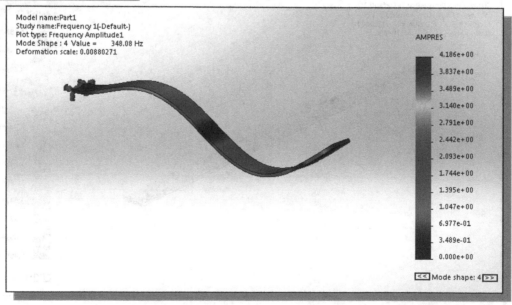

Refinement of the FEA Mesh – Global Element Size 0.15

To confirm the previous calculated FEA results are adequate, we will check for convergence by refining the mesh to using 0.15 as the global element size.

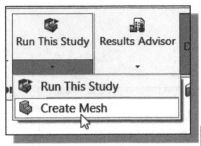

1. Choose **Create Mesh** by clicking the icon in the toolbar as shown.

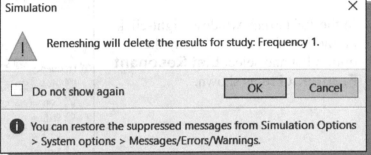

2. Click on the **OK** button to proceed with deleting the old mesh and create a new mesh of the FEA model.

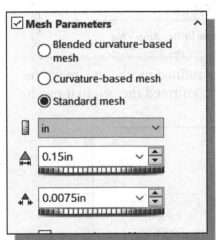

3. Enter **0.15** inch as the *Global Element size*.

4. Click on the **OK** button to accept the *Mesh* settings and create a new mesh.

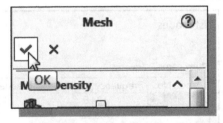

❖ The refined mesh consists of **12254** nodes and **6709** solid elements, which is about four times more elements than the original mesh and it is a very dense mesh.

Mesh Details	
Study name	Frequency 1 (-Defa
Mesh type	Solid Mesh
Mesher Used	Standard mesh
Automatic Transition	Off
Include Mesh Auto Loops	Off
Jacobian points	4 points
Element size	0.15 in
Tolerance	0.0075 in
Mesh quality	High
Total nodes	12254
Total elements	6709
Maximum Aspect Ratio	4.638
Percentage of elements with Aspect Ratio < 3	98

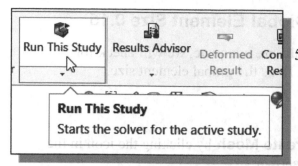

5. Click on the **Run This Study** button to start the *FEA Solver* to calculate the results.

6. In the *FEA Study* window, right-click on the *Results* item to display the option list and select **List Resonant Frequencies** as shown.

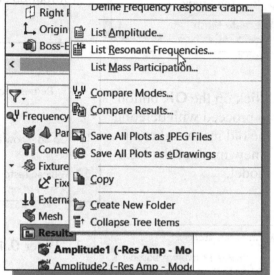

❖ Note that the FEA calculated natural frequencies of the refined mesh are almost the same as the results of the original mesh. These results confirmed the original results are adequate.

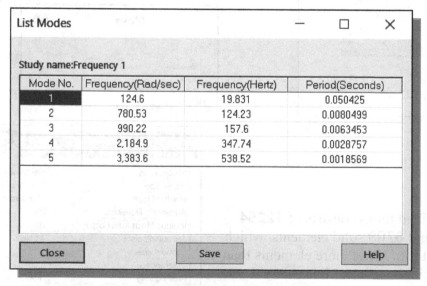

List Modes

Study name:Frequency 1

Mode No.	Frequency(Rad/sec)	Frequency(Hertz)	Period(Seconds)
1	124.6	19.831	0.050425
2	780.53	124.23	0.0080499
3	990.22	157.6	0.0063453
4	2,184.9	347.74	0.0028757
5	3,383.6	538.52	0.0018569

Close   Save   Help

❖ Note that comparing these results to the original analysis, page 14-5, showed the modal frequencies had not changed with the more refined elements, which validates the performed FEA analysis.

Add an Additional Mass to the system

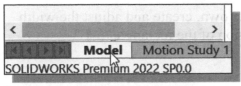

1. Switch to the *CAD model* by clicking on the **Model** tab near the bottom of the graphics window.

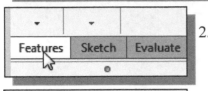

2. In the *Command Manager*, select the **Features** tab to display the feature toolbar.

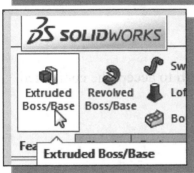

3. Click the **Extruded Boss/Base** icon, in the *Features* toolbar, to create a new extruded feature.

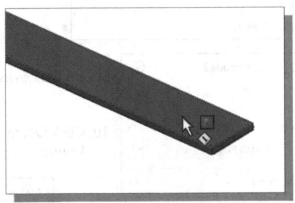

4. Select the **top surface** of the beam model to align the sketching plane.

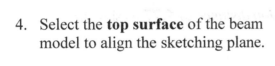

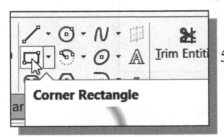

5. In the *Sketch* toolbar, select **Rectangle**.

6. On your own, create a rectangle with the top, bottom and right edges aligned to the corresponding edges of the solid model as shown.

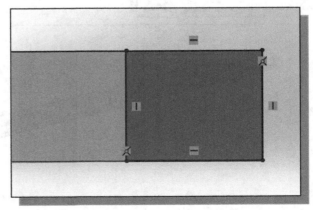

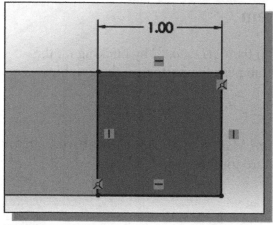

7. On your own, create and adjust the width dimension of the rectangle to **1"**.

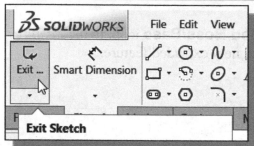

8. Click **Exit Sketch** to accept the completed sketch.

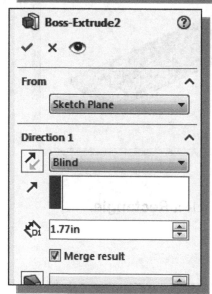

9. Set the *extrusion distance* to **1.77** as shown.

10. Click **OK** to accept the settings and complete the feature.

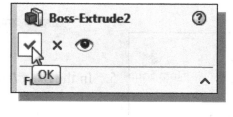

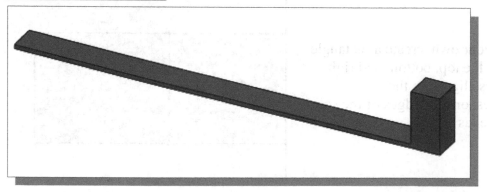

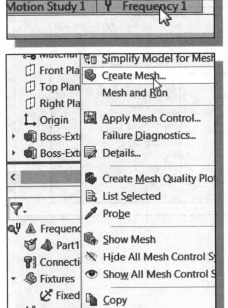

11. Switch to the *FEA Study* by clicking on the **Frequency 1** tab near the bottom of the graphics window.

12. The warning symbol, in front of the mesh item in the study1 window, indicates the change in the CAD model has invalidated the old mesh. Right-click once on the **Mesh** item to display the option list and select **Create Mesh** as shown.

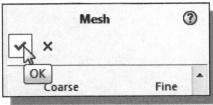

13. Enter *0.15* inch as the *Global element size.*

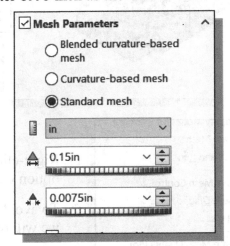

14. Click on the **OK** button to accept the *Mesh* settings.

15. On your own, perform the **Modal Analysis**.

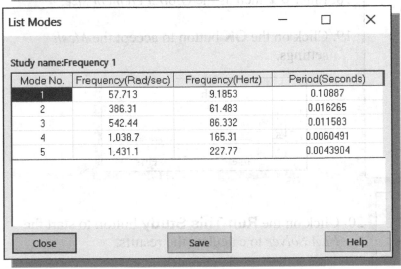

List Modes

Study name:Frequency 1

Mode No.	Frequency(Rad/sec)	Frequency(Hertz)	Period(Seconds)
1	57.713	9.1853	0.10887
2	386.31	61.483	0.016265
3	542.44	86.332	0.011583
4	1,038.7	165.31	0.0060491
5	1,431.1	227.77	0.0043904

Close   Save   Help

➢ Note the first natural frequency is calculated as **9.185Hz**, which is similar to the result from our preliminary calculation, page 14-5. We will also examine the effect of shifting the additional mass so it is centered at the right edge of the beam.

16. On your own, modify the 2D section and update the additional mass as shown.

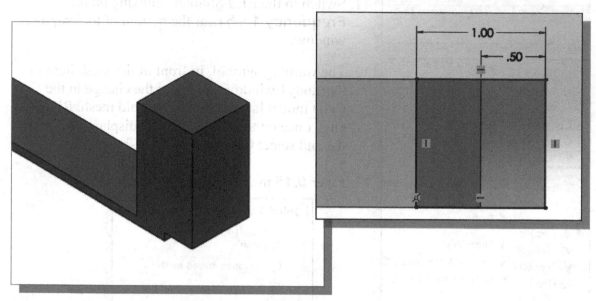

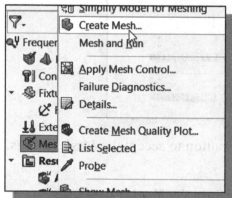

17. Right-click once on the **Mesh** item to display the option list and select **Create Mesh** as shown.

❖ To avoid generating a large number of elements, we will reset the element size to 0.2.

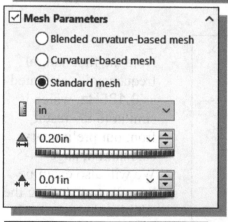

18. Enter *0.2* inch as the *Global element size*.

19. Click on the **OK** button to accept the *Mesh* settings.

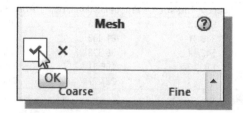

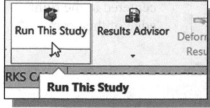

20. Click on the **Run This Study** button to start the *FEA Solver* to calculate the results.

21. On your own, perform the **Modal Analysis** and **view the results**.

Mode No.	Frequency(Rad/sec)	Frequency(Hertz)	Period(Seconds)
1	55.411	8.8189	0.11339
2	373.39	59.427	0.016827
3	513.44	81.716	0.012238
4	1,031.4	164.15	0.0060922
5	1,361.9	216.75	0.0046137

Study name:Frequency 1

List Modes

> Note the calculated first natural frequency is very close to the result of our preliminary analysis.

One-Dimensional Beam Frequency Analysis

You are also encouraged to create and compare FEA modal analyses of the straight beam problem using one-dimensional beam elements.

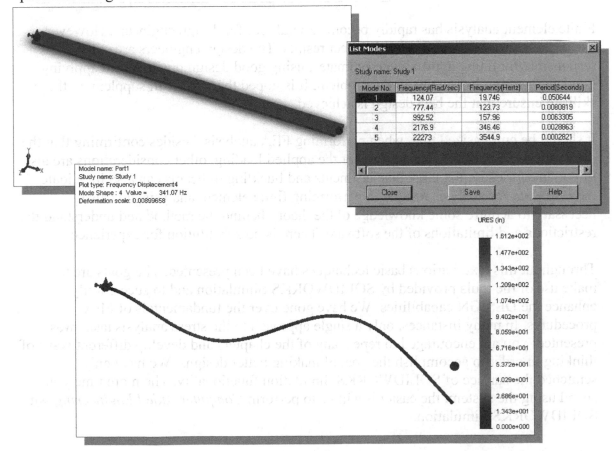

Model name: Part1
Study name: Study 1
Plot type: Frequency Displacement4
Mode Shape : 4 Value =   341.07 Hz
Deformation scale: 0.00899658

List Modes

Study name: Study 1

Mode No.	Frequency(Rad/sec)	Frequency(Hertz)	Period(Seconds)
1	124.07	19.746	0.050644
2	777.44	123.73	0.0080819
3	992.52	157.96	0.0063305
4	2176.9	346.46	0.0028863
5	22273	3544.9	0.0002821

URES (in)

1.612e+002
1.477e+002
1.343e+002
1.209e+002
1.074e+002
9.402e+001
8.059e+001
6.716e+001
5.372e+001
4.029e+001
2.686e+001
1.343e+001
0.000e+000

Conclusions

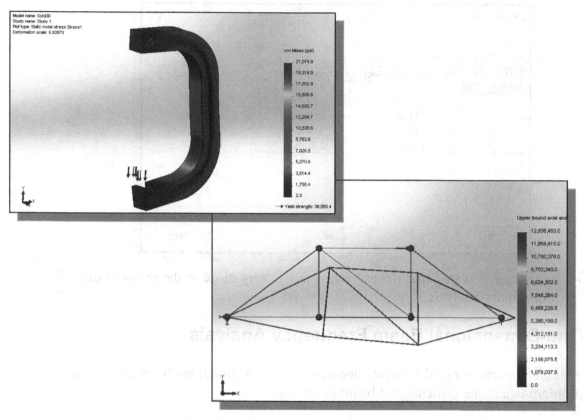

Finite element analysis has rapidly become a vital tool for design engineers. However, use of this tool does not guarantee correct results. The design engineers are still responsible for doing approximate estimates, using good design practice, and applying good engineering judgments to the problem. It is hoped that FEA will supplement these skills to ensure that the best design is achieved.

It should be emphasized that, when performing FEA analysis, besides confirming that the designs remain in the elastic regions for the applied loading, other considerations are also important; for example, large displacements and buckling of beams can also invalidate the *linear statics* analysis results. In performing finite element analysis, it is also necessary to acquire some knowledge of the theory behind the method and understand the restrictions and limitations of the software. There is no substitution for experience.

Throughout this text, various basic techniques have been presented. The goals are to make use of the tools provided by SOLIDWORKS Simulation and to successfully enhance the DESIGN capabilities. We have gone over the fundamentals of FEA procedures. In many instances, only a single approach to the stress analysis tasks was presented; you are encouraged to repeat any of the chapters and develop different ways of thinking in order to accomplish the goal of making better designs. We have only scratched the surface of SOLIDWORKS Simulation functionality. The more time you spend using the system, the easier it will be to perform *Computer Aided Engineering* with SOLIDWORKS Simulation.

Questions:

1. What is the main purpose of performing a modal analysis on a system?

2. What are the relations between resonant frequencies and the natural frequencies of a system?

3. In vibration analysis, any physical structure can be modeled as a combination of what objects?

4. A cantilever beam can be modeled as a simple mass-spring system by examining the deflection at the tip of the beam; what is the equivalent spring constant for such a system?

5. In performing a modal analysis of a cantilever beam, would changing the material properties affect the natural frequencies of the system?

6. In SOLIDWORKS Simulation is it required to have a load applied to the system in order to perform a Modal analysis?

7. In SOLIDWORKS Simulation can we perform a modal analysis without applying a constraint?

8. For the cantilever beam with a mass attached at the end, when the attached mass is relatively big, we can ignore the weight of the beam as a simplified approximation. Calculate the first natural frequency of the tutorial problem using this approximation and compare the results obtained.

Exercises:

The first four natural frequencies and mode shapes of uniform cross section beams, with different types of supports at the ends, are shown in the below table:

Simply supported ends	C = 1.57	.500 C = 6.28	.333 .667 C = 14.1	.25 .50 .75 C = 25.2
Fixed ends	C = 3.56	.500 C = 9.82	.359 .641 C = 19.2	.278 .5 .722 C = 31.8
Free ends	.224 .776 C = 3.56	.132 .500 .868 C = 9.82	.094 .356 .644 .906 C = 19.2	.0734 .277 .500 .723 .927 C = 31.8
Fixed-hinged	C = 2.45	.560 C = 7.95	.384 .692 C = 16.6	.394 .529 .765 C = 28.4

$$f_n = C \sqrt{\frac{EI}{m_b L^3}}$$

f_n = natural frequency in cycles/sec
C = constant from above table

1. Determine the **natural frequencies** and **mode shapes** of a simply supported beam. The beam is made of steel and has the dimensions of 14.5" x 1/8" x 1".

2. Determine the **natural frequencies** and **mode shapes** of a fixed-hinged beam. The beam is made of steel and has the dimensions of 14.5" x 1/8" x 1".

3. Modal analyses can also be performed using 1D beam elements; for the above two problems perform beam modal analyses and compare the results.

Appendix - Answers to Selected Exercises

Ch. 1 Exercise 1:

1. For the one-dimensional 3 truss-element system shown, determine the nodal displacements and reaction forces using the direct stiffness method.

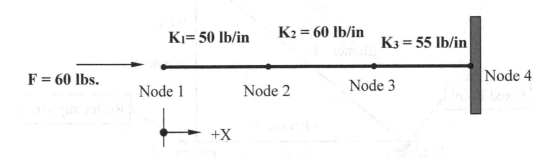

$$\begin{Bmatrix} F_1 \\ F_2 \\ F_3 \\ F_4 \end{Bmatrix} = \begin{bmatrix} K_1 & -K_1 & 0 & 0 \\ -K_1 & (K_1+K_2) & -K_2 & 0 \\ 0 & -K_2 & (K_2+K_3) & -K_3 \\ 0 & 0 & -K_3 & +K_3 \end{bmatrix} \begin{Bmatrix} X_1 \\ X_2 \\ X_3 \\ X_4 \end{Bmatrix}$$

$$\begin{Bmatrix} 60 \\ 0 \\ 0 \\ F_4 \end{Bmatrix} = \begin{bmatrix} 50 & -50 & 0 & 0 \\ -50 & 110 & -60 & 0 \\ 0 & -60 & 115 & -55 \\ 0 & 0 & -55 & 55 \end{bmatrix} \begin{Bmatrix} X_1 \\ X_2 \\ X_3 \\ 0 \end{Bmatrix}$$

X1= 3.29 inch, X2= 2.09 inch, X3=1.09 inch
F4 = -60 lbs.

Ch. 2 Exercise 1:

1. Given: two-dimensional truss structure as shown.

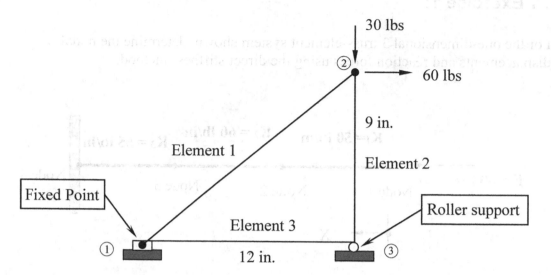

Material: Steel, diameter ¼ in.

Find: (a) Displacements of the nodes.
         (b) Normal stresses developed in the members.

Answers: X2= 1.301E-03 inch, Y2= -4.591E-04 inch
            Stress in Element 1 = 1530 psi
            Stress in Element 2 = -1530 psi
            Stress in Element 3 = 0 psi

Ch. 3 Exercise 2:

2. Given: Two-dimensional truss structure as shown (All joints are pin joints).

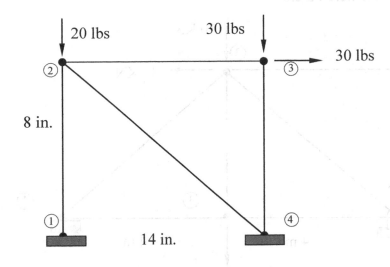

Material: Steel, diameter ¼ in.

Find: (a) Displacements of the nodes.
      (b) Normal stresses developed in the members.

Answers: X2= 4.276E-04 inch, Y2= -1.554E-05 inch
            X3= 7.133E-04 inch, Y3= -1.632E-04 inch
            Stress in Element 12 = -58.3 psi
            Stress in Element 23 = 612 psi
            Stress in Element 24 = 705 psi
            Stress in Element 34 = -612 psi

Ch. 4 Exercise 2:

2.   Material: Steel,
          Diameter: 3 cm.

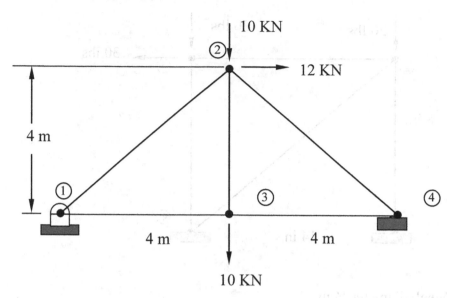

Answers: X2= 9.326E-04 m, Y2= -1.252E-03 m
                X3= 4.526E-04 m, Y3= -1.535E-03 m
                X4= 9.052E-04 m, Y4= 0 m
                Stress in Element 12 = - 8.001E+06 Pa
                Stress in Element 23 = 1.414E+07 Pa
                Stress in Element 13 = 2.263E+07 Pa
                Stress in Element 24 = -3.200E+07 Pa
                Stress in Element 34 = 2.263E+07 Pa

Ch. 5 Exercise 1:

1. Material: Steel pipe,
    NPS 1½", Schedule 40

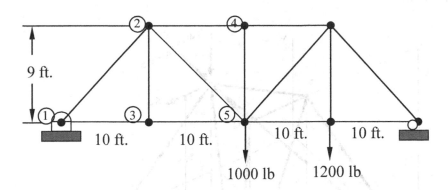

Answers: X2= 3.169E-03 inch, Y2= -5.481E-03 inch
               X3= 7.244E-04 inch, Y3= -5.481E-03 inch

               Stress in Element 12 = -1495 psi
               Stress in Element 23 = 0 psi
               Stress in Element 13 = 1111 psi
               Stress in Element 24 = -2223 psi
               Stress in Element 25 = 1495 psi

Ch. 6 Exercise 2:

2. All joints are ball-joints, and joints D, F, F are fixed to the floor.

Material: Steel
Diameter: 1 in.

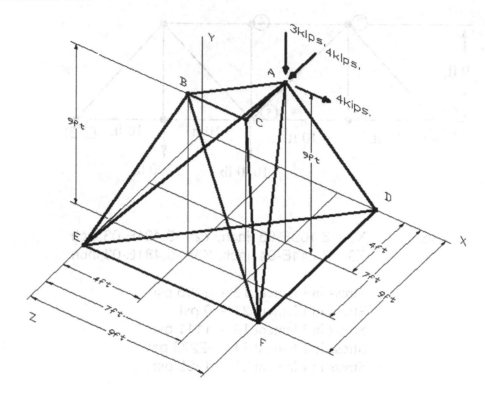

Answers: XA= 1.033E-01 inch, YA= 3.290E-03 inch, ZA= 7.776E-02 inch

Stress in Element AB = 0 psi
Stress in Element AC = 0 psi
Stress in Element BC = 0 psi
Stress in Element AD = 3317 psi
Stress in Element AE = 5870 psi
Stress in Element AF = -1285 psi

Ch. 7 Exercise 1:

1. Material: Steel, Diameter 2.5 in.

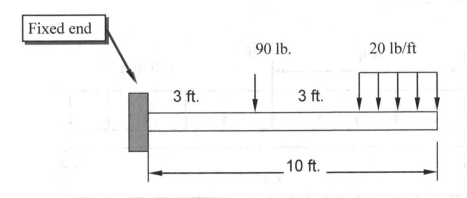

Reactions at the fixed end:
Rx = 0, Ry = 170 lbs, M = 910 ft - lb.

Ch. 8 Exercise 2:

2. A Pin-Joint and a roller support.
    Material: Aluminum Alloy 6061 T6

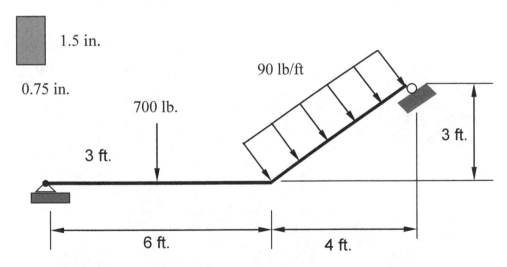

Reactions at the pin-joint support:
Rx = 59.7 lbs, Ry = 620 lbs.

Reaction at the roller support:
R = 549.5 lbs

Ch. 9 Exercise 2:

2.   Material: Steel
             Diameter 2.0 in.

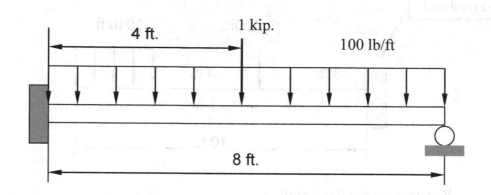

Reaction at the roller support:
             R = 612.5 lbs

INDEX

Notes: